Prognostics and Health Management of High-Speed Railway Traction Power Substations

Zhengyou He • Sheng Lin • Ding Feng • Qi Wang

Prognostics and Health Management of High-Speed Railway Traction Power Substations

Springer

Zhengyou He
School of Electrical Engineering
Southwest Jiaotong University
Chengdu, Sichuan, China

Sheng Lin
School of Electrical Engineering
Southwest Jiaotong University
Chengdu, Sichuan, China

Ding Feng
School of Electrical Engineering
Southwest Jiaotong University
Chengdu, Sichuan, China

Qi Wang
School of Electrical Engineering
Southwest Jiaotong University
Chengdu, Sichuan, China

ISBN 978-981-95-4537-7 ISBN 978-981-95-4538-4 (eBook)
https://doi.org/10.1007/978-981-95-4538-4

Jointly published with Southwest Jiaotong University Press
The print edition is not for sale in China (Mainland). Customers from China (Mainland) please order the print book from: Southwest Jiaotong University Press.

This Springer imprint is published by the registered company Springer Nature Singapore Pte Ltd.
The registered company address is: 152 Beach Road, #21-01/04 Gateway East, Singapore 189721, Singapore

Preface

With the rapid development of its theories and technologies, prognostics and health management (PHM) has been widely applied in multiple fields such as aerospace, military equipment, electronic equipment, naval ships, large machinery, and complex systems. The core of PHM is the integration of advanced sensors, utilizing various intelligent algorithms and inference models to monitor, assess, predict, and manage the health condition of the system. This involves providing timely predictive warnings prior to a system fault, conducting effective and rapid diagnostics upon fault occurrence, and assessing the system's health condition and availability level by integrating multiple sources of data, information, and resources. Ultimately, this leads to the formulation of appropriate maintenance and repair measures, as well as system safety assurance strategies.

Compared to traditional fault diagnosis, PHM represents a conceptual and methodological shift that moves from sensor-based diagnostics to prognostics based on intelligent systems. Additionally, PHM extends the traditional fault diagnostic timeline to include periods before faults occur. As the sole power source for high-speed trains, the traction power supply system's operational safety is crucial for ensuring the safe, reliable, and efficient operation of high-speed railways. The traction power supply system of high-speed railways involves various spatial and temporal scales, dynamics, and randomness, making problem descriptions and model establishment more complex. Traditional fault handling and maintenance work is inefficient in managing the demanding operational maintenance tasks of traction power supply systems, leading to significant pressure on railway maintenance workers, and it is no longer adequate as high-speed railways enter a comprehensive operational maintenance phase. This transition poses new and substantial challenges to the safe, reliable, and efficient operation of high-speed railways. By introducing PHM technology into the high-speed railway traction power supply system, it is possible to fully utilize existing multisource data from monitoring and other sources. By integrating multisource status information of the system, PHM can provide timely predictive warnings before system and equipment faults occur, allowing for comprehensive control over the system's health condition and availability assessment. This ensures that the high-speed railway traction power supply system remains in optimal operational status, reducing or even preventing faults. Based on the assessment results, appropriate maintenance and repair strategies are formulated to ensure the safe and stable operation of China's high-speed railway traction power supply system from both availability and economic perspectives.

Since 2011, the author has been engaged in research on PHM for high-speed railway traction power supply systems, supported by multiple grants. These include the National Natural Science Foundation's High Speed Railway Fundamental Research Joint Fund (HSR Joint Fund) nurturing the project *Research on Latent Fault Early Warning and Rapid Diagnostic Recovery Methods in High-Speed Railway Power Supply Systems* (U1134104), the key project *Research on the Evolution Mechanism of Service State and System Health Diagnostics for Key Traction Power Supply Equipment in High-Speed Railways* (U1234203), and several China Railway Corporation's science and technology research and development programs. Notable among these are *Research on Safe Service Technology for Traction Power Supply—Study on Service Performance of High-Speed Railway Traction Power Supply Systems* (2013J010-A), *Research on Safe Service Technology for Traction Power Supply—Key Technology Research for Operation and Maintenance of High-Speed Railway Traction Power Supply Systems* (2014J009-A), and *Research on Safe Service Technology for Traction Power Supply—State Monitoring and Fault Early Warning Technology Research of Intelligent Traction Power Supply Equipment* (2016J010-A). This book represents a systematic summary and discussion of PHM in high-speed railway traction power supply, building on a consensus, scientific research results, and engineering practice experience.

The book is organized into six chapters. It starts with a systematic discussion on the theoretical framework of PHM for high-speed railway traction power supply systems, focusing on the concepts and development of PHM. The text then provides a detailed overview of prognostics for high-speed railway traction power supply equipment and systems, enriched with case studies. It goes on to integrate the service performance and degradation laws of equipment, the fault evolution process, and the intelligent early warning system for safe operation to perform health assessments of the high-speed railway traction power supply system, including traction substations and catenaries. Following this, the book assesses the availability and risk of the system from perspectives of operating characteristics, relay protection, and multi-factor integration, supported by case analysis. Finally, the book delves into decision-making and optimization research for maintenance strategies of traction power supply equipment and the whole system, covering degradation modeling, prognostics, risk assessment, and multi-attribute decision-making.

During the writing of this book, the author consulted numerous pioneers in power systems and scholars from various related institutions, to whom he expresses sincere gratitude. Special thanks are extended to master's students Xiaojun Sun, Ao Zhang, Zhen Wang, Xiyang Chen, Mingzhou Luo, Xiaohong Lin, Tong Li, Qianqian Yang, Ruidong Fan, Di Huang, and Lin Long for their diligent research efforts in the field of PHM in the high-speed railway traction power supply system. Additional gratitude is owed to graduate students Zhen Cui, Qinyang Yu, Sheng Bi, Xinchang Chen, Kaipu Zhang, and Xiaorui Zhang for their assistance in proofreading and typesetting this book. Acknowledging the limits of his expertise, the author invites readers to provide criticism and corrections, as there may be deficiencies in the text.

Chengdu, Sichuan, China Zhengyou He
Chengdu, Sichuan, China Sheng Lin
Chengdu, Sichuan, China Ding Feng
Chengdu, Sichuan, China Qi Wang

Contents

Chapter 1
Introduction

In recent years, China's high-speed railway has experienced rapid development. By the end of 2019, the operational mileage of high-speed railways in China surpassed 35,000 km, representing over 70% of the global high-speed railway mileage and securing the top position worldwide. In response to the fast-paced domestic growth of high-speed railways, the State Council approved the revised *Medium-and Long-Term Railway Network Plan* in July 2016. This plan outlines milestones for the expansion of China's high-speed railway network: reaching 30,000 km by 2020, extending to 38,000 km by 2025, and eventually exceeding 45,000 km, more than doubling the current operational mileage. The vision aims to construct a modern high-speed railway network featuring the *eight vertical and eight horizontal* main routes, coupled with interwoven regional connecting lines and a comprehensive integration of intercity railways.

Additionally, the *Belt and Road Initiative* has provided Chinese high-speed railways with a significant opportunity for international expansion. Beyond considerable domestic construction, the development of China's high-speed railways now spans over 20 countries across Asia, Europe, the Americas, and Africa.

As high-speed railways advance towards large-scale and networked operations both domestically and internationally and strive for extensive operational mileage within a relatively short construction period, the post-construction peak will mark the beginning of a comprehensive operational maintenance era. The traction power supply system, as the sole power source for high-speed trains, is pivotal in ensuring the safe, reliable, and economical operation of high-speed railways. With increasing operational mileage, service time, and equipment performance degradation, the safety risks associated with the high-speed railway traction power supply system will become more pronounced, and the demand for maintenance and repair will grow daily. It is crucial to ensure the safe and reliable operation of thousands of high-speed trains over tens of thousands of kilometers, improve the utilization rate of traction power supply equipment, reduce or even prevent faults, alleviate the maintenance workload of frontline operational staff, and lower maintenance and repair costs. These measures are essential to maintain the optimal operational status of the

Z. He et al., *Prognostics and Health Management of High-Speed Railway Traction Power Substations*, https://doi.org/10.1007/978-981-95-4538-4_1

high-speed railway traction power supply system and facilitate a smooth transition through the comprehensive operational maintenance phase in China.

As the primary energy source for electric locomotives on high-speed railways, the safety and availability of the traction power supply system are paramount for the seamless operation of the entire network. A significant portion of abnormal operation and delays in high-speed railways can be traced back to faults within this system. When faults occur, locomotives lose power and cease operation, which disrupts passengers' travel plans, affects railway transportation schedules, and can lead to substantial safety production losses. These issues can adversely impact the reputation of the railway department. Moreover, the high-speed railway traction power supply system is characterized by its complex structure and demanding operational conditions. For instance, catenary components are numerous and extensively distributed along railway lines, exposed to the outdoors, subject to high-speed impacts from locomotive pantographs, and lack backup systems. All these factors make the traction power supply system particularly prone to faults.

Traditionally, high-speed railway power supply systems rely on post-fault handling approaches. This reactive method is often slow in restoring function, resulting in significant economic losses and negative social repercussions. For example, soon after its opening, the Beijing-Shanghai high-speed railway encountered several power supply equipment failures, with fault resolutions taking hours and leading to major train delays. Similarly, within the first 3 months of operation, the Wuhan-Guangzhou high-speed railway faced five incidents of severe delays due to equipment failures, as reported by the media. These incidents highlight that the onsite maintenance operations are poorly adapted to the high-speed railways' operational characteristics. They fail to implement maintenance practices tailored to the unique service environment of the traction power supply equipment and do not adequately address the substantial challenges that arise during the transition of large-scale network-based high-speed railway lines to their comprehensive operational maintenance, both domestically and internationally. The primary deficiencies in the current fault handling, maintenance, and repair of the high-speed railway traction power supply system can be summarized in three main areas:

1. The evaluation of the operational status of equipment and systems is often inaccurate and incomplete. Current practices in monitoring the operational status of equipment include checking specific feature indicators of key traction power supply equipment, such as oil chromatography for traction transformers and gas pressure for gas-insulated switchgear (GIS). However, these methods tend to rely on single feature indicators without considering multiple or all indicators collectively; they focus on the status of individual pieces of equipment without accounting for the interdependencies that impact the system's overall performance; they typically make binary judgments of *normal* or *fault*, which fails to capture the gradual degradation trend from *normal* to *fault*, including various intermediate stages; and they often use only current values for simple threshold comparisons, without integrating historical data or predicting future trends, thereby failing to comprehensively and accurately grasp the operational status of the equipment and systems.

2. The methods used for fault diagnosis and handling are fundamentally flawed. The traction power supply system predominantly relies on a post-fault response approach, where fault diagnosis and remediation only occur after a fault has already led to equipment damage, railway outage, or even casualties. Fault judgments are often determined solely by overt manifestations (fault consequences) rather than a reasoned estimation derived from actual service performance degradation patterns or operational availability criteria. This reactive approach results in severe consequences that include not only damage to valuable equipment and high maintenance costs but also extend to incalculable adverse impacts such as power outages, casualties, and adverse social reputations. Additionally, in the aftermath of accidents, the origin of faults is difficult to determine, making it challenging to trace and backtrack the fault process, clarify accident responsibilities, and systematically collect, classify, summarize, and refine fault events. This lack of systematic analysis hinders the accumulation of accident experience that could be used to prevent future faults.
3. The maintenance and repair model for the traction power supply system is significantly outdated. Currently, the system still relies on a mix of periodic maintenance, manual inspections, and emergency repairs. This approach is inherently reactive, often resulting in both insufficient and excessive maintenance. The development of maintenance strategies lacks specificity; some are based on operating procedures, maintenance guidelines, and technical standards of power system equipment, while others apply existing protocols from conventional railways without adequately considering the structural characteristics and service conditions of the high-speed railway traction power supply system, hindering the scientific formulation of maintenance strategies. This results in inefficient and even misguided guidance in the implementation of maintenance activities. The execution of maintenance often involves guesswork and speculation, with maintenance personnel relying on experience or even luck to resolve faults. While such practices may temporarily fix issues, this unscientific approach to maintenance can create future operational risks for the equipment. Consequently, this passive maintenance and repair model leads to low maintenance efficiency, high economic costs, and substantial labor intensity and workload.

The aforementioned shortcomings mean that the current fault handling and maintenance work cannot efficiently manage the demanding operational maintenance tasks of the traction power supply system, nor can it effectively alleviate the immense work pressure on railway workers. It is no longer well-equipped to address the various challenges faced as high-speed railways enter the comprehensive operational maintenance phase, posing severe threats and significant challenges to the safe, reliable, and efficient operation of high-speed railways.

In response to these issues, this book focuses on the theme of prognostics and health management of the high-speed railway traction power supply system. It provides an in-depth analysis of prognostics during abnormal operations, health condition assessment during normal operations, and long-term availability and risk assessments of the system. Building on this foundation, the book establishes theories and develops technologies for maintenance and repair, aiming to fully address the deficiencies in fault handling and maintenance and repair. This enhances the service capability of equipment and the operational efficiency of the system, saves a substantial amount of operational maintenance costs, effectively reduces the workload and work pressure of operational maintenance personnel, and ensures a smooth transition for high-speed railways both domestically and internationally into the comprehensive operational maintenance phase.

This chapter serves as the introduction to this book, briefly presenting the development of PHM the concepts and technologies of PHM, and the established PHM technological framework for high-speed railway traction power supply systems.

1.1 The History and Current Applications of Prognostics and Health Management

1.1.1 History of PHM Technology

With the rapid advancement of information and automation technologies, the complexity and integration of various large-scale systems have significantly increased, leading to more pronounced system maintenance support challenges. In this context, the PHM technology as a convenient, reliable, and cost-effective support approach has been increasingly studied, emerging as a crucial development that continues to evolve and strengthen. Systems based on this technology are increasingly being applied across diverse sectors, including aerospace, military equipment, ships, and power systems, showcasing substantial developmental potential and promising application prospects.

The developmental trajectory of PHM technology reflects the evolution from a reactive approach to faults and abnormalities to a proactive and preventative management strategy. The evolution of PHM technology can be categorized into five distinct stages:

1.1.1.1 Availability Analysis Stage

The origins of PHM technology date back to the mid-twentieth century, around World War II. During this period, many complex systems such as weapon systems, communication systems, and aviation electronics exhibited low availability levels. As these issues became increasingly apparent, coupled with the advent of planned space research programs, the foundational development of availability theory was initiated. At this stage, researchers utilized traditional data collection methods to conduct availability analyses based on system availability data. This foundational work continuously evolved, improving and perfecting system designs to enhance performance and meet availability requirements under operational conditions and extreme environments. This stage represents the nascent phase of PHM technology development.

1.1.1.2 Fault Analysis Stage

As aerospace systems grew in complexity, the likelihood of faults due to inadequate design, manufacturing errors, maintenance mistakes, and unforeseen events also increased. This scenario in the 1970s prompted researchers to explore new methods for monitoring system states and preventing abnormal operating conditions. This led to the development of onboard critical fault response methods. Subsequently, technologies emerged for diagnosing fault sources and causes, eventually giving rise to prognostics methods. Prognostics technology, which can employ either physical or intelligent models, integrates various collected data to assess and predict the future status of systems and components and estimate their remaining service life. Prognostics represents a significant milestone in the development of PHM technology, marking an early indication of its evolving capabilities.

1.1.1.3 System Monitoring Phase

In the early 1990s, Vehicle Health Monitoring (VHM) emerged as a pivotal management model within the National Aeronautics and Space Administration (NASA). A notable feature of this phase was the use of sensors and software to monitor the health condition of space transportation tools. This period marked the first use of the term "health" to describe the operating status of mechanical equipment. The main advancements of this stage were the sophisticated application of sensor technology, data transmission technology, and data processing technology for system status monitoring. These developments not only supported the safe operation of systems but also laid the groundwork for subsequent practical innovations.

1.1.1.4 System Health Management Phase

Following the advancement of health monitoring theories, it became evident that monitoring alone was insufficient to guarantee equipment safety. A critical challenge was formulating management strategies by optimally analyzing the collected data. As a result, the term "management" began to replace "monitoring". This shift marked the first integration of maintenance decisions with health management, significantly enriching PHM technology and signaling its rapid development and gradual maturation.

1.1.1.5 Integrated System PHM Phase

In the 1990s, NASA introduced the concept of Integrated System Health Management (ISHM), where the term "integrated" emphasized addressing the distinctions between system-level issues and various subsystem-level challenges. Previously, different systems mainly focused on addressing specific fault issues within their respective domains without considering a systematic, comprehensive perspective. This phase emphasized adopting a systemic approach, fully leveraging PHM technology to ensure comprehensive equipment protection.

By the end of the twentieth century, the United States initiated research on the F-35 Joint Strike Fighter (JSF), integrating various diagnostic and prognostic technologies. This research led to the establishment of an architectural system that combined component, subsystem, and system-level status monitoring strategies, marking the inception of PHM technology. Subsequently, with the collaborative efforts of researchers from various countries and government support, PHM technology systems across different fields were developed, promoting the further maturation and advancement of PHM technology.

1.1.2 Current PHM Technology Application

1.1.2.1 Aerospace Field

PHM technology was initially applied in the aerospace sector, starting with availability and fault analysis in the 1970s. Since then, the rapid advancements in sensing technologies, predictive algorithms, and computational capacities have broadened its application from power systems to mechanical structures and electromechanical equipment within the aerospace industry. This evolution has seen PHM technology become more integrated, intelligent, and capable, supported by an ever-improving technical standards system. To date, developed Western countries have equipped their aerospace equipment with comprehensive PHM systems. These systems engage in extensive information collection, fusion, and analytical processing, facilitating real-time monitoring and maintenance management decisions.

One of the most notable applications of PHM in civilian aviation is Boeing's development of the Airplane Health Management (AHM) system. This system has been adopted by multiple international airlines for commercial aircraft, significantly enhancing safety and operational efficiency. It is estimated that the use of the AHM system has led to a reduction of approximately 25% in costs associated with flight delays and cancellations. Additionally, the AHM system is capable of identifying frequent faults and predicting trends in avionics systems, aiding airlines in maintaining long-term reliable operations. NASA, in collaboration with ARINC, developed the *Aircraft Condition Analysis and Management System* (ACAMS), similar in function to PHM systems, and tested it on the B757 aircraft, earning a U.S. patent for this innovation. Furthermore, NASA implemented the *Integrated*

System Health Management (ISHM) system, developed by Qualtech, on the Space Shuttle to monitor its health status and to diagnose and troubleshoot system faults, thereby reducing the incidence of system failures.

The capabilities of PHM technology to acquire a system's technical condition through real-time monitoring, predict functional failures, and execute preventive maintenance decisions not only shorten maintenance time but also reduce the life-cycle maintenance costs of aircraft and enhance system availability and safety. This robust support system paves the way for future aircraft to achieve rapid and accurate maintenance assurance.

1.1.2.2 Weapons and Equipment Field

In the 1970s and 1980s, complex weapons and equipment exhibited issues such as low testability, prolonged fault diagnosis times, high false alarm rates from built-in tests (BIT), high operational and maintenance costs, and insufficient maintenance manpower. These issues drew significant attention from military and industrial sectors in the United States, the United Kingdom, and other countries. In response, the U.S. military and industry implemented various independent measures targeting ATE, technical data, BIT, and testability among other diagnostic elements. However, these efforts did not yield satisfactory results. In-depth research identified the root of the problem as the isolated operation of each diagnostic element without a comprehensive analysis; moreover, apart from testability and BIT, other diagnostic elements were often designed only after the main equipment design was nearly completed. To address these support issues, the U.S. Department of Defense issued military standards and directives emphasizing the use of integrated logistic support to effectively resolve support challenges for weapons and equipment.

During this period, diagnostic issues became a bottleneck in implementing integrated logistic support. The National Defense Industrial Association (NDIA), formerly known as the American Society for Industrial Security (ASIS), first proposed the concept of integrated diagnostics in 1983. This concept involved synthesizing various elements that contribute to the diagnostic capabilities of weapons and equipment. It received recognition and strong advocacy from the U.S. military. Starting from the mid-1980s, the U.S. Department of Defense implemented a series of integrated diagnostics research programs and, in April 1991, issued the military standard MIL-STD-1814, designating integrated diagnostics as an effective means to enhance the diagnostic capabilities and mission readiness of new-generation weapon systems, as well as to reduce operational and support costs. The integrated diagnostic strategy was applied to new-generation equipment developed from the mid to late 1980s, such as F-22, SSN-21, and M1A2, as well as the then-in-development F-35 (JSF). After the 1990s, developed European countries like the UK and France, along with Russia, followed the U.S. approach by advocating for the adoption of integrated maintenance modes in their weapon systems. These systems were designed to maximize fault detection and isolation capabilities, thereby enhancing mission readiness and reducing life-cycle costs of the equipment. After the Cold War, with reductions in U.S. military budgets and the aging of weapon systems, both the U.S. military and the industrial sector recognized that the main factor hindering the widespread application of integrated diagnostics was the absence of an open system approach tailored specifically for it. In 1999, the Office of the Secretary of Defense (OSD) initiated the *Open Systems Integrated Diagnostics Demonstration* (OSAIDD) study program in the United States. This program explored the feasibility of a unified, generic method to implement integrated diagnostics functionalities aimed at reducing costs, increasing interoperability, and accelerating the adoption of new technologies. Through in-depth studies and demonstration validations of typical cases with varying testing and diagnostic features from both military and civilian sectors, the project ultimately proposed an information-based integrated diagnostics open architecture.

Currently, PHM technology is widely applied in helicopters in countries such as the UK, USA, Canada, the Netherlands, Singapore, South Africa, and Israel, featuring an integrated application platform known as *Health and Usage Monitoring Systems* (HUMS). The U.S. Department of Defense's new generation HUMS—JAHUMS—exhibits comprehensive PHM capabilities along with an open and flexible system architecture. To date, over 180 U.S. Army helicopters have been equipped with HUMS, including the AH-64 Apache, UH-60 Black Hawk, and CH-47 Chinook. The mission readiness of U.S. Army helicopters equipped with HUMS has increased by 10%. The Army has issued airworthiness certificates or maintenance permits to aircraft equipped with HUMS and has approved the installation of HUMS in 750 Apache helicopters. The UK Ministry of Defense has also reached an agreement with Smiths Aerospace to develop a system that integrates health and usage monitoring systems with cockpit voice and flight data recorders (HUMS/CVFDR) capability for 70 future Westland Lynx helicopters, with deliveries starting in 2011.

PHM has become a core technology for the new generation of domestic and international weapon systems to achieve state-based maintenance and autonomous support. It is a highly promising dual-use technology for both military and civilian applications in the twenty-first century, aimed at enhancing the five characteristics (availability, maintainability, testability, supportability, and safety) of complex systems and reducing life-cycle costs.

1.1.2.3 Maritime Domain

A ship is a complex systems engineering project, and within this domain, PHM focuses on one or several core subsystems, such as the propulsion system and power system. PHM primarily performs the following functions in this domain:

1. Through real-time monitoring of the state parameters and health conditions of critical equipment within ship systems, it facilitates fault diagnosis and predicts potential fault timings to eliminate safety hazards, ensuring the safe and stable operation of ship systems.

2. Based on the fault state or health condition of the equipment, this approach predicts the fault development trend and the remaining life of the ship systems and equipment, providing a solid foundation for formulating maintenance strategies.
3. By developing maintenance strategies that eliminate unnecessary scheduled maintenance and maintenance costs, it significantly reduces the economic costs of ship operation.
4. It furnished data support for the optimization design of ship systems and equipment, grounded on parameters obtained throughout the prognostics, and health management processes.

Currently, from a methodological perspective, maritime PHM can be segmented into several approaches based on physical models, empirical models, and data-driven methods.

Maritime PHM technology based on physical models utilizes precise physical or mathematical models of the monitoring targets to conduct prognostics, and health management of ship systems or equipment. For example, using a failure mode-based ship equipment fault information processing system to analyze and evaluate diagnostic compensation measures and improvements for ship system faults, and designing fault diagnostic observers based on the T-S fuzzy model of the ship propulsion system.

Maritime PHM technology based on empirical models leverages expert experience and prior knowledge, predominantly including methods based on expert systems and fault trees. These methods are primarily used for fault diagnosis of critical ship equipment such as diesel engines, electric propulsion systems, and related electronic equipment.

Data-driven maritime PHM technology effectively addresses the dependency on models and experience, hence it is currently attracting widespread attention and research. This approach requires historical data from the monitoring targets under various operating conditions, including methods based on statistical analysis and artificial intelligence algorithms. These methods have shown excellent performance in ship system prognostics, and health management.

Some scholars advocate for the integration of two or all three types of models—physical, empirical, and data-driven—to fully utilize the advantages of each approach and achieve higher precision in PHM functionalities.

Currently, marine PHM systems developed by various countries have been put into practical use. The German company MAN B&W has implemented fault diagnosis for large two-stroke marine diesel engines by monitoring state parameters through sensors and integrating a knowledge base. The Norwegian company KYMA has developed the *Marine Performance Monitoring* system based on ship performance parameters. This system evaluates the main engine condition by comparing various operating parameters with deviations from normal operating status. Additionally, companies like Siemens, ABB, and GE have also conducted extensive research and development in the areas of ship prognostics, and health management.

In China, research related to marine PHM technology began in the 1980s, and several systems have since been deployed. Examples include the CJBW20 and

CY8800 ship monitoring systems developed by Shanghai Ship and Shipping Research Institute, the marine rescue ship state monitoring and fault diagnosis system developed by Wuhan University of Technology, and the QY-GZZD-01 diesel engine fault diagnosis expert system developed by Shanghai Marine Diesel Engine Research Institute.

1.1.2.4 Power System Domain

In the context of rapidly developing information technology and automation technology, the integration and complexity of various large systems have increased significantly, giving rise to PHM technologies. In the power system sector, there is considerable research on the health diagnosis of equipment such as generators, transformers, and circuit breakers, which are key primary equipment. However, studies on health diagnosis at the system level of electrical grids are less frequent. In 2004, Professor Ju first proposed the concepts and theoretical framework of power system health diagnosis based on the foundation of power system fault diagnosis, which includes the concept of diagnosability. He also introduced the methodology of health diagnosis, which involves establishing a 'health record', implementing dynamic tracking, diagnosing the 'health' level based on changes in feature quantities, and promptly identifying and alerting to any 'pathology.' This concept laid a solid theoretical foundation for the study of health management in power grids. Subsequently, in *Evolution of the Power Dispatching Control Center — Real Time Power Security Early Warning System*, Sun et al. proposed using 'triage' and 'consultation' to provide real-time safety warnings for the power grid. They designed the functional framework and system architecture of the power grid real-time safety warning system. Later, ideas for real time fault diagnosis of power gird based on the fingerprints identify algorithm of power flow emerged. Mirrasoul J et al. used digital modeling and statistical diagnostic techniques for early detection studies on underground distribution systems. Hu and Zhang from Yanshan University introduced the concept of power system health diagnosis and proposed using the AHP-Eigenvalue method combined with fuzzy theory to assess the health condition of power grids. They constructed a power system health diagnostic indicator system from three aspects: historical foundation, online operation, and environmental impact, and incorporated fault indicators to make the indicator system more comprehensive.

Currently, research on prognostics in the power system domain, both internationally and domestically, focuses on two main aspects. The first aspect targets electrical equipment such as transformers, circuit breakers, and power cables. This involves using online monitoring techniques combined with the equipment's historical condition and operational circumstances to objectively assess the equipment's status and detect faults in advance. For example, by establishing a grey prognostics model, prognostics and forecasts can be made regarding the operating status and trends of power transformers. The second aspect involves predicting and analyzing potential cascading faults in large power grids. However, existing prognostics methods have several limitations: they mostly predict faults in individual pieces of equipment, and when monitoring data is insufficient, the prognostics results are unreliable. These methods do not consider uncertainties such as sensor errors and data transmission losses. Power systems exhibit distributed or network structures, ranging from equipment to subsystems and from primary to secondary systems. Within these structures, fault information exhibits varying degrees of association and interactive coupling, with failure manifestations including multiple faults, common-cause faults, correlated faults (subordinate or induced failures), and coupled failures. Therefore, from a holistic system perspective, the future development trend of prognostics in power systems involves collecting multi-point information and establishing a multi-input uncertainty prognostics model to enhance the credibility of prognostics.

1.2 Concepts and Technologies of PHM

As PHM technology continues to develop and refine across various fields, researchers have gradually formed a unified understanding of its concepts and implications. This involves utilizing various advanced sensors for data acquisition regarding equipment conditions, operating environments, and online or historical operating states. Through various intelligent reasoning algorithms (such as physical models, neural networks, data fusion, fuzzy logic, expert systems, etc.), PHM technology analyzes and assesses the health condition of systems and equipment. It predicts faults before they occur and combines various available resource information to provide a series of maintenance and support measures, both theoretically and technically.

From an operational maintenance philosophy perspective, PHM represents an innovation and transformation from breakdown maintenance to regular inspections, and further to proactive prevention. In terms of conceptual content, PHM encompasses specific tasks such as state monitoring, prognostics, lifespan prognostics, and operational maintenance decision-making. From a technical (implementation) standpoint, PHM involves aspects of both prognostics and health management. Specifically, prognostics entails assessing and diagnosing the current health condition of equipment based on historical and current data, enabling abnormal monitoring, fault localization, and state assessment of the equipment. It also combines information about future operating environments and planned functions to predict changes in equipment performance, remaining useful life, and fault probability of occurrence. Based on the analysis results from PHM involves dynamically adjusting the operational conditions of the equipment and devising appropriate operational maintenance strategies to ensure the equipment fulfills its designated tasks.

Currently, research in PHM has developed three types of mature and commonly used approaches: model-driven methods, data-driven methods, and statistical availability-based methods.

1.2.1 Model-Driven Method

This method utilizes knowledge of a product's lifecycle loads and failure mechanisms for prognostics. Prognostics are influenced by numerous interacting factors both during the data processing phase and in the final outcomes. These factors include the type of failure, the form of failure models, the form of failure criteria, and the methods used for prognostics. This method typically requires domain-specific knowledge for physical fault modeling, meaning the mathematical model of the target system is known or solvable. Prognostics using a physics-of-failure model require input from indirectly measured relevant energy parameters of the target, such as temperature, humidity, pressure, operating voltage, and current. Supported by the model, this method can achieve high prognostics accuracy. However, it is often challenging to establish precise mathematical models for complex dynamic systems, which limits the applicability of this approach. Typical model-driven approaches include Fault Mode, Mechanisms and Effects Analysis (FMMEA), Kalman filtering, particle filtering, and methods based on expert knowledge. The process is illustrated in Fig. 1.1.

In the prognostics of some equipment, the physics-of-failure model is often used due to the difficulty of monitoring performance parameters. This approach involves various failure mechanisms and different forms of failure models in equipment prognostics. Therefore, clarifying the impact of each factor is particularly important for constructing an effective prognostics technical framework. The form of failure models plays a foundational role in influencing prognostics. Commonly used physics-of-failure model include lifespan models and performance degradation models:

1. Lifespan Models are applicable to physical failure mechanisms related to wear-out failure. They describe the anticipated service life of equipment under given operating environmental conditions. When the equipment's working time reaches

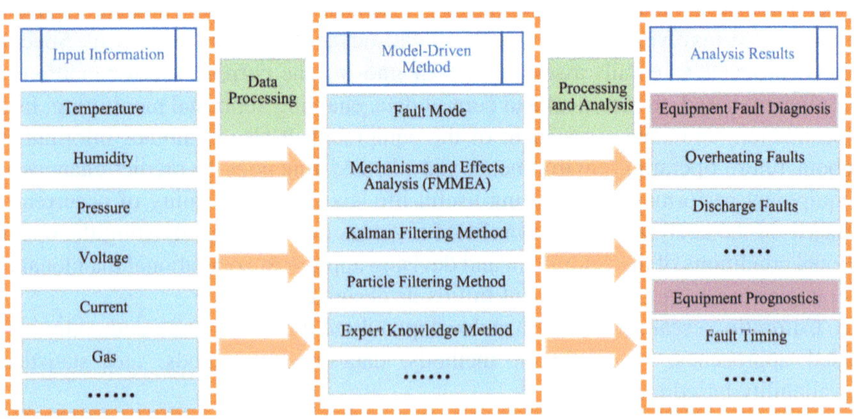

Fig. 1.1 Model-Driven Diagnostic Process

its service life, the equipment fails. However, in practical use, the working environment is generally not constant, which requires a mathematical model for accumulating life loss—i.e., damage accumulation in failure criteria.

2. Performance Degradation Models are also applicable to failure mechanisms related to wear-out failures. These models describe the pattern of performance parameter degradation over time under specific operating environments and conditions. When one of these performance parameters exceeds the specified range, the equipment fails. In cases where the operating environment of the equipment is not constant during use, it is necessary to integrate changes in performance parameters to determine the service life of the equipment at the point of failure.

1.2.2 Data-Driven Method

In many practical situations, the mathematical models of complex systems may be unavailable or difficult to solve. In such cases, installing various sensors at key locations of the study subject or conducting multiple experiments and organizing data to form information becomes an effective way to understand the operational patterns of the system. The data-driven approach utilizes data collected from these sensors or experiments to build models and perform analysis and prognostics through machine learning. Since it relies solely on available data for health assessment and prognostics, this method has wide applicability and can achieve a certain level of accuracy.

In the era of industrial big data, conducting comprehensive data-driven research on the application of artificial intelligence technologies, including expert systems, uncertainty reasoning, machine learning, and intelligent optimization computing, in the comprehensive state data mining and analysis of equipment can provide critical decision support for the operation and maintenance management of equipment. The application research of AI analysis techniques in equipment state-based maintenance primarily focuses on three core tasks: processing and mining of multi-source heterogeneous data, comprehensive and accurate operation state assessment and prognostics, and integrated maintenance decision modeling and optimization. These tasks are further subdivided into seven typical application scenarios: data cleaning, text mining, image recognition, state assessment, prognostics, state prognostics, and maintenance decision optimization. The decision-making process for power equipment driven by data is shown in Fig. 1.2.

Currently, the application of data-driven AI analysis technologies faces numerous challenges. At this stage, the storage forms of relevant equipment status data vary widely, and the quality is uneven, often plagued by data duplication, anomalies, and missing data. Frequently, data originate from different departments and information management platforms, with no capability for cross-platform data interaction and sharing. This leads to data silos, which significantly increase the difficulty of obtaining complete data samples and severely affect the efficiency and accuracy of the assessment and prognostics stages.

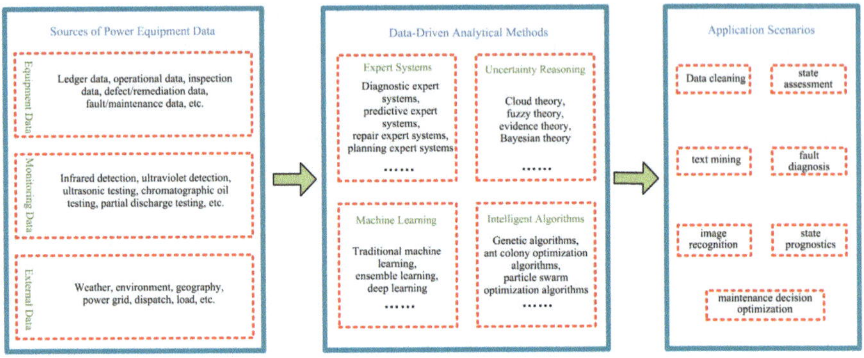

Fig. 1.2 Data-Driven Decision-Making Process for Power Equipment

1.2.3 Statistical Availability-Based Method

Due to various constraints, it is sometimes difficult to obtain mathematical models of systems and sufficient status data. However, in cases where sufficient historical failure data have been accumulated, statistical availability-based methods can be utilized to prognostic. These methods model the life distribution, structural characteristics, and operating status of the object from a availability perspective and infer fault information. Given the potential for incomplete data, the accuracy of this method can vary significantly. To mitigate this, a confidence level is often added to the prognostics results to indicate the degree of accuracy.

Availability management encompasses a comprehensive integration of various aspects, including equipment planning, manufacturing, installation, commissioning, facility operation, maintenance, repair, and production management. It evaluates the integrated level of all these aspects. In specific systems, equipment availability management serves as a foundational type of management, drawing information directly from the production front line, which forms the basis for availability statistical indicators and the conduct of availability analysis.

A availability assessment model is a process that leverages existing data to handle unexpected situations and establish connections between equipment and inter-systems. Employing artificial intelligence assessment methods to design availability models for systems and their equipment facilitates rapid identification and resolution of issues during the assessment process. Analytical methods frequently used in assessments typically involve data analysis and statistical computations based on mathematical knowledge to address potential faults. Typical statistical availability-based methods include D-S theory, Bayesian network methods, and fuzzy statistical methods, among others. The process for statistical availability analysis is illustrated in Fig. 1.3.

Availability assessment aims to obtain effective system operational state evaluation results. Clearly defining the purpose of the assessment can determine the

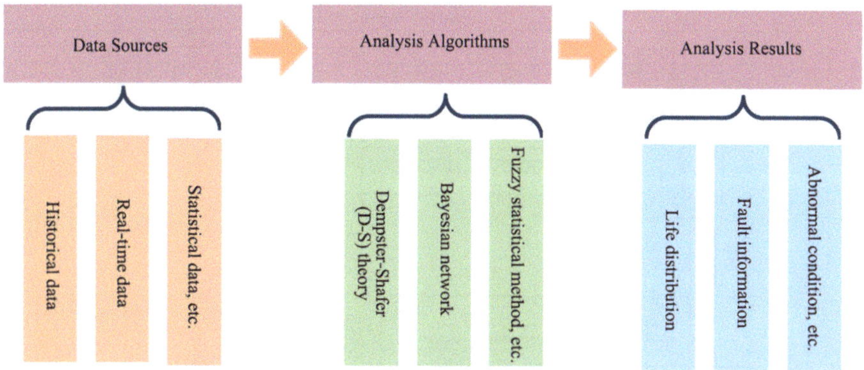

Fig. 1.3 Statistical Availability Analysis Process

method of evaluation. If the assessment method is correct, it ensures reduced risks during the assessment process. Comprehensive assessments guarantee that the results are irreplaceable in terms of authenticity and effectiveness. To meet application standards for systems and equipment, system planning should aim to ensure adequacy, reduce unnecessary complications, and guarantee that the system can handle emergencies timely and effectively.

Currently, whether for power systems or traction power supply systems, the application of PHM technology is still in its early stages. Related research primarily focuses on two aspects: one is the predictive warning for early and concealed faults, and the other involves studies on availability, risk, and safety assessments. In-depth studies on monitoring, assessing, and managing the health condition of traction power supply equipment and systems are relatively scarce and still require further development.

1.3 PHM Technology Framework for High-Speed Railway Traction Power Supply System

1.3.1 Structure and Function of the High-Speed Railway Traction I Power Supply System

In high-speed railways, the traction power supply system serves as the sole power source for the high-speed electric multiple unit. It receives electrical energy from the power system and supplies the electric multiple unit with electricity that conforms to specific voltage and frequency standards after transformation, phase change, or frequency conversion through the traction electric power transmission line. China's high-speed railways utilize single-phase industrial frequency (50 Hz) AC at a voltage level of 25 kV. The power supply method commonly employs an autotransformer (AT) feeding scheme, which offers several advantages. This scheme can

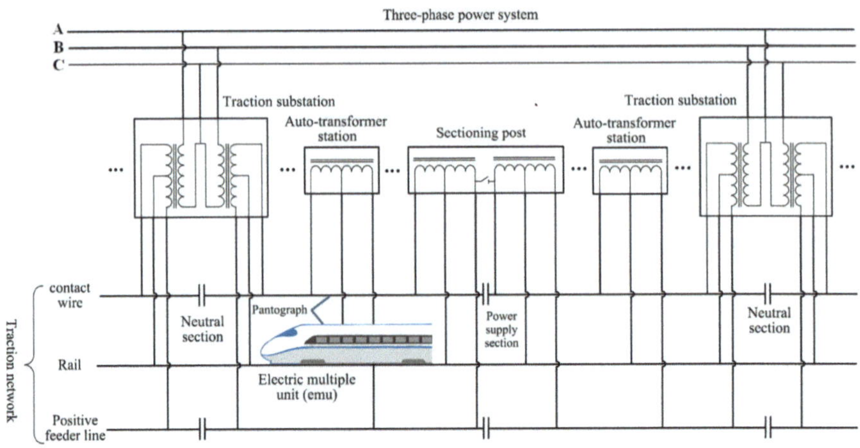

Fig. 1.4 Schematic Diagram of the High-Speed Railway Traction Power Supply System Structure

significantly increase the supply voltage, reduce the impedance of the traction network, extend the power supply distance, and effectively minimize voltage and energy losses on the traction network. It is particularly suitable for operating conditions involving high speeds, heavy loads, and other scenarios that demand high currents.

Under the auto-transformer feeding scheme, the structure of the high-speed railway traction power supply system is illustrated in Fig. 1.4. Typically, the traction power supply system sources electrical energy from a three-phase power system with a voltage level of 220 kV, which is 330 kV in the northwest region and 110 kV in a few special areas. The power is then stepped down by traction transformers located within the traction substation to two-phase 2 × 27.5 kV industrial frequency alternating current, which is supplied separately to the left and right power supply arms of the traction network, each carrying single-phase voltage. The phases are separated by electric phase splitting in the middle, with the contact wire voltage on the traction network at 27.5 kV and the positive feeder line voltage at −27.5 kV. The system is rail-grounded and connected to the neutral point of the traction transformer. Along the traction network, an auto-transformer station or sectioning post is established every 10–15 km. These stations contain an AT with a transformation ratio of 2:1, creating a connection between the contact wire and the positive feeder line, with the neutral point connected to the rail. This configuration generates a 25 kV voltage between the contact wire (which contacts the pantograph of the electric multiple unit) and the rail (which contacts the electric multiple unit's wheels), supplying electric power to the electric multiple unit. Beyond the basic AT functionality, sectioning posts also divide the power supply zones and facilitate cross-zone power supply through the switching operation of the switchgear.

In the high-speed railway traction power supply system, the traction substation serves as the critical interface between the three-phase power system and the single-

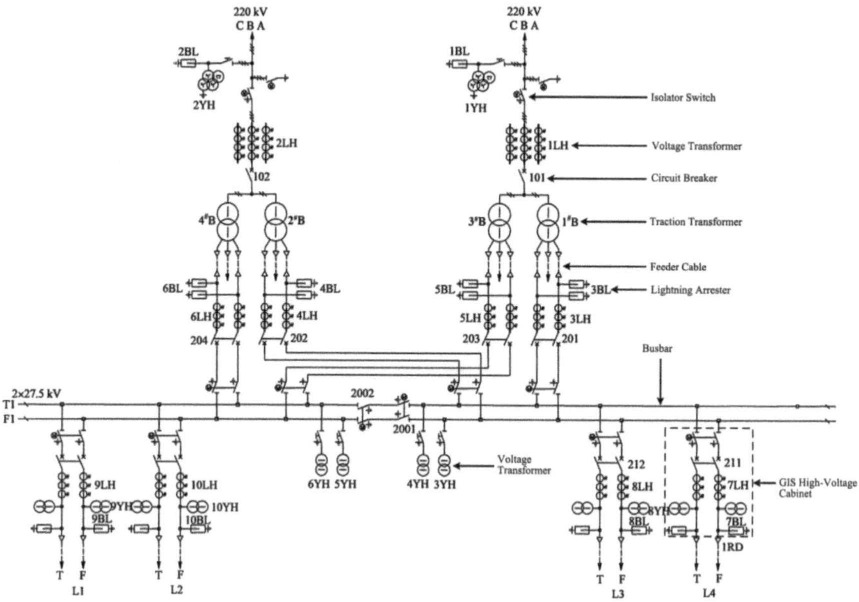

Fig. 1.5 Typical Traction Substation Main Wiring Diagram

phase traction power system. It plays a vital role in voltage transformation and power transmission and is a key component of the system. A typical main wiring diagram of a traction substation is depicted in Fig. 1.5. The high-voltage power supply equipment and current-carrying conductors, generally referred to as primary equipment, include traction transformers, circuit breakers, isolators, GIS high-voltage cabinets, voltage transformers, current transformers, lightning arresters, feeder cables, and busbars. As illustrated in Fig. 1.5, power is received at the traction substation from a 220 kV three-phase power system via two incoming lines. These lines are connected to two traction transformers (each comprising two single-phase transformers configured in a Vx wiring setup), reducing the voltage to 27.5 kV. The electrical energy is then transmitted via single-phase feeder cables to a collecting busbar, where it is further distributed and transferred through feeder cables to the contact wires and positive feeder lines of the up and down line traction network. The high-voltage incoming side, both sides of the traction transformers, and the 27.5 kV side are equipped with circuit breakers and isolators to enable normal load switching and to coordinate with relay protection systems for tripping operations during short circuits. Voltage transformers and current transformers are installed at various measuring points within the substation to measure high voltage, meter electric energy, and provide the electrical quantities necessary for various relay protection systems and automatic equipment. Lightning arresters, essential for suppressing overvoltages, are installed throughout the substation. Additionally, in many high-speed railway traction substations, equipment such as circuit breakers, isolators, current transformers, lightning arresters, and busbars are encapsulated within

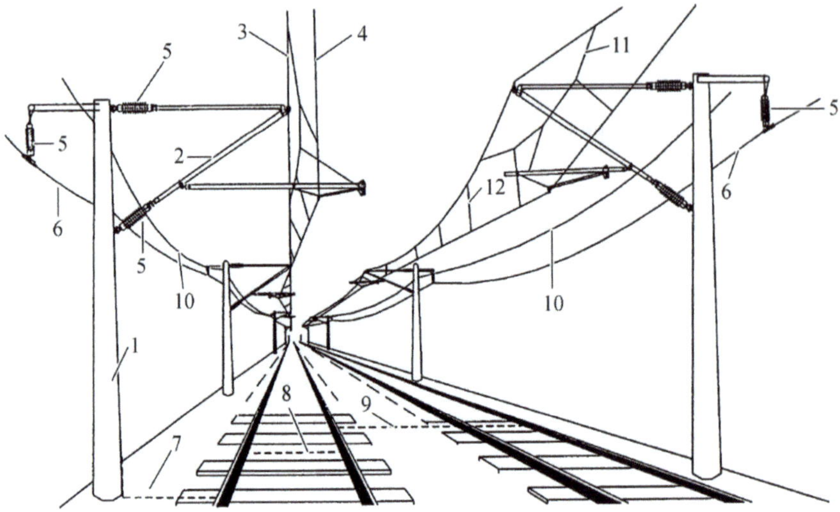

1-Pole 2-Arm 3-Catenary Wire 4-Contact Wire 5-Insulator 6-Feeder Line 7-Pole Foundation 8-Rail Connection 9-Up and Down Line Rail Connection 10-Return Line 11-Elastic Suspension Rope 12-Dropper

Fig. 1.6 Basic Structure of the Catenary

grounded metal casings filled with sulfur hexafluoride (SF_6) gas to provide phase-to-phase and ground insulation. This arrangement forms a GIS, which helps to reduce the footprint, decrease construction workloads, and enhance availability.

Along electrified railways, transmission lines are installed at intervals along the tracks, with the catenary system designed to provide sliding contact to pantographs under various conditions. Due to economic and technical reasons, the catenary in the traction power supply system is singular and lacks a backup, making it a critical component of the system. Figure 1.6 illustrates a typical high-speed railway catenary structure, employing an elastic chain-type suspension. The diagram includes four main parts: pole and foundation, support and positioning equipment, contact suspension, and electrical auxiliary equipment. The entire line's catenary is composed of various spans, with a center misalignment at the center of the anchor section, and a ratchet compensation equipment at the end of the anchor section. Transitions between adjacent anchor sections are completed by anchor section joints. The main components of the catenary are listed below.

1. Pole and Foundation.

   The pole and foundation provide the main vertical support, tracking, and positioning for the catenary, capable of withstanding the total load of the contact suspension support equipment and suspending the overhead contact wire to its designated position and height. Due to the catenary poles enduring repeated actions from fixed and non-fixed loads over time, combined with possible errors in design and construction and the imperfection of craft standards, there are some safety hazards associated with the catenary pole foundations.

2. Contact Suspension.

   The contact suspension primarily comprises the contact wire, dropper, catenary wire, elastic suspension rope, compensation equipment, and their connecting parts. Its function is to transmit electrical energy through the contact wire to the locomotive's pantograph, which is in contact with it, and then from the pantograph to the traction motor, enabling the operation of the electric locomotive. The contact suspension subsystem generally consists of wire-like components, which should possess sufficient mechanical strength and integrity, as well as strong wear resistance.
3. Support and Positioning equipment.

   The role of the support and positioning equipment is to support and position all components of the contact suspension above the track and maintain them in the required position, transferring all mechanical loads to the pole. The supporting equipment primarily includes arms, soft spans, and hard spans, among other forms. The structure of the supporting equipment should be adaptable to various locations, as lightweight and durable as possible, with sufficient mechanical strength, and convenient for construction and maintenance. The positioning equipment consists of a positioning tube, locators, and their connecting parts. The function of the positioning equipment is to spatially position the contact wire, determining its height and draw-out value. The support and positioning equipment is a key subsystem for maintaining the good status of the catenary, decisively impacting the safety of pantograph-catenary operations and the quality of the current collection.
4. Power Supply Auxiliary Equipment.

   Power supply auxiliary equipment includes sectional insulators, isolators, lightning protection equipment, electromagnetic interference protection equipment, grounding equipment, additional conductors, etc. These components enhance the safety and flexibility of the catenary power supply.

1.3.2 Classification and Operating Characteristics of High-Speed Railway Traction Power Supply Equipment

1.3.2.1 Primary Equipment

In high-speed railway traction power supply systems, the primary equipment located in the traction substation is directly used for transmitting, converting, and distributing electrical energy. This equipment converts the power system's 220 kV or 110 kV three-phase alternating current into the 27.5 kV single-phase alternating current required by the traction power supply system and transmits it to the catenary to power high-speed electric multiple unit. Primary equipment includes traction transformers for energy conversion, current transformers, circuit breakers, isolators for connecting or disconnecting electrical power, lightning arresters for suppressing overvoltage, and feeder cables and busbars that carry current.

1. Traction Transformer

   The traction transformer is a pivotal equipment within a traction substation, primarily responsible for stepping down the voltage from high to low. It converts the 220 kV or 110 kV high voltage supplied by the power system into a voltage suitable for the operation of the electric multiple unit, typically 27.5 kV. Most traction transformers used in high-speed railways are oil-immersed transformers, which include components such as the core, coils, oil tank, bushings, oil pillow, explosion vent pipe, oil purifier, radiator, breather, thermometer, and gas relay.

   Compared to power transformers, traction transformers exhibit unique characteristics due to their interaction with multiple physical fields such as electric, magnetic, temperature, and stress fields, as well as environmental factors that influence their service performance. They frequently endure shocks from traction loads and catenary short-circuit currents, which highlight their specificity. These specific characteristics include: (1) Load Imbalance: The imbalance in the load between the two power supply arms of the traction network can lead to significant differences in temperature rise and aging rates across different windings. (2) Load Shock: Situations like sudden load changes, frequent cyclical variations, and demands exceeding load capacity can impose immense electromagnetic stress, potentially damaging the windings, core, and insulation. (3) Load Non-Linearity: Train inverters generate a substantial amount of harmonics, causing additional losses in the core and windings of the traction transformer, thereby exacerbating the aging of the core and windings. Overall, in high-speed railways with high power, high speeds, and high traffic density, the operating environment for traction transformers is more severe than in conventional railways.

2. Current Transformer

   The current transformer is a crucial equipment in the traction power supply system, linking the primary and secondary systems. It converts high voltage and large currents from the primary system into lower voltages and smaller currents that are easier to manage and measure, which are then supplied to measuring instruments, protection equipment, and automatic control equipment. The accurate and reliable operation of current transformers is vital for the monitoring, measurement, protection, and control of the traction power supply system, significantly impacting the safety and availability of its operations.

   Currently, traction substations commonly use electromagnetic induction transformers, which are susceptible to various issues such as insulation moisture absorption, partial discharge, core steady-state saturation, drift deviation, and secondary circuit anomalies due to aging, improper configuration, or environmental factors. These complications can lead to secondary side output waveform distortion and deviations in measurement values, posing hidden dangers to the safe operation of the traction power supply system. The abnormalities associated with current transformers accumulate and change over time. However, due to the multitude of influencing factors and the uncertainty of their evolution, these might not immediately impact the normal operation of the system. Nonetheless, under trigger events or stress conditions, they can potentially provoke protection misoperations or other issues, leading to fault escalation.

3. Circuit Breaker

A circuit breaker, also known as a high-voltage switch, is a crucial equipment in a traction substation. It is designed to open and close high-voltage circuits under both no-load and load conditions. Moreover, in the event of a system fault, it collaborates with protection equipment and automatic systems to rapidly isolate the fault current. This action helps to minimize the extent of power outages and prevent the escalation of accidents.

High-voltage circuit breakers can be categorized based on the type of arc extinguishing medium used. These include compressed air circuit breakers, oil circuit breakers, vacuum circuit breakers, and SF_6 circuit breakers. Among these, SF_6 circuit breakers have gained widespread application in high-speed railway traction substations due to the excellent insulation performance and arc extinguishing capabilities of SF_6 gas. Despite the superior performance of SF_6 circuit breakers compared to other types, faults do occur due to complex operating conditions, frequent operations, and factors related to manufacturing and installation quality, which directly impact system safety and power supply availability. Mechanical failures and SF_6 gas leaks account for a significant proportion of all faults. Currently, high-speed railway traction power supply systems often adopt a strategy of performing preventive testing during scheduled maintenance to understand the operating characteristics of circuit breakers. This approach not only consumes considerable manpower and financial resources, increasing the lifecycle costs of the equipment, but frequent operations and excessive disassembly for maintenance can also reduce the operational availability of the circuit breakers, leading to certain negative impacts.

4. Isolator Switch

The isolator switch is the most widely used and distributed electrical equipment in traction substations and is operated most frequently. Its main function is to segment lines, open and close small load currents, provide clear disconnection points for maintenance operations, and work in conjunction with circuit breakers to facilitate changes in operational modes.

The isolator switches on the incoming side are generally installed outdoors, often facing the challenge of various severe weather conditions. Combined with the multiple operating modes of the traction power supply system and frequent switch operations, isolator switches, which are made of metal components from the frame and transmission parts to bearings and contacts, are prone to corrosion and even fracture. Components such as connecting rods, pins, springs, bolts, nuts, flanges, bases, brackets, and mechanism boxes are generally severely corroded. Corrosion can cause rotation and transmission connections to jam, preventing the switch from opening and closing properly. This severely reduces the mechanical strength of the frame and transmission parts, making them prone to deformation and damage. Additionally, increased operational force may cause damage and breakage to insulators. Corrosion at the contacts of isolator switches can lead to poor contact and, combined with the impact characteristics of the traction load, can cause the conductive parts to heat up and even burn out. Heating and burning of the conductive circuit in isolator switches are relatively common. Generally, switches exceed the specified temperature rise when the actual load reaches 60% of the rated value.

5. Feeder Cable

   The feeder cable serves as a connection between the traction substation and the traction network, responsible for transmitting electrical power. For instance, when a high-speed railway traction substation or sectioning post is located under an overpass, or there are buildings or urban areas between the traction depot and the catenary that make it inconvenient to use overhead bare conductors for power transmission, 27.5 kV high-voltage single-core cross-linked polyethylene (XLPE) feeder cables are often used to replace overhead bare conductors for transmitting electrical energy.

   Compared to typical power system cables, high-speed railway feeder cables are quite unique due to their use in single-phase AC power transmission. When AC flows through a single-core cable, an alternating magnetic field is generated around the cable. This magnetic field intersects with the cable circuit, inducing an EMF on the cable's metal sheath. Improper grounding can lead to sheath circulating currents. The magnitude of the induced EMF is determined by the current value, the arrangement of the cables, and the cable length. The longer the cable and the larger the current, the greater the induced voltage and sheath circulating currents will be. Excessively high sheath currents not only result in energy loss, reducing transmission efficiency and capacity, but in severe cases, they can also endanger personal and equipment safety, affecting the normal operation of the railway system. Due to the current level of manufacturing processes and some issues present in onsite cable laying, cable faults are inevitable. Once a fault occurs, it not only affects the normal operation of the railway system but in severe cases, can cause equipment damage and even endanger personal safety. Since cables are laid underground, locating and repairing faults after they occur can also be resource-intensive.

6. Lightning Arrester

   Lightning-induced trips in the catenary can increase the complexity of managing the entire high-speed railway traction power supply system, necessitating measures to reduce the number of lightning-induced trips and improve power supply availability. Installing line lightning arresters is one of the measures to reduce the number of lightning-induced trips on the line.

   The catenary, situated almost directly above the trains, experiences significantly stronger vibrational disturbances compared to power system overhead lines. To prevent damage to the equipment and components on the catenary from harming the train equipment and passengers inside during high-speed travel, the safety and availability of these equipment and components must be exceptionally high. Therefore, in the design of lightning arresters for catenary use, apart from meeting the traditional requirements for electrical performance and mechanical properties of line lightning arresters, it is crucial to address issues such as the overall strength under high-intensity vibration stress loads, long-term installation stability, and the product's explosion-proof performance.

7. Busbar

   The busbar is a conductor that connects various voltage-level electrical equipment within the traction substation and serves as the connection conductor for transformers and corresponding distribution equipment. It is usually made of rectangular or circular cross-section bare wires or stranded wires. The primary role of the busbar is to collect, distribute, and transmit electrical energy.

   In traction substations, the busbar carries the traction load current. Generally, busbars themselves rarely fault, but the equipment connected to them can easily form fault coupling. For example, flashovers caused by dirty surfaces on busbar insulators and circuit breaker bushings, faults occurring in voltage transformers mounted on busbars, and faults in current transformers between the busbar and the circuit breaker can lead to damages to the support insulators of the circuit breaker or isolator during switching operations. In China, high-speed railway traction power supply systems commonly use switchgear enclosures to house busbars, where the size of the switchgear and the insulation margins are relatively small. Coupled with poor environmental conditions at railway sites, the failure rate of switchgear due to arc short circuits is very high. Any faults can damage the 27.5 kV busbar, directly impacting the operation of the high-speed railway.

1.3.2.2 Secondary Equipment

In the high-speed railway traction power supply system, secondary equipment controls, monitors, and protects the primary equipment directly used in the traction substation for transmitting, converting, and distributing electrical energy. This ensures their normal and safe operation. The secondary equipment mainly includes protection equipment that rapidly responds to abnormal and fault states, measuring and control equipment for executing measurement and control tasks, fault ranging equipment for fault localization, and communication equipment for achieving communication and liaison functions.

1. Protection Equipment

   In the traction power supply system, the protection equipment should selectively and rapidly respond to abnormal and fault states when a fault occurs in the system. It automatically switches circuits, signals, or triggers the circuit breaker to disconnect the fault line or equipment, ensuring the safety of trains, equipment, and passengers. The protection equipment installed in the traction substation is mainly divided into main transformer protection, traction network feeder protection, and other protections.

   In China's high-speed railway traction power supply system, the autotransformer feeding scheme is prevalently utilized. The traction transformers primarily employ several protective measures including differential protection, low voltage start overcurrent protection, overload signal protection, touch shell protection, and gas protection. Moreover, transformers with a grounded neutral point are equipped with zero-sequence overcurrent ground protection. The feeder protection configuration typically includes primary protection and backup protection, featuring three-stage impedance protection, instantaneous overcurrent protection, and incremental current protection at the traction substation; one-stage impedance protection at the incoming direction of the switching station and two-stage impedance protection at the feeder direction; and two-stage impedance protection in both forward and reverse directions at each sub-district station.

Due to the harsh environment surrounding the traction substation, along with quality issues of the protection equipment or misalignments between the action logic settings of the protection equipment and actual operational modes, the protection equipment may experience failures to operate and misoperation faults. Improper manual adjustments by personnel, leading to errors in protection settings, can also cause the protection equipment to fail to operate or misoperate. Moreover, protection equipment may also suffer from equipment failures induced by lightning.

2. Measurement and Control Equipment

The measurement and control equipment is crucial for measuring and collecting various electrical and non-electrical quantities in the traction power supply system, such as voltage, current, active power, reactive power, active energy, reactive energy, protection actions, main transformer temperature, and system frequency. It works in conjunction with the protection equipment to monitor and control the operating conditions of the equipment, forming an integral part of the comprehensive automation system of the traction power supply system.

Currently, the more commonly used measurement and control equipment utilizes a 32-bit CPU hardware platform, supporting advanced measurement and control functions. High-precision data acquisition systems based on DSP technology accurately collect operating parameters of substation equipment, perform protection calculations, and promptly and accurately execute commands issued from the station level. High-precision hardware components on each unit module, reasonable wiring, and high standards of electromagnetic compatibility performance significantly enhance the availability of the hardware system. This makes it suitable for operation in harsh industrial environments, overcoming the inherent flaws of traditional relay protection such as singular protection, poor accuracy, complex wiring, and communication difficulties. At present, the measurement and control equipment in the traction power supply system mainly include transformer protection measurement and control equipment, compensatory protection measurement and control equipment, feeder protection measurement and control equipment, and for AT traction power supply features, AT measurement and control equipment.

3. Fault Ranging Equipment

The primary function of the fault-ranging equipment is to automatically measure the distance to the fault point when a fault occurs in the traction power supply line. This capability is crucial for promptly detecting and eliminating faults, saving on fault patrolling costs, and improving fault clearance efficiency.

Taking the commonly used autotransformer feeding scheme in high-speed railways as an example, the AT fault distance measurement equipment is mainly installed at the traction substation, switching station, sub-district booth, and autotransformer station under the AT power supply mode. The main electrical quantities measured include bus voltage and the current absorbed by the AT neutral point. To ensure the safe operation of electrical equipment, the AT fault distance measurement equipment is equipped with a bus voltage PT wire break detection function and provides sufficient inputs, outputs, and AC circuits to meet the special requirements of certain traction substations. The fault-ranging equipment must possess four characteristics: availability, accuracy, economy, and convenience. Availability ensures that the measuring equipment does not fail to operate or misoperate. Accuracy is the most critical requirement; under the constraints of technology and economics, an absolute ranging error of no more than 300 meters is considered very ideal. Economy refers to the equipment having a high performance-to-cost ratio. Convenience means that in debugging, usage, and maintenance, the equipment should automatically produce ranging results, reducing the workload of personnel.

4. Communication Equipment

   The communication equipment includes two types: general communication equipment and remote communication equipment.

   The general communication equipment primarily handles the exchange of information between various substations and the dispatch center. It is suitable for all types of traction substations, sectioning posts, and switching stations. The general communication equipment utilizes a high-performance communication apparatus developed with a 32-bit microprocessor. It allows the integration of third-party equipment into the comprehensive automation system. The main interfaces available include Ethernet, RS232, RS485, and RS422.

   The remote control management equipment primarily facilitates the remote communication tasks of the comprehensive automation system. It collects telemetry and remote signaling information from the system to be transmitted to the dispatch center and relays remote control and remote adjustment commands issued from the dispatch center. The remote communication unit and the comprehensive automation system typically use an optical network or an Ethernet connection with an electrical interface, allowing flexible adaptation to different remote control channels, such as analog remote control channels, digital remote control channels, and optical interface remote control channels. Communication equipment faults generally include issues due to design reasons or subsequent operational divisions causing single-channel operation at stations; traction channel interruptions caused by backup channels forming a loop with fault channels due to personnel operational errors; and problems with the remote unit due to unexpected power outage accidents or software upgrade faults.

1.3.3 Degradation Process Within the Full Life Cycle of Traction Power Supply Equipment

The high-speed railway power supply system consists of a series of primary and secondary equipment. The genesis and development of faults in equipment or subsystems inevitably undergo a time process. What may appear as random faults are bound by inherent regularities, and even sudden faults have a period of emergence and development, albeit very brief.

Drawing an analogy between the traction power supply system and a biological organism, just as an organism progresses from the onset of a disease to death through a definite time process, during which the disease continuously gestates and develops, so does the health condition of the organism follow a deteriorating trajectory. This is similar to the description in *Bian Que Visiting Duke Huan of Cai* by Han Fei during the Warring States period, where Duke Huan progresses from having "no illness" to ailments "in the loins," "in the flesh," "in the stomach and intestines," "in the bones and marrow," eventually leading to death. In pathological terms, the health condition of an organism goes through a degradation process from "healthy" to "sub-healthy," "mildly ill," "moderately ill," and "severely ill." Similarly, traction power supply equipment undergoes a degradation process in fault genesis and performance, experiencing a development process in health conditions analogous to those of a biological organism.

Taking the current transformer as an example, a newly commissioned current transformer is in a healthy condition and functions effectively, accurately measuring voltage and current at critical locations in the traction power supply system. As it ages, the current transformer may degrade into a sub-healthy or mildly ill status. During this stage, it might develop defects such as oil leakage or moisture ingress, or face issues like unstable screen grounding or secondary circuit wiring errors. These issues can lead to deviations in the current transformer's measurements. Although it still retains the function of measuring electrical quantities, the performance of this function declines (measurement error). It is necessary to provide early warning and identification for current transformers with defects or potential issues to prevent such measurement deviations from causing misoperations in protection systems. If left unaddressed, the performance of the current transformer will continue to deteriorate, progressing into a moderately or severely ill status. At this stage, the defects or potential issues will manifest certain functional consequences, such as oil leaks causing defects in insulation that lead to the destruction of the current transformer's insulation, or unstable screen grounding leading to decreased insulation performance. These functional consequences mean that the current transformer can no longer meet the requirements for measuring electrical quantities. If the fault continues to develop, it will eventually lead to an electrical breakdown explosion of the current transformer, posing a severe threat to the traction power supply system. This necessitates a power outage for fault handling across the entire facility, leading to significant operational losses.

Taking the feeder cable as an example, a newly laid cable line that has passed acceptance testing generally operates well, being in a healthy condition and meeting all requirements for powering the traction network. However, as the service life increases, the electrical and physical performance of the cable will degrade, leading to aging and a gradual shift into a sub-healthy or mildly ill status. At this point, the cable material may have developed defects such as water treeing, insulation aging, or external insulation corrosion, or it may encounter issues such as exceeding the grounding grid resistance standards, poor cable grounding, or single-end grounding with the other end left unconnected. Despite these issues, the cable can still continue to perform its power transmission function, although its overall performance has declined, necessitating the detection and early warning of potential hidden faults in the cable line. If these issues are allowed to progress, the performance of the cable line will continue to deteriorate, entering a moderately or severely ill status. At this stage, the defects or potential issues in the cable will manifest certain functional consequences, such as local overheating caused by the progression of water treeing or discharges to the ground or other metals caused by external insulation corrosion. These functional consequences mean that the cable can no longer fulfill its normal power transmission function. Continued deterioration could lead to severe accidents such as cable insulation breakdown, cable fire, or joint explosion, causing power outages in the traction network and line shutdowns, severely impacting the safe and stable operation of the traction power supply system.

Finally, considering the traction transformer, a newly commissioned traction transformer is in a healthy condition with excellent performance, meeting all operational requirements. As operating time progresses, the various components of the traction transformer gradually age, and performance metrics decline, transitioning into a sub-healthy or mildly ill status. At this stage, the traction transformer may develop defects such as moisture in the insulation oil, a drop in oil level, or long-term heating of the insulating oil. Additionally, issues such as loosening of core fasteners or insufficient spring pressure in the tap changer may arise. These defects or potential issues are not severe enough to prevent the transformer from operating, but the overall performance of the transformer has decreased, necessitating inspections and early warnings for traction transformers with defects or potential issues. If these conditions are left unaddressed, the performance of the transformer will continue to deteriorate, progressing into a moderately or severely ill status. At this stage, the defects or potential issues will manifest certain functional consequences, such as coil inter-turn short circuits, phase-to-phase short circuits, or tap changer contact discharges. These functional consequences will impact the normal operation of the transformer. If the fault continues to develop, it could lead to severe faults such as transformer burnout, heavy gas action, or tap changer burnout, causing significant impact and losses to the traction power supply system.

From the examples provided, we can summarize the evolution process from safety to failure of traction power supply equipment, as illustrated in Fig. 1.7. The "defects" mentioned in the examples can be referred to as potential faults, and the "hidden dangers" as concealed faults. By mastering the fault development laws and failure mechanisms of the high-speed railway power supply system or its components, it is possible to predict potential faults and identify concealed faults. Once the

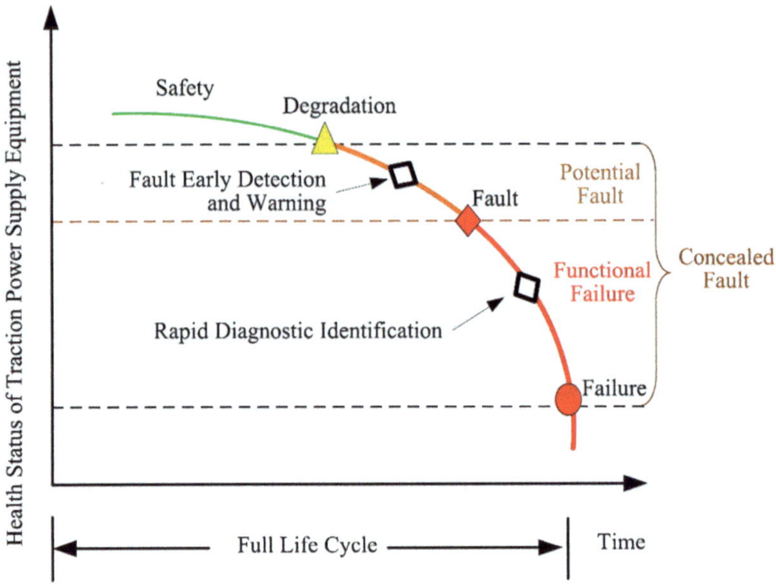

Fig. 1.7 Schematic Diagram of the Operational Status Process of Traction Power Supply Equipment

probability of occurrence of potential faults has been predicted, and the concealed faults in the subsystems and equipment of the high-speed railway power supply system have been identified, it is necessary to assess the impact of these latent faults on the overall safety and availability of the system, i.e., to conduct a health condition assessment (health monitoring and diagnosis) of the high-speed railway power supply system. This enables the timely implementation of early warning measures to reduce the probability of power outage accidents.

The service status evolution process of traction power supply equipment, as depicted in Fig. 1.7, shows that equipment faults generally do not occur abruptly. Initially, when equipment is newly commissioned, it is in a healthy condition, operating safely to fully realize its expected functions. As the service life of the equipment extends, its performance gradually declines, entering a sub-healthy or mildly ill status. At this stage, the equipment may harbor potential faults or concealed faults, which are often difficult to detect, necessitating fault early warning to facilitate timely detection. If these latent faults are not identified in time, the equipment's performance will further degrade, advancing to a moderately or severely ill status. At this point, potential faults transform into functional failures, and concealed faults begin to manifest functional consequences, meaning the equipment can no longer perform its expected functions. It becomes essential to conduct rapid fault diagnosis to restore normal function through measures such as replacing faulty components and performing remedial maintenance. If still neglected, the equipment will completely fail or even cease operation altogether, severely impacting the system's function and potentially causing a total system shutdown.

1.3.4 PHM Technology Framework for High-Speed Railway Traction Power Supply System

The health diagnosis technology of the high-speed railway traction power supply system utilizes advanced sensor data acquisition to collect feature parameters related to system attributes. These feature parameters are then correlated with useful information using intelligent algorithms and models for detection, analysis, evaluation, and prognostics. By analyzing the historical data of equipment operation, the trends in equipment status operation and transition are identified. Combining current structural characteristics, parameters, and environmental conditions, the current health condition of the equipment is accurately assessed, and reasonable prognostics are made for future operational trends. This process allows for early detection and forecasting of potential faults, providing a reliable basis for discovering faults early and preventing fault-related losses. Risk assessment quantifies the potential impact and loss that possible risk events in the traction power supply system might cause before they occur, using the results of the assessment to carry out targeted maintenance on the traction power supply system. This effectively reduces the operational risks of the traction power supply system, thereby avoiding power supply faults.

Based on the above objectives, a high-speed railway traction power supply system PHM technological framework is developed as illustrated in Fig. 1.8. This technological framework is detailed further to form the main research content of each chapter in this book.

Below is a summary of the contents of each section under this theoretical framework.

1. Research on PHM

   In the short time scales (real-time operation) and medium time scales (limited failure and renewal cycles of equipment), random process theory is applied to model the randomness of potential and concealed fault occurrence and development, as well as the dynamics of equipment health condition degradation. The research focuses on the probability distribution descriptions of potential and concealed faults, and real-time assessment methods of equipment health conditions, aiming for timely prognostics and early warning before a functional failure occurs or health condition degradation exceeds a threshold, and predicting the remaining useful life of equipment within a certain confidence interval. After a functional failure occurs or the health status deteriorates beyond the threshold, rapid diagnosis of faults is conducted using case-based reasoning. On a spatial scale, the equipment-level analysis of a single component and single fault mode is expanded to a system-level analysis of multiple components and multiple fault modes. For the traction power supply system's series and parallel connections, redundancy, cold/hot standby configurations, and multiple fault modes such as degradation fault, shock fault, and concurrent multiple faults, a PHM approach that combines data-driven and physics-of-failure model-based methods is used for integrated modeling to perform comprehensive assessment of the system's health condition and achieve system-level prognostics.

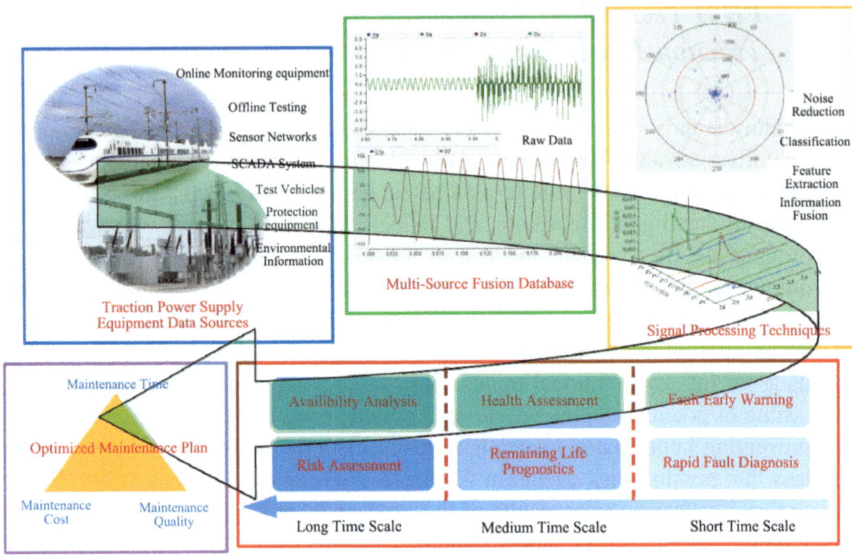

Fig. 1.8 High-Speed Railway Traction Power Supply System PHM Theoretical Framework

2. Full Life Cycle Availability Analysis and Risk Assessment Research

   In the long-time dimension (covering the full life cycle of the traction power supply system), the availability and availability of the system are analyzed from the perspective of random processes. Utilizing the bathtub curve of equipment failure rates, a time series model of system availability indicators is developed to predict future trends of this series, accurately tracking intervals where security risks increase and arranging preventive measures during the turning point to suppress the rise in failure rates. On the spatial scale, several factors are considered: on one hand, the impact of severe weather and natural disasters on the system fault rate and availability is assessed to determine the magnitude of risk; on the other hand, the probability distribution forms of system failure rates and availability for the traction power supply system's chain (serial) and backup (parallel) structures are examined. Traditionally, failure rate models have time as the sole independent variable. This research incorporates the influences of health condition (H) and service environment (E) into the model, creating a comprehensive failure rate model that quantitatively reflects internal and external risks of the system (from H, t, E). Additionally, scale variables are introduced to establish a conditional availability model that analyzes the full life cycle availability of the system.

3. Maintenance Decision and Optimization Research

   Building on the foundation of the previous two areas of research, this segment considers the various dynamics and randomness in maintenance activities. It examines the multi-scale effects of maintenance actions over time—such as the urgency of emergency repairs, the real-time nature of state-based maintenance,

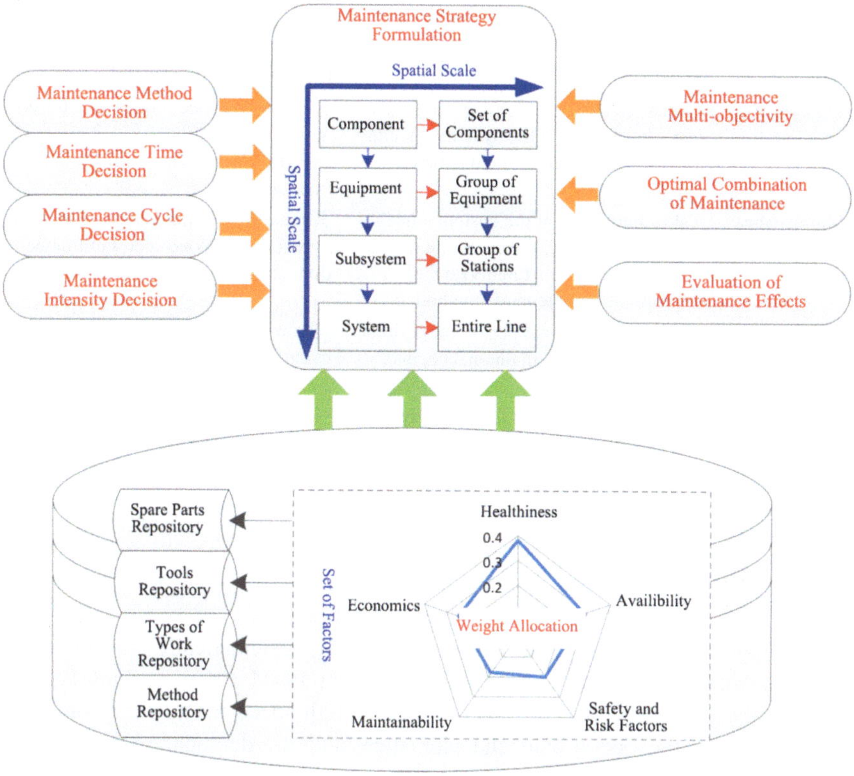

Fig. 1.9 Maintenance Decision and Optimization Research

and the periodicity of scheduled maintenance—and across spatial dimensions—such as component-level, equipment-level, subsystem-level, and system-level maintenance, etc. A multi-factor, multi-scale spatio-temporal joint decision model is established, focusing on decision-making issues related to maintenance time, maintenance cycle, maintenance intensity, and maintenance methods. This involves the application of sensitivity analysis methods, non-linear mathematical programming, and multi-objective intelligent optimization algorithms to study the optimization and optimal combination of maintenance decision variables. This research analyzes the various attributes displayed by the traction power supply system during maintenance activities, assigns appropriate weights to these attributes, forms a set of influencing factors for maintenance, and combines these with the spare parts, tools, and types of work needed for maintenance activities to establish a maintenance repository. Ultimately, this approach provides a comprehensive, customized maintenance and repair solution for each uniquely structured and performing traction power supply system. The proactive maintenance research content is shown in Fig. 1.9 and Table 1.1.

Table 1.1 Four types of maintenance decisions and their meanings

Decision types	Meanings	Example set of decisions
Maintenance time decision	Considers the urgency of maintenance tasks in relation to the system's health condition to schedule maintenance delays	{No maintenance required, deferred maintenance, priority maintenance, expedite maintenance, immediate maintenance}
Maintenance method decision	Takes into account the system's healthiness, availability, safety, maintainability, and economics to choose the appropriate maintenance method	{Breakdown maintenance, scheduled maintenance, concealed fault detection, state-based maintenance}
Maintenance cycle decision	Chooses the optimal period for periodic inspection, repair, maintenance, and care activities	{Short cycle, medium cycle, long cycle}
Maintenance intensity decision	Quantitatively selects the intensity and effectiveness of maintenance activities based on the equipment's service status and performance degradation	{No maintenance, minimal maintenance, partial maintenance, complete maintenance}

1.4 Summary

This chapter begins with the development of PHM, systematically introducing the developmental history and current applications of PHM across various fields. It elaborates on the structure and function of the high-speed railway traction power supply system, the operational characteristics, and the degradation processes of power supply equipment. Based on the high-speed railway traction power supply system PHM technology framework introduced in this section, subsequent chapters will provide detailed descriptions of the research content within each part of this technological framework.

Chapter 2
Fault Diagnosis of High-Speed Railway Traction Power Supply System

The high-speed railway traction power supply system operates in a harsh working environment with a complex system structure, making it susceptible to faults. The increasing burden on frontline maintenance staff necessitates prompt fault detection and analysis by field personnel when a fault occurs, enabling them to implement effective measures to restore system operation. Advances in information technology have equipped the traction power supply system with extensive monitoring and protective measures, enhancing the system's informatization and automation, and thus increasing its complexity. A fault in the traction power supply system generates a vast amount of fault information. Personnel often struggle to handle this overwhelming data, leading to delays in fault resolution. Therefore, it is necessary to adopt effective intelligent fault diagnosis measures to assist personnel in identifying the source of faults, managing accidents rapidly, restoring power or reducing the power outage area, and minimizing the impact of faults.

This chapter provides an overview of fault diagnosis for high-speed railway traction power supply equipment and systems and specifically explores case studies of fault diagnosis in the high-speed railway traction power supply system and its equipment.

2.1 Overview of Traction Power Supply Equipment Failure Diagnosis in High-Speed Railway

Fault diagnosis involves a series of detection, testing, and data analysis processes to check if there is a fault in the system or equipment, and if so, to further determine the approximate location of the fault. The main tasks include fault detection, fault type determination, fault localization, and fault recovery. The development of fault diagnosis technology has spanned nearly three decades. Initially, it was in its primitive stage where experts in the field used sensory information to determine

Z. He et al., *Prognostics and Health Management of High-Speed Railway Traction Power Substations*, https://doi.org/10.1007/978-981-95-4538-4_2

the status of equipment and based their fault judgments on this information. This was followed by a phase based on material life analysis, where, with the development of availability theory, fault diagnosis was performed through analysis and assessment of material life and testing of material performance. The subsequent diagnostic phase involved sensor and computer technology, acquiring more accurate data through the advancement of sensors and using computers to process data for fault diagnosis. The final stage is the intelligent diagnostic phase, wherein artificial intelligence techniques and signal processing algorithms are integrated to perform fault diagnosis, facilitating the diagnosis through knowledge-based inference. From a methodological perspective, fault diagnosis can be divided into three main categories: diagnostic methods based on mathematical analytical models, diagnosis methods based on machine learning models, and diagnosis methods based on experimental testing. Timely identification of the type and location of faults through these diagnostic methods allows for prompt maintenance of equipment, thereby minimizing the impact on the traction power supply system. The traction power supply system, subjected to a harsh working environment and characterized by a complex structure, is prone to experiencing faults across various components. Rapid fault location can be achieved by analyzing circuit breaker trip information and relay protection equipment data. In the fault diagnosis of the traction power supply system, machine learning methods are widely applied to identify faults.

2.1.1 Traction Transformer Fault Diagnosis Methods

Many severe faults in traction transformers originate from the evolution of minor faults. In the early stages of a fault's latency, the transformer shows signs of insulation performance degradation and winding deformation. The main factors affecting insulation performance during the operation of the equipment include discharge and overheating. The decomposition of insulating oil and the aging of insulating materials in transformers, driven by electrical and thermal stress, can produce small amounts of low molecular weight hydrocarbons, H_2, CO_2, and CO. The volumetric proportions of these gases and the ratios among their various components can, to a significant degree, reflect the nature of transformer faults, providing a basis for the application of oil chromatography diagnosis.

2.1.1.1 Fault Diagnosis Method Based on the Three Ratio Method

Similar to common power transformers, traction transformers' faults can be categorized into discharge faults and overheating faults, each affecting the transformer oil gas content differently. For traction transformers, dissolved gases in the oil serve as indicators of faults. Discharge faults primarily affect gases such as hydrogen (H_2) and acetylene (C_2H_2), with secondary impacts on methane (CH_4), ethane (C_2H_6), and ethylene (C_2H_4). Overheating faults primarily affect CH_4 and C_2H_4, with secondary impacts on H_2. Analyzing the principles of dissolved gases in oil and their content features is crucial for diagnosing faults in traction transformers. When an oil-immersed transformer experiences a fault, the gas generation rates for different faults vary among the low molecular weight hydrocarbons, leading to potential inaccuracies if diagnosis relies solely on the content of dissolved gases in the oil. Diagnostic accuracy can be improved by coding based on the relative content of gases according to their mutual ratios. By analyzing the ratios of CH_4, C_2H_6, C_2H_4, C_2H_2, and H_2 produced during transformer faults, we can encode the three ratios of C_2H_2/C_2H_4, CH_4/H_2, and C_2H_4/C_2H_6 according to their respective ranges. These encoded ratios are then used to diagnose the corresponding faults.

Gas Generation Principle

According to the Chinese electric power industry standard DL/722-2014, *Guide to Analysis and the Diagnosis of Gases Dissolved in Transformer Oil*, the dissolved gases in the oil of traction transformers mainly originate from two sources: the decomposition of transformer oil and the decomposition of solid insulation materials. Transformer oil primarily consists of compounds such as alkanes, cycloalkanes (saturated hydrocarbons), and aromatic (unsaturated hydrocarbons). When a fault occurs in a transformer, due to temperature changes at the fault site and the impact of internal discharges, the carbon-carbon (C-C) and carbon-hydrogen (C-H) bonds in the insulating oil and solid materials break. CH_3, CH_2, and CH chemical groups rapidly synthesize H_2 and low molecular weight hydrocarbon gases through complex chemical reactions. CH_4 and C_2H_4 constitute about 80% of the total hydrocarbons, with CH_4 predominating at lower temperatures; when the temperature exceeds 500 °C, H_2 and C_2H_4 increase sharply, though H_2 generally does not exceed 30% while C_2H_4 surpasses CH_4. C_2H_2 typically does not form during low-temperature overheating but becomes significantly more prevalent in transformer oil when the temperature exceeds 800 °C. In overheating faults, the content of C_2H_6 is typically not high. Regarding solid insulation materials, which are primarily composed of fibrous insulation materials like paper, wood blocks, and laminated boards, cellulose mainly consists of a long-chain high polymeric hydrocarbon ($C_6H_{10}O_5$) composed of many glucose monomers. The thermal stability of C-O bonds in cellulose is weaker than that of C-H bonds, leading to polymer breakdown at temperatures above 105 °C, and complete degradation and carbonization at temperatures above 300 °C. Polymer decomposition produces a large amount of CO and CO_2, a small amount of low molecular weight hydrocarbon gases, and furfural and its derivatives. The gas content produced varies depending on the fault state. When solid insulation materials begin to thermally decompose, CO_2 is primarily produced. As the temperature increases, the content of CO increases, gradually surpassing that of CO_2. Water in the oil can react with iron to produce H_2; at higher temperatures, dissolved oxygen in the oil can generate large amounts of H_2 under the catalysis of certain paints on stainless steel, or through the catalytic reaction of stainless steel with oil to produce H_2.

Three-Ratio Method for Fault Diagnosis

In transformers, the primary and secondary gas components produced vary with different fault types, providing a basis for the three-ratio method of transformer fault diagnosis. Table 2.1 illustrates the gas components produced under different fault types in oil-immersed transformers.

The three-ratio method leverages the interdependent relationship between the relative content of low molecular weight hydrocarbons (CH_4, C_2H_6, C_2H_4, C_2H_2) and H_2 produced by the decomposition of insulating oil under fault conditions and varying temperatures. From the five characteristic gases, two gases with similar solubility and diffusion coefficients are selected to form three pairs of ratios: C_2H_2/C_2H_4, CH_4/H_2, and C_2H_4/C_2H_6. These gas ratios are then encoded within specific ranges to perform fault identification. Table 2.2 and 2.3 provide the encoding rules and the fault type determination method, respectively.

Three-Ratio Diagnostic Process

The fault diagnosis using the three-ratio method involves calculating the ratios of C_2H_2/C_2H_4, CH_4/H_2, and C_2H_4/C_2H_6 from CH_4, C_2H_6, C_2H_4, C_2H_2, and H_2 field gas data. Using the ratio values listed in Table 2.2, each ratio's corresponding ID number is determined. Once the ID number is determined, fault types and possible fault locations are judged based on Table 2.3. The diagnostic flowchart is illustrated in Fig. 2.1.

A certain 110 kV substation's #3 main transformer in Shenzhen, manufactured by a transformer factory with the model SFZ10-50000/110, was produced in March 1994 and put into operation in December of the same year. It had been running

Table 2.1 Gas components produced by different fault types

Fault type	Primary gas components	Secondary gas components
Oil overheating	CH_4, C_2H_2	H_2, C_2H_6
Oil and paper overheating	CH_4, C_2H_4, CO, CO_2	H_2, C_2H_6
Partial discharge in oil-paper insulation	H_2, CH_4, CO	C_2H_2, C_2H_6, CO_2
Spark discharge in oil	H_2, C_2H_2	
Arc discharge in oil	H_2, C_2H_2	CH_4, C_2H_4, C_2H_6
Corona discharge in oil and paper	H_2, C_2H_2, CO, CO_2	CH_4, C_2H_4, C_2H_6

Table 2.2 Encoding rules

	Ratio range encoding		
Gas ratio range	C_2H_2/C_2H_4	CH_4/H_2	C_2H_4/C_2H_6
<0.1	0	1	0
≥ 0.1 to <1	1	0	0
≥ 1 to <3	1	2	1
≥ 3	2	2	2

Table 2.3 Fault type determination method

Encoding combination			Fault type judgment	Fault instance
C_2H_2/C_2H_4	CH_4/H_2	C_2H_4/C_2H_6		
0	0	1	Low-temperature overheating (below 150 °C)	Insulation wire overheating
	2	0	Low-temperature overheating (150–300 °C)	Poor contact in tap changers, loosening of line clamp screws, defective joint welding, copper overheating caused by eddy currents, core magnetic flux leakage, local short circuit, poor interlayer insulation, multiple grounding points in the core, etc.
	2	1	Medium-temperature overheating (300–700 °C)	
	0, 1, 2	2	High-temperature overheating (above 700 °C)	
	1	0	Partial discharge	Partial discharges with low energy density in the oil triggered by elevated temperatures and dissolved gases
2	0, 1	0, 1, 2	Low-energy discharge	Continuous spark discharge between lead wires and components without fixed potentials, decomposition of tap lead wires and oil gap flashover, spark discharge in oil between different potentials or between suspended potentials
	2	0, 1, 2	Low-energy combined with overheating discharge	
1	0, 1	0, 1, 2	Arc discharge	Turn-to-turn, interlayer short circuit, phase-to-phase flashover, flashover in oil gaps between tap lead wires, discharges causing casing discharge, coil fusing, tap changer arcing, arc caused by loop currents, discharges from lead wires to other grounded bodies, etc.
	2	0, 1, 2	Arc discharge with overheating	

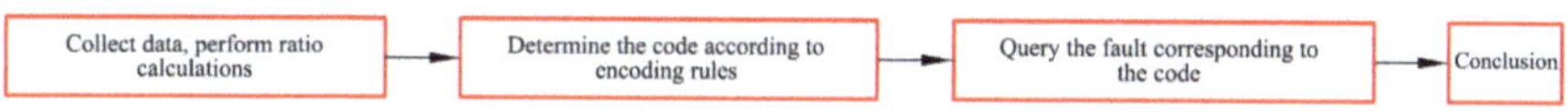

Fig. 2.1 Three-Ratio Diagnostic Process

continuously for about 12 years until March 2006, during which the transformer operated normally. On March 12, 2007, a chemical test was performed on the transformer oil. After multiple sampling experiments to minimize errors, the gas data was as shown in Table 2.4.

Using the three-ratio method, the chromatographic test data from March 12, 2007, were used to calculate the ratios of C_2H_2/C_2H_4, CH_4/H_2, and C_2H_4/C_2H_6. Based on Table 2.2, the three-ratio coding combination at that time was determined to be: 0, 2, 2, as shown in Table 2.5. Consulting the fault type determination method table (Table 2.3), it was preliminarily judged that a high-temperature

Table 2.4 #3 main transformer oil chromatographic test data (μL/L)

Experiment date	H_2	CH_4	C_2H_6	C_2H_4	C_2H_2	CO	CO_2	Total hydrocarbons	Conclusion
2007-03-13	84	562.0	136.2	615.2	Not detected	994	7222	1313.4	Total hydrocarbons exceed the standard
2007-03-13	85	579.6	137.9	632.6	Not detected	1041	7468	1350.1	Total hydrocarbons exceed the standard
2007-03-12	79	552.2	133.7	598.6	Not detected	879	7190	1284.5	Total hydrocarbons exceed the standard
2007-03-12	82	564	136.3	608.7	Not detected	893	7303	1309	Total hydrocarbons exceed the standard
2006-03-14	5	43	10.6	23.1	Not detected	1012	6687	76.7	Qualified
Caution value	150				5			150	

Table 2.5 #3 main transformer fault gas ratios and coding

Ratio item	Ratio value	Coding value
C_2H_2/C_2H_4	0	0
CH_4/H_2	6.9	2
C_2H_4/C_2H_6	4.5	2

Table 2.6 Modified three-ratio method coding rules

	Three-ratio coding		
Gas ratio range	C_2H_2/C_2H_4	CH_4/H_2	C_2H_4/C_2H_6
<0.1	0	1	0
[0.1, 1)	1	0	0
[1, 3)	1	2	1
≥3	2	2	2

Table 2.7 Fault type judgment criteria

Coding combination			
C_2H_2/C_2H_4	CH_4/H_2	C_2H_4/C_2H_6	Fault type judgment
0	0	0	Low-temperature overheating (<150 °C)
	2	0	Low-temperature overheating (150–300 °C)
	2	1	Medium-temperature overheating (300–700 °C)
	0, 1, 2	2	High-temperature overheating (>700 °C)
	1	0	Partial discharge
2	0, 1	0, 1, 2	Low-energy discharge
	2	0, 1, 2	Low-energy discharge with overheating
1	0, 1	0, 1, 2	High-energy discharge
	2	0, 1, 2	High-energy discharge with overheating

overheating fault occurred inside the #3 main transformer, with a hotspot temperature possibly above 700 °C.

A field inspection concluded that the transformer experienced a high-temperature overheating fault at that time.

Modified Three-Ratio Method

China's current standard DL/T722-2014 and the International Electrotechnical Commission (IEC) both recommend the modified three-ratio method, which is derived from the four-ratio method. The C_2H_6/CH_4 ratio in the four-ratio method was not effectively reflecting the temperature range of thermal decomposition, hence the IEC removed it. The Chinese electric power industry standard DL/722-2014, *Guide to Analysis and the Diagnosis of Gases Dissolved in Transformer Oil*, refines the coding combinations and fault types based on domestic practical experience. Tables 2.6 and 2.7 respectively show the coding rules and fault type judgment criteria for the modified three-ratio method.

Limitations of the Modified Three-Ratio Method:

1. The three-ratio method is only effective in confirming the presence of a fault in transformers; it does not provide meaningful information for transformers with normal gas content and no trend of increase.
2. Fault overlap and normal aging can influence the judgments made by the three-ratio method.
3. The coding for the three-ratio method is derived from statistical analysis of typical faults, yet actual cases often present codings not listed in Table 2.6.
4. The classification of fault types has overlapping ambiguities, such as high-energy discharge faults and high-energy discharge with overheating faults, both characterized by high-energy discharge, which can often lead to misjudgments.

Three-Ratio Method Summary

Dissolved gases in oil are used for fault determination based on the content of gases dissolved in transformer oil. Under thermal or electrical stress, the internal insulating oil and solid insulation materials in transformers undergo decomposition, producing low molecular weight hydrocarbon gases, H_2, CO_2, and CO. These gases can dissolve in the oil, and the presence of gas concentrations in the oil allows for the determination of the traction transformer's fault type and location based on the concentrations or ratios of these gases.

The three-ratio method is currently widely applied in the fault diagnosis of traction transformers. It uses the relationships among three specific ratios-CH_4/H_2, C_2H_4/C_2H_6, and C_2H_2/C_2H_4-to determine whether the traction transformer is functioning normally or experiencing discharge faults and overheating faults. The three-ratio method has a low computational load and a simple judgment process, avoids the volumetric effects of oil, and provides relatively accurate diagnostic results, making it extensively applied in engineering.

2.1.1.2 Fault Diagnostic Methods Based on Expert Systems

Currently, large transformers still utilize the three-ratio method for fault analysis and determination. However, due to certain inherent limitations of the three-ratio method and potential overfitting issues when using artificial intelligence methods for fault diagnosis coding, this section introduces a transformer fault diagnosis expert system that combines heuristic methods and field expert experience based on an improved version of the IEC three-ratio method.

Principles of Expert Systems

Expert systems integrate years of diagnostic experience, inference logic, and technical knowledge from several experts within a particular field into a computer to solve specific problems. An expert system consists of a knowledge base, a database, an inference engine, and an interface. The knowledge base is formulated from expert experience as rules that establish relationships between input data and output results, serving as the basis for decision-making. This knowledge base can be continually refined to make the diagnostic results of the expert systems more accurate. Experts infer whether a fault has occurred based on various experimental data stored in the database, including gas chromatographic analysis of oil, coil insulation resistance leakage current, dielectric loss, electrical characteristics of oil, and trace moisture content in the oil. The inference engine connects the database with the knowledge base, retrieving necessary data including the latest and historical data. It then performs inference analysis using the experience and professional knowledge stored in the knowledge base, processing the data in a certain way to produce inference results.

Transformer Fault Diagnosis Expert System

The knowledge base of the transformer fault diagnosis expert system is composed of several parts: a section that stores expert diagnostic knowledge; a section for storing historical archives and raw data; a section that stores equipment performance indicators and health condition references; and a fault set section. The fault set includes a range of common and typical faults associated with transformers, such as core multipoint grounding faults, interturn short circuits, phase-to-phase short circuits, issues arising from moisture ingress, lead faults, and overheating caused by magnetic flux leakage. The overall structure of the expert system is illustrated in Fig. 2.2.

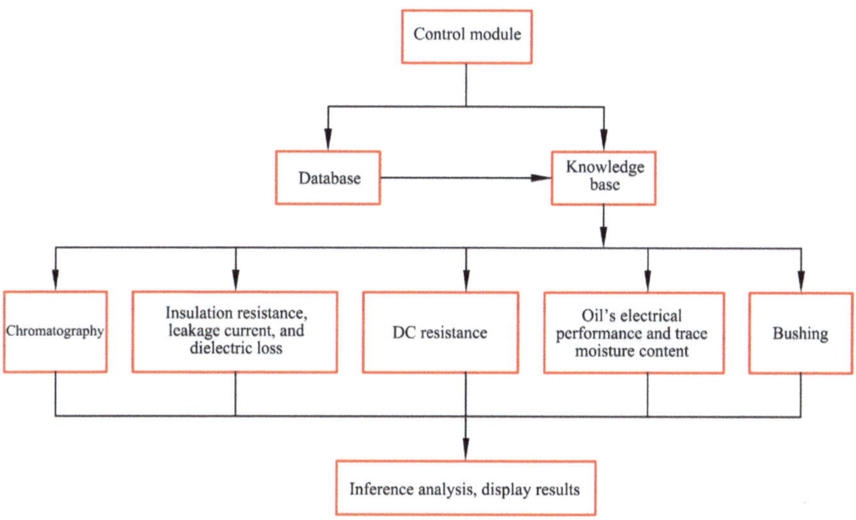

Fig. 2.2 Overall Structure of Expert Systems

Oil chromatographic gas data, with its ease of extraction and analysis and relatively accurate diagnostic capabilities, is widely applied in current transformer fault diagnostics. By integrating the three-ratio method with expert systems, a dynamic rule system for transformer fault diagnosis has been established, combining a vast array of modern heuristic methods and field expert experience. The combined expert system diagnostic process is illustrated in Fig. 2.3, where A, B, and C represent the ratio codings of CH_4/H_2, C_2H_4/C_2H_6, and C_2H_2/C_2H_4, respectively, from the three-ratio method. Expert systems overcome the inherent ambiguity in fault classification associated with the three-ratio method, which directly links the fault types to coding types. Through a meticulous step-by-step classification process, these systems enhance the precision of fault categorization and refine the accuracy of diagnostic outcomes.

The expert system diagnostic process is intuitive, and the inference process is visualized, providing a reasonable explanation capability for its results. Furthermore, it allows for modification of internal rules to update its structure, ensuring timeliness. In transformer fault diagnosis, upon detecting a previously encountered fault type, it is incorporated into the existing rule set; conversely, if it does not conform to the existing rules of the expert system, it is integrated into the knowledge base. By continually removing outdated or undiagnosable rules and adding new ones, expert systems are gradually perfected, leading to more accurate diagnostic results.

Expert Systems Diagnostic Case Study

To validate the effectiveness of this system, a typical case study is analyzed. The actual fault refers to the fault type finally determined at the site; the three ratios are CH_4/H_2, C_2H_4/C_2H_6, and C_2H_2/C_2H_4, and the results for these ratios are obtained according to the encoding rules shown in Table 2.2. The result from the three-ratio method indicates whether the traditional diagnosis was accurate; the expert system's judgment result is based on the accuracy of the judgment proposed by the expert system.

From the judgment results of 20 samples in Table 2.8, the traditional three-ratio method had a diagnostic accuracy rate of 65%, while the proposed expert system diagnostic framework achieved an accuracy rate of 85%, which is a significant improvement over the traditional method. Among these, 17 groups received judgment results completely consistent with the actual faults; the 19th and 20th samples had slight deviations in diagnostic results from the actual faults due to the ambiguity of coding boundaries; only the 12th sample failed to diagnose high-energy discharge.

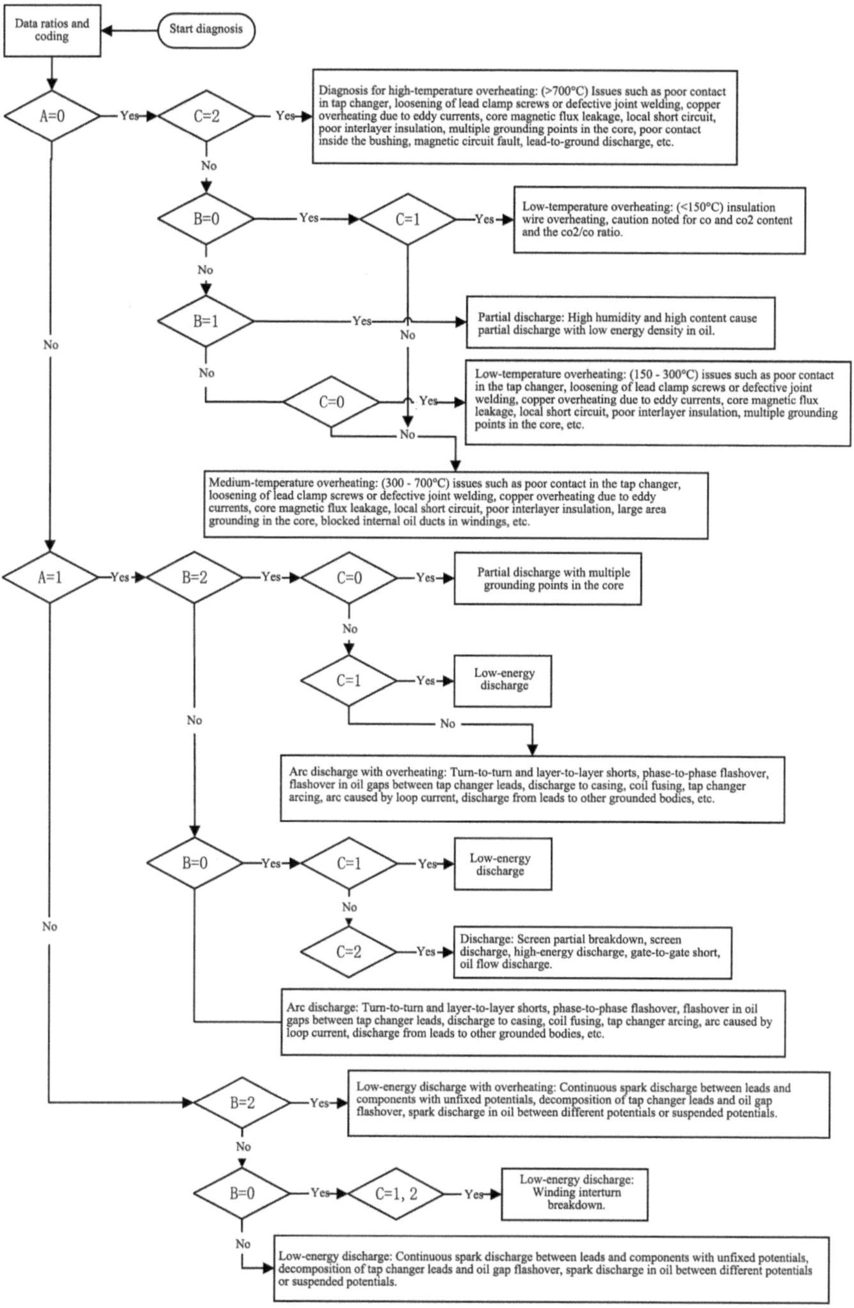

Fig. 2.3 Expert System Diagnostic Process for Transformer Faults

Table 2.8 Fault data samples and three-ratio coding

Index	H_2	CH_4	C_2H_4	C_2H_6	C_2H_2	Actual fault	Three-ratio coding	Expert systems method judgment result
1	29	26.3	27	1.8	82.4	Winding interturn breakdown	202	T
2	65.8	21.26	7.85	4.44	24.5	Low-energy discharge	201	T
3	42	97	600	157	0	Poor contact inside the bushing	022	T
4	4.1	15.8	42.6	19.1	0.53	Blocked oil duct between turns	021	T
5	100.4	150.8	129.68	110.3	1.02	Core magnetic flux leakage	021	T
6	115.9	75.77	45.5	86.78	1.67	Local short circuit and poor insulation	000	T
7	41	240	297	104	0	Core large area grounding	021	T
8	57	225	781	133	16	Magnetic circuit fault	022	T
9	99.39	28.86	16.07	6.82	52.3	Low-energy discharge	201	T
10	56	13	22	2.4	22.13	Screen partial breakdown	102	T
11	32.4	5.5	12.6	1.4	13.2	Screen discharge	102	T
12	220	340	480	42	14.	High-energy discharge	022	F
13	68.72	9.05	11.38	1.71	34.88	Low-energy discharge	202	T
14	225.1	215.9	605.5	88.7	1849	Low-energy discharge	202	T
15	100.4	50.8	159.68	20.31	13.85	Core magnetic flux leakage or local short circuit	002	T
16	99.47	64.17	200.85	25.81	14.72	Core magnetic flux leakage	002	T
17	101.2	98.12	202.63	87.50	6.35	Medium-temperature overheating	001	T
18	93	68.99	20	43	0	Medium-temperature overheating at 510 °C	000	T
19	61.88	88.76	83.81	76.14	0.14	Low-temperaturc overheating	021	F
20	27.7	75.78	335.18	88.52	0.51	Lower low-temperature overheating	022	F

2.1.1.3 Bayesian Network-Based Fault Diagnosis Method

The occurrence of transformer faults is a slow process, which can be divided into two stages from the onset of the fault to the point where the transformer can no longer operate.

1. Initially, the fault progresses slowly over time, during which the transformer fault gradually worsens but can still maintain normal operation.
2. When a fault reaches a critical point, it intensifies rapidly, causing the transformer to become inoperable. Additionally, transformer faults are often not isolated incidents; they frequently occur concurrently with multiple faults. Timely diagnosis during the period when abnormal indicators appear, but before the transformer stops operating, can prevent many unnecessary losses and identify other existing fault modes. Bayesian networks are effective tools for expressing knowledge of uncertainty and for inference; their capability for causal probability derivation can be utilized for transformer fault diagnosis. Based on the abnormal indicators produced when a transformer fault occurs, Bayesian network inference can diagnose the type of fault and facilitate timely repairs.

Bayesian Network Diagnostic Principles

A Bayesian network models each abnormal indicator and fault mode as nodes within a DAG. By considering the prior probabilities and conditional probabilities between nodes, Bayesian networks solve the most probable explanation (MPE) inference problem. This allows for the inference of other fault modes based on known abnormal indicators and fault modes through the network.

A Bayesian network is a directed acyclic graph. For example, in Fig. 2.4 depicting a Bayesian network, nodes X_1, X_2, X_3, X_4 represent random variables, and directed edges represent conditional dependencies between nodes. X_1, X_2, X_3 are parent nodes with prior probabilities $P(X_1)$, $P(X_2)$, $P(X_3)$, and X_4 is a subnode with a corresponding probability given its parent nodes: $P(X_4 \mid X_1, X_2, X_3)$.

The concept of conditional independence within Bayesian networks significantly simplifies the joint probability distribution, as illustrated in Eq. (2.1).

$$P(X_1, X_2, X_3, X_4) = \prod_{i=1}^{4} P(X_i \mid \pi(X_i)) = P(X_1)P(X_2)P(X_3)P(X_4 \mid X_1, X_2, X_3) \tag{2.1}$$

where, $\pi(X_i)$ denotes the parent nodes of $X_i (i = 1, 2, 3, 4)$.

In practical scenarios, determining eight conditional probabilities for node X_4 can be challenging. If each parent node independently affects the subnode, then the relationship is expressed by Eq. 2.2:

Fig. 2.4 Bayesian Network Schematic

X_1	$P(X_1)$	X_2	$P(X_2)$	X_3	$P(X_3)$
0	0.2	0	0.02	0	0.4
1	0.8	1	0.98	1	0.6

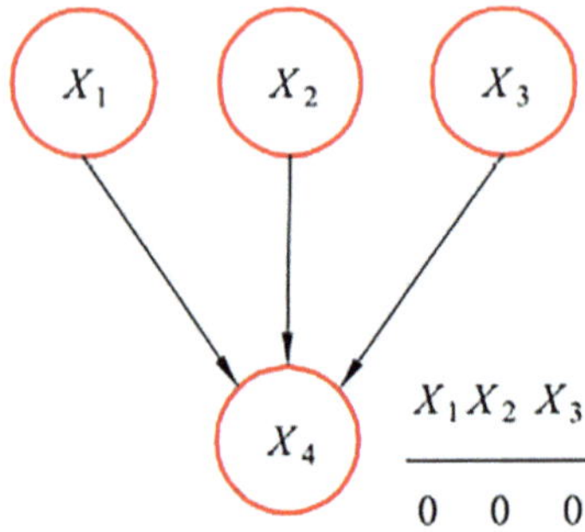

X_1	X_2	X_3	$P(X_4=1\mid X_1,X_2,X_3)$
0	0	0	0.02
1	0	0	0.12
0	1	0	0.38
1	1	0	0.41
0	0	1	0.1
1	0	1	0.3
0	1	1	0.5
1	1	1	0.9

$$\mathrm{P}(X_4 = \alpha \mid X_1, X_2, X_3) = \sum_{\alpha_1^* \alpha_2^* \alpha_3 = \alpha} \mathrm{P}(\xi_1 = \alpha_1 \mid X_1) p(\xi_2 = \alpha_2 \mid X_2) p(\xi_3 = \alpha_3 \mid X_3) \tag{2.2}$$

where, α_i represents a specific state in the state space; $*$ is the basic composition operator; ξ_i is the contribution of X_i to X_4; $\mathrm{P}(\xi_\mathrm{i} \mid X_\mathrm{i})$ is the contribution probability distribution of X_i to X_4 ($i = 1, 2, 3$).

In the transformer fault diagnosis process, the basic composition operator is the logical Or, and X_4 node is a noisy-Or node, characterized by causal independence, as shown in Fig. 2.5.

However, in the case of a transformer fault, even when all fault modes have not occurred, abnormal indicators may still arise. Therefore, a leaky node is introduced on the foundation of the noisy-Or node to represent the impact of other unconsidered factors on the outcome, as depicted in Fig. 2.6. The fault diagnosis model based on the Bayesian network is established at this node.

The Bayesian network's MAP problem involves, given known conditions $E = e$, determining the most probable set of state combinations for other events that are not part of the evidence, denoted as hypothesis variables H. A combination of states of H forms a hypothesis, denoted as h. Among all state combinations, seeking the one with the highest posterior probability defines the MAP problem, as shown in Eq. (2.3):

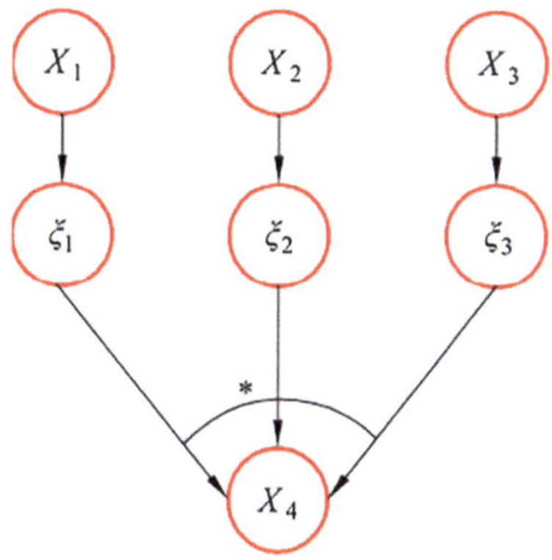

Fig. 2.5 Diagram of Causal Independence

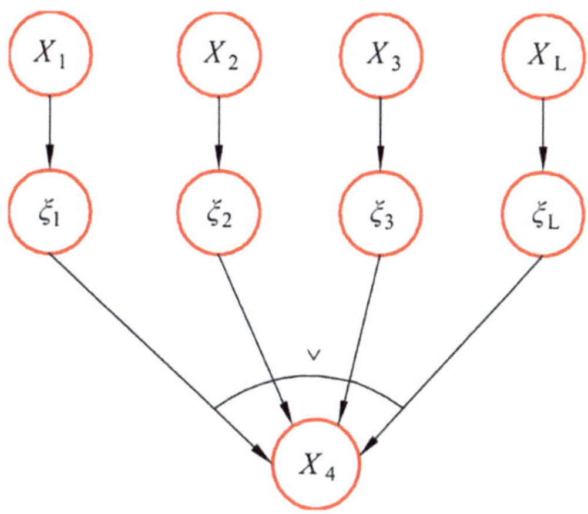

Fig. 2.6 Diagram of the Noisy-OR Node

$$h^* = \arg\max{}_h \mathrm{P}(H = h \mid E = e) \tag{2.3}$$

In transformer fault diagnostics, known abnormal indicators are treated as evidence variables, while various fault modes are treated as hypothesis variables. Through MAP inference, the fault mode combination with the highest posterior probability is output as the diagnostic result.

Development of a Bayesian Network for Transformer Fault Diagnosis

By selecting adverse conditions, fault mode, and abnormal indicators as the three-tier structure of the Bayesian network. According to the standards of the State Grid Corporation, three types of adverse conditions are adopted for assessment: short-term emergency load, output short circuit, and lightning impact. These common adverse conditions are also chosen for analysis. Under the condition of a short-term emergency load, transformers generate thermal stress, leading to insulation aging.

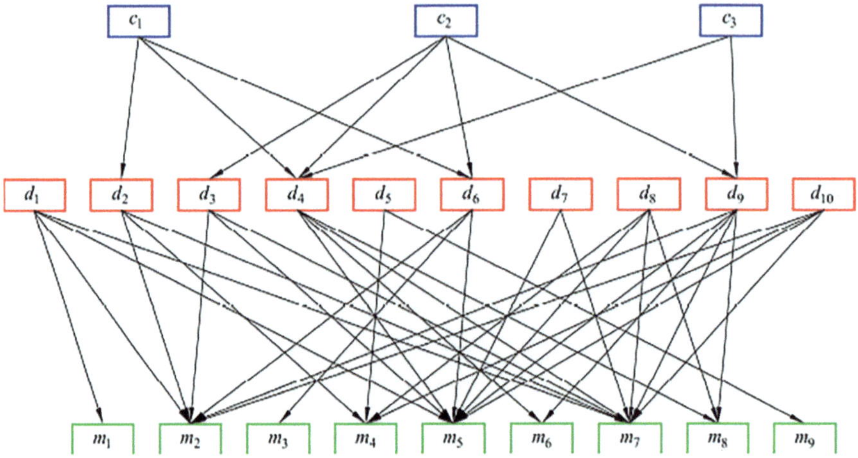

Fig. 2.7 Three-layer Bayesian Network

Additionally, this can cause turn insulation damage, interturn short circuits, as well as tap changers and lead faults. When subjected to nearby or output short circuit adverse conditions, it is prone to winding deformation and inter-turn short circuit. It may also lead to magnetic flux leakage heating or magnetic shielding discharge overheating, turn insulation damage, interturn short circuits, as well as tap changers and lead faults among other fault modes. Under the adverse conditions of thunder and lightning overvoltage impacts, turn insulation damage, interturn short circuit, and winding deformation are the most likely fault modes to occur due to the impact of the invading wave. When a fault mode is present, detecting corresponding indicators reveals abnormal indicators associated with that fault mode. Hence, there is a strong causal relationship between adverse conditions, fault mode, and abnormal indicators, though they are not one-to-one corresponding. A network diagram is used to represent the relationships among them, as shown in Fig. 2.7. Symbols in the diagram are explained in Table 2.9.

For all subnodes, a leaky noisy-Or node is used, and it is only necessary to determine the conditional probability of each parent node to the subnode. Due to the varying conditions of substations in different regions, the prior probabilities of adverse conditions are adjusted accordingly. Here, for illustrative purposes, we assume the conditional probabilities for unconsidered factors from the first to the second layer (leaky node) affecting fault mode as 0.5, and from the second to the third layer (leaky node) affecting abnormal indicators as 0.01. The conditional probability table is shown in Table 2.10.

This structure helps establish a Bayesian network model for diagnosing transformer faults. Inputs include detected abnormal signs, while outputs involve undetected abnormal indicators and fault modes.

Table 2.9 Explanation of symbols in Fig. 2.7

	ID number	Event type
Adverse conditions c	c_1	Overload
	c_2	Output short circuit
	c_3	Lightning impact
Fault mode d	d_1	Core multi-point grounding and local short circuit
	d_2	Insulation aging
	d_3	Magnetic flux leakage heating or magnetic shielding discharge overheating
	d_4	Turn insulation damage and interturn short circuit
	d_5	Insulation moisture absorption
	d_6	Tap changer and lead fault
	d_7	Floating discharge
	d_8	Screen discharge
	d_9	Winding deformation and inter-turn short circuit
	d_{10}	Discharge in oil
Abnormal indicators m	m_1	Core grounding current
	m_2	Three-ratio coding shows overheating fault signs
	m_3	Three-phase unbalance coefficient of winding DC resistance
	m_4	Water content in transformer body oil
	m_5	Three-ratio coding shows discharge fault characteristics
	m_6	Winding ratio deviation
	m_7	Partial discharge
	m_8	$\varphi(CO)/\varphi(CO_2)$
	m_9	Absorption ratio or polarization index of winding

Case Analysis

Case One: Known three-ratio coding features indicate overheating faults, abnormal $\varphi(CO)/\varphi(CO_2)$ ratio, and abnormal three-phase unbalance coefficient of winding DC resistance. Evidence $E = (m_2, m_3, m_8, -m_5, -m_7)$ is inputted into the Bayesian network. Inference results suggest the occurrence of d_2, d_6, m_2, m_3, m_8. Observed abnormal indicators include three-ratio coding showing overheating fault features, $\varphi(CO)/\varphi(CO_2)$, and the three-phase unbalance coefficient.

Case Two: Before a factory transformer accident, no output short circuit occurred, and the lightning arrester acted once. Oil chromatographic analysis results, presented in Table 2.11, show that the transformer's absorption ratio and polarization index, three-phase unbalance coefficient of winding DC resistance, and winding ratio deviation are all within normal ranges.

Analysis Date From the oil chromatography data using the three-ratio method, according to Tables 2.2 and 2.3, the transformer is diagnosed with discharge faults. Current evidence $E = (c_3, m_5, -c_2, -m_2, -m_3, -m_6, -m_8, -m_9)$ leads to the inference that c_3, d_4, m_5, m_7 occur, with no other nodes affected. The transformer experiences a lightning strike; fault mode includes turn insulation damage and

Table 2.10 Bayesian network's conditional probability table

Node information		ID number	Conditional probability table
Adverse conditions *c*	Overload	c_1	$P(c_1) = 0.05$
	Output short circuit	c_2	$P(c_2) = 0.05$
	Lightning impact	c_3	$P(c_3) = 0.05$
Fault mode *d*	Core multi-point grounding and local short circuit	d_1	$P(d_1) = 0.2271$
	Insulation aging	d_2	$Pc_1 = 0.6; P_L = 0.5$
	Magnetic flux leakage heating or magnetic shielding discharge overheating	d_3	$Pc_2 = 0.6; P_L = 0.5$
	Turn insulation damage and interturn short circuit	d_4	$Pc_1 = 0.6; Pc_2 = 0.6; Pc_3 = 0.6; P_L = 0.5$
	Insulation moisture absorption	d_5	$P(d_5) = 0.0506$
	Tap changer and lead fault	d_6	$Pc_1 = 0.6; Pc_2 = 0.6; P_L = 0.5$
	Floating discharge	d_7	$P(d_7) = 0.0798$
	Screen discharge	d_8	$P(d_8) = 0.1421$
	Winding deformation and inter-turn short circuit	d_9	$Pc_2 = 0.6; Pc_3 = 0.6; P_L = 0.5$
	Discharge in oil	d_{10}	$P(d_{10}) = 0.0717$
Abnormal indicators *m*	Core grounding current	m_1	$Pd_1 = 0.90; PP_L = 0.01$
	Three-ratio coding shows overheating fault signs	m_2	$Pd_1 = 0.818; Pd_2 = 0.219; Pd_3 = 0.713; Pd_6 = 0.674; Pd_9 = 0.149; Pd_{10} = 0.20; P_L = 0.01$
	Three-phase unbalance coefficient of winding DC resistance	m_3	$Pd_6 = 0.87; PP_L = 0.01$
	Water content in transformer body oil	m_4	$Pd_2 = 0.267; Pd_5 = 0.718; Pd_8 = 0.416; Pd_{10} = 0.60; PP_L = 0.01$
	Three-ratio coding shows discharge fault characteristics	m_5	$Pd_1 = 0.189; Pd_3 = 0.289; Pd_4 = 0.515; Pd_6 = 0.231; Pd_7 = 0.863; Pd_8 = 0.879; Pd_9 = 0.681; Pd_{10} = 0.70; P_L = 0.01$
	Winding ratio deviation	m_6	$Pd_4 = 0.80; Pd_9 = 0.80; P_L = 0.01$
	Partial discharge	m_7	$Pd_1 = 0.30; Pd_3 = 0.35; Pd_4 = 0.90; Pd_7 = 0.90; Pd_8 = 0.90; Pd_9 = 0.75; Pd_{10} = 0.90; PP_L = 0.01$
	$\varphi(CO)/\varphi(CO_2)$	m_8	$Pd_2 = 0.816; Pd_4 = 0.681; Pd_8 = 0.759; Pd_9 = 0.721; PP_L = 0.01$
	Absorption ratio or polarization index of winding	m_9	$Pd_5 = 0.75; PL = 0.01$

Table 2.11 Oil chromatographic analysis results

Analysis date	CH_4	C_2H_4	C_2H_6	C_2H_2
2009-06-26	6.99	10.87	0	3.72
	H_2	CO	CO_2	Total hydrocarbons
	13.57	229	2344	21.58

interturn short circuit, with abnormal indicators showing three-ratio coding indicating overheating faults and partial discharge. A field inspection confirmed abnormal partial discharge, aligning with damage found on the medium-voltage side insulation, consistent with inference results.

2.1.2 *Switchgear Fault Diagnostic Methods*

Switchgear is electrical equipment used for breaking and making conductive circuits, playing a crucial role in connecting and disconnecting circuits, and cutting off and isolating faults. Major types of switchgear include circuit breakers, isolators, load switches, and fuses. Reliable operation of switchgear ensures the stable functioning of systems, making fault diagnosis of switchgear significantly vital. Currently, fault diagnosis of switchgear is often conducted through vibration signals, circuit breaker operation current signals, and gas decomposition products.

2.1.2.1 Fault Diagnostic Methods Based on Vibration Signals Principles of Diagnosis

Diagnostic Principles Based on Vibration Signal

Vibration signals are commonly used detection signals in switchgear components like isolators and circuit breakers. These signals contain rich information about the mechanical condition, resulting from the superposition of multiple vibration impacts produced by the movement of various mechanisms and components within the switchgear. Any change in the mechanical condition of the switchgear can lead to alterations in the movement state of its components, consequently changing the vibration signal. By analyzing changes in vibration signal feature information, the type of fault can be identified. Thus, this signal can reflect subtle changes in mechanical structures. As long as the vibration propagation path remains constant, the same type of vibration signal will have consistent features. Extensive research has shown that the same type of switchgear can produce similar vibration signals during operation, and different types of mechanical failures in the same type of switchgear can be detected through these vibration signals.

Firstly, feature extraction should be performed on the collected vibration signals to facilitate subsequent fault identification. Time-frequency analysis is the most widely used method for feature extraction in vibration signals. Mechanical failures will cause changes in the time and frequency characteristics of the vibration signal. The time-frequency analytical method involves analyzing the vibration signal from both time and frequency domains simultaneously. In this paper, wavelet packet transform is used to perform time-frequency analysis on the vibration signals, thereby obtaining feature information.

Wavelet packet transform decomposes the signal into a series of frequency bands with equal bandwidth, allowing for localized time-frequency analysis of the signal. The process of wavelet packet transform can be viewed as the decomposition of a signal into high and low frequency bands through high-pass and low-pass filters. The signal components post-decomposition can continue to be processed with wavelet packet transform to determine the distribution of vibration signals within specified frequency bands.

Information about abrupt changes is often reflected in the envelope, and the envelope can also provide valid information from the switchgear vibration signals. By calculating the information entropy vector of the envelope signals for each frequency band, this vector is used as feature information to reflect changes in the switchgear circuit breaker. Finally, the feature signals are classified using a BP neural network to obtain the diagnostic results for the switchgear faults.

Thus, fault diagnosis of switchgear based on vibration signals is completed through signal collection, feature extraction, and fault identification, effectively identifying mechanical failures in isolators, circuit breakers, and other switchgear components.

Fault Diagnostic Methods Based on Vibration Signals

Further Details on Wavelet Transform in Fault Diagnosis. The switchgear vibration signal $f(t)$ after noise reduction is subjected to a wavelet packet transform to obtain the frequency band components of the vibration signal. Let g_k be the orthogonal wavelet function $\psi(t)$, with corresponding high-pass filter coefficients, and h_k be the coefficients of the orthogonal scaling function $\varphi(t)$ corresponding to the low-pass filter, where $g_k = (-1)^k h_{1-k}$. Both filter coefficients satisfy the two-scale equation:

$$\begin{cases} \varphi(t) = \sqrt{2}\sum h_k \varphi(2t-k) \\ \psi(t) = \sqrt{2}\sum g_k \varphi(2t-k) \end{cases} \tag{2.4}$$

Let $u_0 = \varphi(t)$, $u_1 = \psi(t)$, which can be obtained through a recursive relationship:

$$\begin{cases} u_{2n+1}(t) = \sqrt{2}\sum h_k u_n(2t-k) \\ u_{2n}(t) = \sqrt{2}\sum g_k u_n(2t-k) \end{cases} \tag{2.5}$$

Define $u_{2n}(t)$ as a set of wavelet packets determined by the scaling function $\varphi(t)$. Thus, the process of wavelet packet transform of the vibration signal can be seen as dividing the vibration signal into high and low frequency bands. These two signal components can continue to undergo further decomposition using wavelet packet transform. After m layers of wavelet transform, the vibration signal can be decomposed into 2^m frequency bands with signal components $f_i(t)$, where $i = 1, 2, \cdots, 2^m$.

Switchgear vibration signals are non-stationary, and their transient information often appears on the signal's envelope. The Hilbert transform is commonly used in the fault diagnosis of switching equipment to extract the envelope. For a signal component $f_i(t)$ on a certain frequency band, its Hilbert transform $f(t)$ is defined as follows:

$$\widehat{f}_i(t) = \frac{1}{\pi} \int_{-\infty}^{+\infty} \frac{f_i(\tau)}{t-\tau} \mathrm{d}\tau \tag{2.6}$$

The amplitude function of the signal component, i.e., the envelope signal $A_i(t)$, is:

$$A_i(t) = \sqrt{f_i^2(t) + \widehat{f}_i^2(t)} \tag{2.7}$$

Divide the envelope signal along the time axis into N segments and calculate the energy $E_{i,\ n}$ in each segment:

$$E_{i,n} = \int_{t_n}^{t_{n+1}} A_i(t)^2 \mathrm{d}t \tag{2.8}$$

where $n = 1, 2, \cdots N$; t_n, t_{n+1} are the start and end times of the N time segment, respectively. Normalize the energy $E_{i,\ n}$ in each time segment as follows:

$$q_{i,n} = E_{i,n} / \sum_{n=1}^{N} E_{i,n} \tag{2.9}$$

According to basic entropy theory, define the feature entropy of the vibration signal component $f_i(t)$ as follows:

$$H_i = -\sum_{n=1}^{N} q_{i,n} \log(q_{i,n}) \tag{2.10}$$

Calculate the feature entropy for each signal component to obtain the wavelet packet-feature entropy vector $T = [H_1, H_2, \cdots, H_{2m}]$ of the switchgear vibration signal $f(t)$.

By collecting wavelet packet-feature entropy vectors of the circuit breaker under various statuses as a training set, train a BP neural network through the backpropagation algorithm. When diagnosing a fault in an unidentified vibration signal, similarly compute its wavelet packet-feature entropy vector and input it into the trained BP neural network to complete the fault diagnosis of the switchgear.

Fault Diagnosis Case Study Based on Vibration Signal

Using a specific model of a minimum oil circuit breaker as a case study, simulate two types of fault signals: base screw loosening (Fault 1) and insufficient lubrication (Fault 2), while also collecting vibration signals under normal conditions. The steps for fault diagnosis are as follows:

1. After noise reduction, perform wavelet packet transform on the vibration signals $f(t)$ under each status using the Daubechies10 wavelet as the base function. The transform has three layers, resulting in eight different frequency band signal components $f_i(t)$, $i = 1, 2, \cdots, 8$.
2. Apply the Hilbert transform to each signal component $f_i(t)$ to obtain the envelope $A_i(t)$.
3. Divide the envelope of each signal component under normal conditions into 15 segments of equal energy, and apply the same segmentation approach to the envelopes of the signal components of the signals to be diagnosed.
4. Calculate the feature entropy vectors $\boldsymbol{T} = [H_0, H_1, H_2, H_3, H_4, H_5, H_6, H_7]$ for each status. The calculation results for the signals to be diagnosed are shown in Table 2.12.
5. A BP neural network was constructed with an input layer containing 8 neural nodes, a hidden layer with 10 neural nodes, and an output layer with 3 neural nodes. After completing its training, the diagnostic signals shown in Table 2.12 were input into this network for recognition and fault diagnosis. The output results, with three types of encodings corresponding to three different circuit breaker statuses, are displayed in Table 2.13. The case demonstrates that based on vibration signals, it is possible to effectively diagnose mechanical failures in switchgear.

2.1.2.2 Fault Diagnosis Based on Closing and Opening Coil Current

Principle of Fault Diagnosis Based on Closing and Opening Coil Current

The closing and opening coils are crucial components in circuit breakers and other types of switchgear. As current flows through these coils, it generates magnetic flux

Table 2.12 Vibration signal entropy

Status	H_0	H_1	H_2	H_3	H_4	H_5	H_6	H_7
Norm	1.2499	1.1801	1.0906	1.2250	1.2533	1.2829	1.0699	1.1721
Fault 1 signal 1	1.1320	0.8010	0.6139	0.6341	0.8988	1.0187	0.5067	0.7057
Fault 1 signal 2	1.1890	0.7447	0.6049	0.5399	0.8382	0.9012	0.5209	0.6334
Fault 1 signal 3	1.1889	0.7864	0.7173	0.5685	0.9045	1.0152	0.5420	0.6697
Fault 2 signal 1	1.1640	0.8963	0.8925	0.8353	1.0798	1.2816	0.7964	0.8892
Fault 2 signal 2	1.1276	0.8089	0.7304	0.7653	0.9802	1.1781	0.6418	0.8820
Fault 2 signal 3	1.0883	0.6340	0.4677	0.5491	0.7663	0.8360	0.4072	0.6585

Table 2.13 Fault diagnostic result

Circuit breaker status	Actual value	Expected value	Diagnostic result
Normal	(0.9359, −0.0277, 0.0863)	(1, 0, 0)	Normal
Fault 1 signal 1	(−0.0388, 0.8457, 0.2766)	(0, 1, 0)	Fault 1
Fault 1 signal 2	(−0.0147, 0.8984, 0.1148)	(0, 1, 0)	Fault 1
Fault 1 signal 3	(−0.0942, 0.9258, 0.0951)	(0, 1, 0)	Fault 1
Fault 2 signal 1	(−0.0289, −0.0927, 0.9600)	(0, 0, 1)	Fault 2
Fault 2 signal 2	(−0.0585, −0.0692, 0.9185)	(0, 0, 1)	Fault 2
Fault 2 signal 3	(−0.0405, −0.1157, 0.9120)	(0, 0, 1)	Fault 2

Fig. 2.8 Circuit Breaker and Gate Coil Current

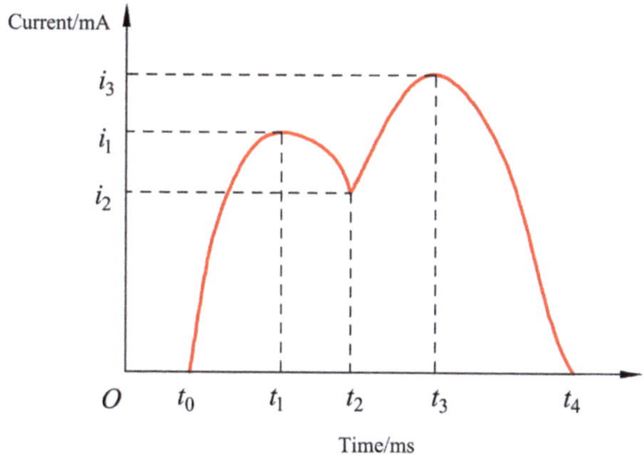

that causes the core within the coil to experience electromagnetic attraction, initiating action and thereby enabling the control of the switchgear to interrupt the current. Conversely, when the current in the coil is interrupted, the magnetic flux within the coil ceases, causing the core to lose electromagnetic attraction. The core then returns to its original position due to the action of a spring, and the switchgear resumes the current flow. The closing and opening coil currents vary depending on the mechanical conditions of the switchgear, allowing the use of different current features to distinguish the operating statuses of the switchgear.

Taking the circuit breaker closing coil current waveform as a case study, this waveform can be segmented into four stages, as illustrated in Fig. 2.8:

First Stage (t_0 to t_1): Rest-to-Motion Stage. During this stage, the core remains stationary, not yet commencing movement, although the magnetic flux in the coil continuously increases, causing the current to rise exponentially. The circuit breaker receives the closing command at time t_0, triggering the electromagnet coil to begin exhibiting an exponentially increasing current. By time t_1, the electrical force on the core has sufficiently increased to overcome the static frictional resistance, prompting the core to begin moving.

Second Stage (t_1 to t_2): Core Operating Phase. In this phase, the core is propelled by force, leading the current to gradually decrease until it stops moving at time t_2.

During this phase, the current progressively declines, and the current at time t_2 denotes a minimum value within the coil current waveform, marking the moment the core contacts the tripping mechanism and ceases movement. The interval from t_1 to t_2 represents the duration it takes for the core to commence movement and engage the tripping mechanism. The current magnitude at the moment the core contacts the tripping mechanism, denoted as i_2, is contingent upon the resistance the core encounters during its movement.

Third Stage (t_2 to t_3): Core Motion Stop Phase. During this stage, the core is stationary again, and the current starts to rise exponentially. As the closing operation continues, or after a sufficient duration, the current stabilizes at a steady-state value.

Fourth Stage (t_3 to t_4): Coil Current Interruption Phase. At time t_3, the auxiliary switch is activated, generating an arc between the switch and the contact. This arc results in a rapid increase in arc voltage, which in turn causes a sharp decline in coil current. By time t_4, the coil current reduces to zero, completing the interruption of the coil circuit.

In the four stages of the closing coil current waveform. First Stage (t_0 to t_1): Rest-to-Motion Phase of the core, which is crucial for overcoming static frictional resistance. Analyzing this stage helps diagnose potential issues such as sticking or binding of the core, by examining static frictional resistance, applied voltage, coil resistance, and inductance encountered during this phase. Second Stage (t_1 to t_2): Core Action Phase, where one can identify issues like sticking or excessive resistance encountered by the moving core. Third Stage (t_2 to t_3): Core Motion Stop Phase, where measurements such as t_2, t_3, i_2, and i_3 are critical for diagnosing issues like DC resistance of the wire break, jitter during the core's movement, coil voltage, and the timing of the mechanism's operation. Fourth Stage (t_3 to t_4): Core Stop Phase, where the core remains stationary. By analyzing measurements from this phase, including i_3, t_3, and t_4, one can evaluate the breaking performance of the auxiliary switch.

Thus, by monitoring critical values (t_0, t_1, t_2, t_3, t_4, i_1, i_2, i_3) and their changes across different stages, it is possible to effectively diagnose faults based on the closing and opening coil currents.

Fault Diagnosis Method Based on Closing and Opening Coil Current

The diagnostic approach based on closing and opening coil current starts by capturing the complete waveform, ensuring the initial point of the coil current, denoted as t_0, aligns with zero on the time axis. Subsequent steps involve determining the two peak times, t_1 and t_3, as well as the trough time t_2 and the endpoint time t_4, thereby capturing the current values at these times: two peak currents i_1 and i_3, and the trough current i_2. This data collection results in a complete feature set of the coil current, represented as $T = [t_0, t_1, t_2, t_3, t_4, i_1, i_2, i_3]$, enabling effective fault diagnosis.

Initially, the advantages of the wavelet transform, particularly in detecting abrupt changes and reducing noise, are utilized to capture the coil current signal $x(t)$. An appropriate wavelet function, $\varphi(t)$, is selected, and the wavelet transform is performed with specific translation τ and scaling α factors, as outlined in eq. (2.11).

$$WT_x(a,\tau)=\frac{1}{\sqrt{a}}\int_{-\infty}^{+\infty}x(t)\varphi^*\left(\frac{t-\tau}{a}\right)\mathrm{d}t \tag{2.11}$$

Following wavelet transformation, a soft thresholding method is applied to the entire current signal, $x(t)$, to reduce noise. Post noise reduction, wavelet analysis is used to detect abrupt changes, pinpointing the start t_0 and end t_4 times of the coil current. The high-frequency wavelet coefficients D_0 of the noise-reduced signal are then extracted to facilitate the detection of these abrupt changes.

To accurately detect abrupt change points, the high-frequency coefficients undergo normalization, resulting in processed coefficients D_1:

$$D_1=[D_0-min(abs(D_0))]/[max(abs(D_0))-min(abs(D_0))] \tag{2.12}$$

These coefficients are then subjected to clipping processing to eliminate disturbances, enhancing the accuracy of detection. After clipping, the coefficients are termed D_2,

$$D_3=\begin{cases} D_2, D_2>\lambda \\ 0, D_2<\lambda \end{cases} \tag{2.13}$$

where λ is the threshold. This processing helps determine the initial and final abrupt change points, corresponding to the times t_0 and t_4, thus extracting a complete coil current signal $x'(t)$ from t_0 to t_4. The peak and trough times within $x'(t)$, t_1, t_2, and t_3, are identified, along with their corresponding current values i_1, i_2, and i_3, forming the coil current feature vector $\boldsymbol{T}=[\ t_0,\ \ t_1,\ \ t_2,t_3,t_4,i_1,i_2,i_3]$.

Finally, apply the K-means algorithm to classify the obtained feature vector. The K-means algorithm is a commonly used classification algorithm that divides n objects into K clusters, ensuring high similarity within clusters and maximum dissimilarity between clusters. Input the coil current feature vector for each status into the K-means algorithm, which then classifies these vectors into K categories, each representing a different switchgear status, completing the fault diagnosis based on closing and opening coil current.

Fault Diagnosis Case Based on Closing and Opening Coil Current

In a specific instance involving a sulfur hexafluoride circuit breaker of a designated model, coil currents during the closing operation were recorded under six distinct operational statuses: normal condition, electromagnet sticking, electromagnet loosening, high operating voltage, low operating voltage, and buffer sticking. Initially, a soft thresholding method was employed to reduce noise in the current signal.

Table 2.14 Mean values of coil current features under various statuses

Status	i_1	i_2	i_3	t_1	t_2	t_3	t_4
Normal condition	1.5778	1.1368	2.3288	10.1041	14.4774	36.7971	41.6986
Electromagnet sticking	1.6593	1.2015	2.2743	12.6840	17.9422	39.9823	44.8808
Electromagnet loosening	1.6501	1.1042	2.3033	9.7122	15.4704	37.7572	42.6268
High operating voltage	1.6954	1.2485	2.5548	9.5188	13.3525	35.8163	41.3420
Low operating voltage	1.3761	0.9427	1.9107	11.2736	16.6737	38.7175	42.8635
Buffer sticking	1.5536	1.1278	2.2818	9.9485	14.4568	34.3108	39.6053

Following this noise reduction, the current signal underwent decomposition into three layers using the Daubechies2 wavelet base function. The critical information of the current signal was predominantly concentrated in the high-frequency component of the first decomposition layer. The high-frequency coefficients D_0 from the first layer of the wavelet transformation, denoted as t_0, were normalized and clipped using a wavelet threshold of 0.2. This process enabled the extraction of the complete coil current waveform, identifying feature points i_1, i_2, i_3, t_1, t_2, t_3, and t_4 as illustrated in Fig. 2.8, facilitating fault diagnosis. Table 2.14 presents the mean values of these feature points under the six operational statuses after multiple measurements.

The coil current feature vectors, $T = [\ t_0,\ t_1,\ t_2,\ t_3, t_4,\ i_1,\ i_2,\ i_3\ ]$, corresponding to these six statuses were input into the K-means algorithm. Each status included 30 sets of feature vectors with K = 6. The classification results, as shown in Table 2.15, revealed only one instance where a high operating voltage was incorrectly classified as buffer sticking, achieving a fault diagnosis accuracy rate of 99.4%. This outcome exemplifies the effectiveness of fault diagnosis based on closing and opening coil currents in accurately identifying switchgear issues.

2.1.2.3 Fault Diagnosis Based on Gas Decomposition Products

Principle of Fault Diagnosis Based on Gas Decomposition Products

With the progression of industrial technology, SF_6 switchgear has largely supplanted traditional oil-filled equipment and now holds a predominant position within switchgear systems. In the early stages of SF_6 switchgear fault diagnosis, conventional electrical testing methods proved inadequate for detecting internal insulation defects. Field experience has shown that by analyzing SF_6 gas decomposition products, latent faults within the equipment can be detected timely and effectively, thereby aiding in fault localization.

When internal insulation faults occur in SF_6 switchgear, or when conductive particles in the gas lead to ground discharges, various discharge types such as arc, spark, or corona can occur. During arc discharges in the fault area, the primary gases produced are SOF_2 and SO_2. Spark discharges additionally generate gases like

Table 2.15 Closing coil current fault diagnostic results

Type	Test sample			Accuracy rate (%)
	Test sample count	Number of misclassifications	Number of correct classifications	
Normal condition	30	0	39	100
Electromagnet sticking	30	0	40	100
Electromagnet loosening	30	0	35	100
High operating voltage	30	1	29	100
Low operating voltage	30	0	30	96.7
Buffer sticking	30	0	30	100
Total	180	1	179	99.4

SO_2F_2, SO_2F_{10}, and S_2OF_{10}, while corona discharges mainly produce SO_2 and HF, with a higher ratio of SO_2F_2 to SOF_2 compared to spark discharges. When the switchgear experiences a gap partial discharge, only trace amounts of SO_2, HF, and H_2S are produced.

Issues such as poor connection quality of conductive poles can increase contact resistance of conductors, leading to a rise in temperature at the fault point. Experiments have demonstrated that solid insulation materials react with SF_6 gas under high gas pressure and temperatures above 190 °C, and at even higher temperatures, the insulation materials may decompose directly. This type of fault primarily generates decomposition products such as SO_2, HF, H_2S, and sulfonyl fluoride; during internal faults in switchgear, SF_6 gas also decomposes into components such as CF_4, SF_4, and SOF_2. The presence of moisture and oxygen in the gas compartment can cause these substances to react again, forming stable compounds like SO_2 and HF.

During discharges and the thermal decomposition process, SF_6 not only produces stable decomposition products but also, under certain conditions, insulation materials decompose, generating complex gases. Thus, detecting the decomposition products of gases within SF_6 switchgear can significantly facilitate the fault diagnosis of the switchgear.

Fault Diagnostic Methods Based on Gas Decomposition Products

Building on the theory of SF_6 gas decomposition, fault diagnosis of SF_6 switchgear involves detecting components such as HF, H_2S, and SO_2 in SF_6 gas decomposition products. Adhering to standards in GB 8905-1996 and IEC 60480-2004, these measurements provide a straightforward and reliable means of diagnosing faults in switchgear.

The electrochemical sensor method is employed for the detection of gas components, noted for its simplicity, resistance to environmental disturbances, and rapid monitoring capability. This makes it a popular choice in practical engineering applications. Common sensors using this method can detect gases such as HF, H_2S, and SO_2. Despite its widespread use, this method has limitations, including an inability to detect gases like CF_4, a restricted detection range, and susceptibility to issues such as temperature drift and short service life. Regular calibration is necessary to maintain detection accuracy.

Moreover, gas chromatography is extensively used in the fault diagnosis of gas decomposition products. This method offers precise and reliable detection of sulfides, electronegative compounds, and halogen compounds in gases. Due to its high sensitivity, gas chromatography is a primary tool for gas detection in laboratory settings.

Fault Diagnosis Case Based on Gas Decomposition Products

Case One: After routine maintenance, a GIS circuit breaker passed all operational tests. However, the tripping of the line protection during the closing operation of its isolator resulted in a line outage. Preliminary analysis based on the symptoms of the incident suggested a failure in the circuit breaker. A sampling tube and a portable gas chromatograph were utilized to analyze the decomposition products in the affected gas compartment. The test results for phase A's gas compartment, detailed in Table 2.16, indicated that the levels of CF_4 and SO_2 significantly exceeded the thresholds established by IEC 60480-2004, pointing to an internal arc discharge fault within the equipment.

Upon opening and examining the faulty GIS circuit breaker, fragments indicative of contact resistance issues upon closing and a substantial amount of white powder decomposition product were discovered scattered inside the GIS. This confirmed that a severe arc discharge had occurred in phase A's circuit breaker, and the contact resistance upon closing had been compromised, corroborating the findings from the SF_6 gas decomposition products test.

Case Two: A substation experienced bus differential protection tripping twice during operation, initially diagnosed as resulting from an insulation discharge fault in a circuit breaker. Gas detection tubes were employed to analyze the gas compartment of the circuit breaker, with results shown in Table 2.17. The detected levels of SO_2, HF, and H_2S all exceeded the limits specified by IEC 60480-2004.

Table 2.16 Gas detection results for phase A of a circuit breaker

Gas component	Content (μL/L)	Reference value (μL/L)
CF_4	4264	≤1000
SO_2	10,000	≤5
H_2S	>1600	–

Table 2.17 Gas detection results for a circuit breaker

Gas compartment number	SO_2 content (μL/L)	HF content (μL/L)	H_2S content (μL/L)
Phase A	0	0	0
Phase B	>1500	>900	>300
Phase C	0	0	0

After performing on-site dismantling and maintenance of the circuit breaker, it was discovered that a breakdown had occurred between the uniform pressure cap and the casing at the inner end of phase B's tank. Discharge traces were found at the tank's bottom, and the fourth string of contact resistance upon closing in phase B was shattered, with fragments scattered at the bottom of the tank. The closing resistance end cap showed clear arc discharge traces. No abnormalities were detected in the circuit breakers of Phases A and C. The fault phenomena observed during the disassembly aligned with the results of the decomposition product detection.

2.1.3 Power Cable Fault Diagnosis Method

In the context of urban planning, power cable lines are increasingly utilized, offering superior power resource transmission compared to traditional line types. Prompt location and identification of faults in power cables are essential to resolve issues and minimize the duration of power supply interruptions, thus enhancing supply stability. Common causes of cable faults include mechanical damage, insulation moisture absorption, insulation aging, overvoltage, overheating, and cable quality defects. Cable fault diagnosis typically involves three steps: first, determining the nature of the cable fault; second, selecting appropriate detection and location methods based on the fault's nature, starting with a rough measurement; third, choosing a pinpointing method to accurately determine the fault location.

2.1.3.1 Diagnosis of Cable Fault Nature

After the occurrence of a cable fault, except for visibly apparent issues like cable termination explosions or external damages, most faults are not detectable through routine inspections. Therefore, employing cable fault testing equipment to measure and pinpoint the location of the fault becomes essential. Given the diverse types of cable faults, the detection methods must be tailored to the specific nature of each fault. Thus, accurately determining the nature of the cable fault is crucial before proceeding with fault location efforts.

Operational cable faults can be classified into several categories based on their nature, including ground faults, short circuit faults, broken wire faults, flashover faults, and mixed-type faults.

The preliminary assessment of a fault's nature is often based on the observed phenomena at the time of the fault's occurrence. For instance, if a fault during operation triggers only a ground signal, it might indicate a single-phase ground fault. If the relay protection's overcurrent relay activates and a trip occurs, this might suggest a short circuit fault in the cable. However, such phenomena alone cannot conclusively determine the nature of the fault, necessitating further tests like measuring insulation resistance and conducting continuity tests.

This test assesses the insulation condition and grounding status of the cable. It should be performed using a megohmmeter that is suitable for the cable's voltage level, measuring the insulation resistance both between the cable cores and between the core and ground.

During this test, the cable's end is short-circuited across all three phases. A multimeter is then used at the start of the cable to measure the resistance between the cores. This test helps determine the integrity and resistive properties of the cable cores.

2.1.3.2 Cable Fault Distance Measurement

Cable fault distance measurement utilizes instruments at one end of the cable to pinpoint the fault's location. Common techniques include the bridge detection method and the traveling wave method.

Bridge Detection Method

This method uses a concentrated parameter model of the cable, employing current and voltage measurements from one or more ends to formulate specific equations for fault localization. Its straightforward and practical approach has maintained its popularity over the years. It is primarily categorized into the DC resistance bridge method and the capacitance bridge method.

1. DC Resistance Bridge Method: Ideal for detecting low-resistance faults where the fault resistance is below 100 kΩ, this method involves short-circuiting the fault phase to a non-fault phase. The bridge is then connected across these phases. The resistance on the bridge arm is adjusted until equilibrium is achieved. The fault distance is subsequently calculated using the Wheatstone bridge principles.
2. Capacitance Bridge Method: Similar in principle to the DC resistance bridge method, this technique is used for identifying open circuit faults in cables. It assumes that the capacitance measured during an open circuit fault is proportional to the fault's distance. Using the known capacitance per unit length from cable manuals, the fault distance can be accurately calculated.

The bridge detection method is appreciated for its simplicity and high accuracy. However, it faces significant limitations: (1) It is unsuitable for online cable fault distance measurements. (2) It cannot be applied to high-resistance faults or flashover faults. (3) Accurate application requires detailed knowledge of the cable's parameters. (4) Excessively high requirements for lead contact. Due to these drawbacks, the bridge method is increasingly being replaced by more advanced measurement techniques.

Traveling Wave Method

The traveling wave method is a pivotal technique in cable fault distance measurement. It calculates the fault location by measuring the time it takes for traveling waves to propagate to and from the fault point and the measurement endpoint. Based on the specific implementation and the type of traveling wave generated, this approach can be segmented into three categories: Types A, B, and C.

1. Type A Distance Measurement Method: This method determines the time required for a traveling wave to oscillate between the fault point and the measurement endpoint. The speed of the wave is calculated using empirical formulas, and the distance is derived from the formula $s = vt/2$. Advantages of this method include its straightforwardness, minimal equipment needs, independence from transitional resistance, and potential for high precision. Nonetheless, it has not seen widespread adoption due to persistent challenges, such as achieving the necessary signal sampling rate for precise timing of the traveling wave's arrival at the measurement endpoint. Common Type A methods include the pulse voltage and pulse current techniques.

   (a) Pulse Voltage Method

   This method is suitable for identifying high-resistance and flashover faults in cables. It involves inducing a fault by applying either a direct current high voltage or a pulse high voltage. The distance to the fault is then determined by timing the round trip of the discharge voltage pulse between the measurement endpoint and the fault point. The pulse voltage method excels in directly detecting the instantaneous pulse signal generated by the fault point breakdown, allowing for rapid testing without needing to burn through the faults. However, it poses challenges such as safety concerns, difficulties in distinguishing voltage waveforms, and complex wiring configurations.

   (b) Pulse Current Method

   Operating similarly to the pulse voltage method, this technique measures the distance by detecting the current pulse signal at the moment of cable breakdown, facilitated by a linear current coupler. This setup allows electrical coupling between the instrument and the high-voltage circuit, effectively bypassing the effects of series resistance and inductance. It is characterized by a straightforward measurement line and the ease of identifying the pulse current waveform.

   (c) Type B Distance Measurement Method

   This method employs the initial traveling wavefront generated at the fault point, using a communication channel to determine the fault distance. Its primary advantage is its ability to measure distance by simply noting the time it takes for the first traveling wavefront to reach both ends of the cable, independent of any waves reflected or transmitted from the fault point.

(d) Type C Distance Measurement Method

Utilizing a pulse transmitter, this method sends high-frequency pulses to an offline faulted cable and measures the distance by calculating the time it takes for these pulses to travel to and from the fault point. The low voltage pulse reflection method is a common implementation of the Type C method. Given that Type C cannot perform online fault distance measurements, it is being phased out in favor of other methods.

2.1.3.3 Cable Fault Pinpointing

Cable fault pinpointing involves determining the approximate location of a fault along the cable based on the measurement results. In a localized area, specific techniques are employed to precisely identify the fault point. Once the fault type is determined, a search for the specific fault location is initiated. Popular techniques for fault location search include the acoustic detection method, bridge detection method, pulse detection method, and capacitive current detection method.

Acoustic Detection Method

1. Basics

   The acoustic detection method is a straightforward technique for diagnosing faults in power cables. It identifies the fault location by detecting the sound produced during the cable's discharge process. For power cables laid above ground, the location of the fault can often be directly determined by the sound of the discharge. However, for underground cables, the discharge sound is generally too faint to pinpoint the fault location directly. In such cases, the direction of the cable is analyzed, and sound amplification equipment is utilized to accurately locate the source of the fault sound.
2. Case Analysis

   (a) Fault Condition

   Cable Model: ZLQ20-10-3 × 70. Operating Voltage: 6 kV. Installation Method: Partly directly buried, partly in cable trenches. Total Length: 1000 meters. Operating Time: 15 years. phase C Operating Grounded. Measured Insulation Resistance: RC = 10 kΩ; RA = RC = 800 MΩ

   (b) Fault Location Search

   - The waveform of the impulse flashover method at the front end is depicted in Fig. 2.9, indicating $L_X = 600$ m.
   - The waveform of the impulse flashover method is shown in Fig. 2.10, where $L_X = 600$ m.

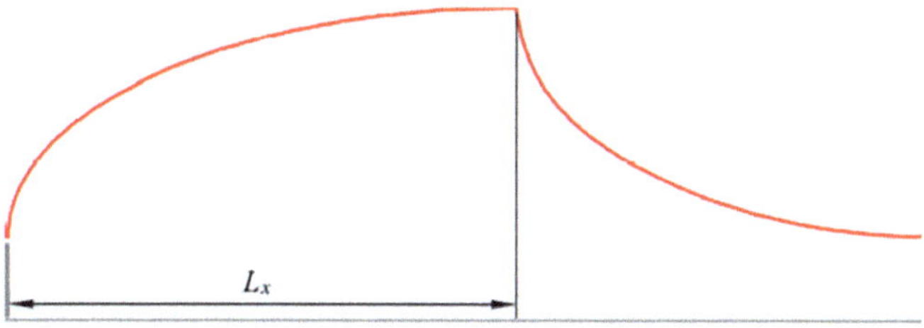

Fig. 2.9 Waveform of the Leading-End Impulse Flashover Method

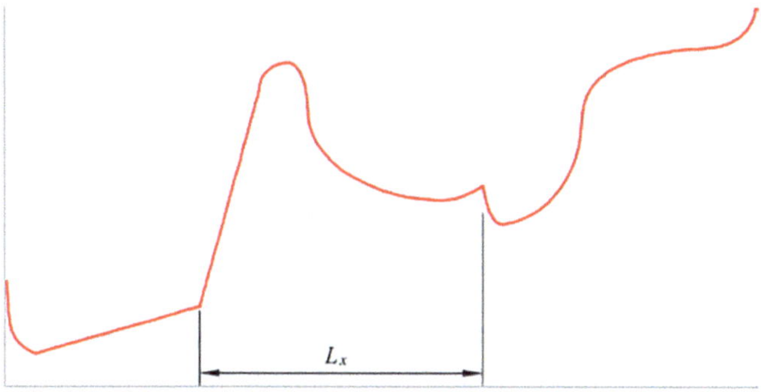

Fig. 2.10 Waveform of the Impulse Flashover Method

Acoustic pinpointing was conducted with an impulse voltage of 21 kV and a discharge frequency of 1/3 to 1/4 per second under favorable discharge conditions. Pinpointing around 600 m, a thorough inspection of the cable path revealed a distinct discharge sound above the cable trench near 603 m, accompanied by a noticeable jitter on the cable cover plate. Excavation confirmed the fault as a blown mid-joint in the cable, likely due to water ingress causing insulation breakdown at the cable head.

- The waveform appears to be quite ideal.
- The effectiveness of the Flashover Acoustic Location method is particularly highlighted in cases involving oil-immersed cables, where the loud discharge sounds facilitate straightforward and precise on-site pinpointing once the fault range is established from the waveform.

Bridge Detection Method

1. Basics

   The bridge detection method utilizes a double-arm bridge to measure the internal resistance of a power cable, which helps determine the cable's length. By analyzing variations in length and resistance values, inconsistencies can be detected, thereby pinpointing the location of faults within the cable. Accuracy in measurements and minimizing the length of cable connections are critical when employing this method for fault detection in power cables.

2. Case Analysis

(a) Fault Condition

One VV-22 3*70 + 1*35 cable, with a length of 339.26 m, is directly buried. Due to changes in construction planning, 75 m of this cable is buried under concrete. The cable has now experienced a short circuit, with the short circuit condition being a three-phase to armor direct short circuit, but the insulation condition of the neutral line remains good.

(b) Fault Location Search

- Internal Resistance Measurement

Load End (Group 1): Resistance between phases A and B: 0.368 Ω. Resistance between phases A and C: 0.371 Ω. Resistance between phases B and C: 0.220 Ω. Internal resistance: 0.152 Ω. Power Supply End (Group 2): Resistance between phases A and B: 0.440 Ω. Resistance between phases A and C: 0.442 Ω. Resistance between phases B and C: 0.277 Ω. Internal resistance: 0.153 Ω

- Data Analysis

The comparative analysis between the two sets of resistance measurements reveals variations in short circuit conditions across the phases, with the most significant shorts occurring between phases B and C, and less severe between AB, and AC. The data deviates from theoretical expectations, showing no linear correlation in short circuit resistance.

Based on these findings, a strategic cut is made at the power supply end-considering usability, safety, ease of recovery, and maximization of length-to facilitate a new measurement scenario. The new short circuit point is identified at 16.74 m with the following resistance measurements in the new three-phase short circuit scenario (Group 3): Resistance between phases A and B: 0.159 Ω. Resistance between phases A and C: 0.160 Ω. Resistance between phases B and C: 0.160 Ω. Internal resistance: 0.152 Ω

- Short Circuit Point Calculation

Upon analyzing the data from Groups 1 and 2, the resistance in the BC phase was the lowest, indicating minimal contact resistance and, consequently, the smallest error in this phase. Therefore, the BC phase resistance is utilized as the primary metric for calculating the short circuit point, with the resistances from the other two phases used as supplementary data.

Data Calculation from Group 3

BC phase Resistance at 16.74 m = (The resistance between phases B and C is 0.160 Ω) − (internal resistance 0.152 Ω) = 0.008 Ω

Resistance per meter for BC phase = 0.008/16.74 = 0.00047789725 Ω/m

According to the calculations, a value of 0.0004779 Ω/m is actually used.

Calculate the short circuit point using the first set of BC phase short circuit resistance:

(0.220 − 0.152)/0.0004779 = 142.29 m

Calculate the short circuit point using the second set of BC phase short circuit resistance:

(0.277 − 0.153)/0.0004779 = 259.46 m

Sum of the two sets of data:

142.29 + 259.46 = 402.75 m (calculated length)

The calculated length of 402.75 m differs from the actual length of 339.26 m. The discrepancy is due to contact resistance, with an error distance of 402.75 − 339.26 = 63.49 m.

Comparing the data, it is evident that the contact resistance at the load end is less than at the supply end. Although the specific contact resistance difference for the BC phase cannot be measured, it can be inferred that the calculated short circuit point should be corrected towards the supply end.

Actual result: From the load end, concrete was excavated starting at the 110.54 m position (under the road) towards the supply end, and the short circuit point was found at 113 m.

Pulse Detection Method

1. Basics

   The pulse detection method is a prevalent technique in diagnosing faults in power cables, encompassing various approaches such as the low voltage pulse method, pulse voltage method, pulse current method, and secondary pulse method. This method employs a pulse generator to emit a pulse wave. At the node where a fault occurs, a reflected pulse is generated. The speed and time interval of this reflected pulse are recorded to determine the approximate location of the fault within the cable. Further waveform analysis of the reflected pulse assists in identifying the specific type of cable fault, providing critical information for fault resolution.
2. Case Analysis

   (a) Fault Condition

   In March 2015, a ground fault was detected on a 10 kV cable line within a residential area. The cable, composed of cross-linked C_2H_4 for its insulation material, is a 10 kV branch cable commissioned in 2011. The total length of the cable is 342 m, with an intermediate joint located 53 m from a fire pole where the cable is vertically installed. The remainder of the line is directly buried in a horizontal layout. To identify the fault type, the cable head near the fire pole served as the test point where a 2500 V megohmmeter was used to test the ground resistance of phases A, B, and C. Test results: The grounding resistance of phase A is approximately 1 MΩ, and the insulation resistance of phases B and C is nearly infinite. Based on this feature, it is preliminarily determined to be a high resistance ground fault in phase A. To

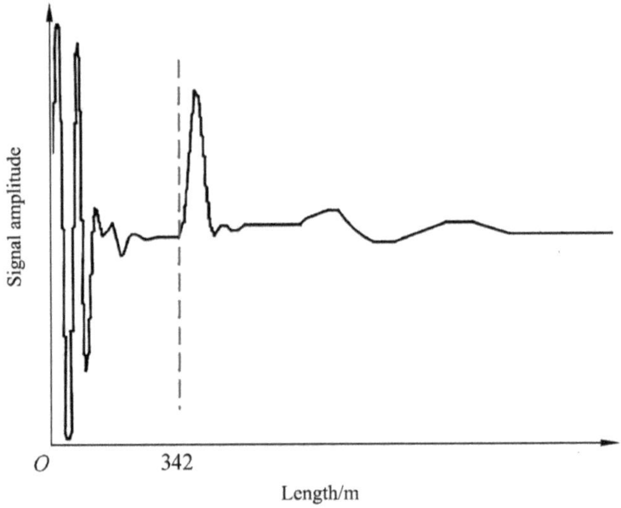

Fig. 2.11 Testing Waveform of the Entire Cable Length

further confirm this fault type, DC withstand voltage tests were conducted on phases A, B, and C. The test results showed that phase A broke down at 18 kV, while phases B and C still passed at 26 kV. The DC withstand voltage test results further confirmed that the fault was a high resistance ground fault in cable phase A.

(b) Fault Location Search.

- Utilizing the PSHZC cable fault tester, the total length of the cable was measured using the low voltage pulse current method, with the wave speed set at 160 m/μs. This test produced a complete waveform, confirming the cable's total length of 342 m and verifying good continuity, thereby excluding any broken wire faults. The corresponding waveform is depicted in Fig. 2.11.
- Initial confirmation of fault location. A DC high-voltage signal generator was employed to conduct a flashover test between phases A and the steel armor, setting the output voltage to 18 kV and the storage capacitor to 2 μF. The resulting waveform, illustrated in Fig. 2.12, pinpointed the fault approximately 53 m from the test end.
- Accurate determination of fault location. The result given in Fig. 2.12 only provides an approximate range. To precisely determine the fault distance, the secondary pulse method was used on cable phase A. After adjusting the output voltage to 28 kV, the cable was discharged, generating a waveform as shown in Fig. 2.13. The test results confirmed the fault distance to be 53 m, consistent with the initially determined fault distance.

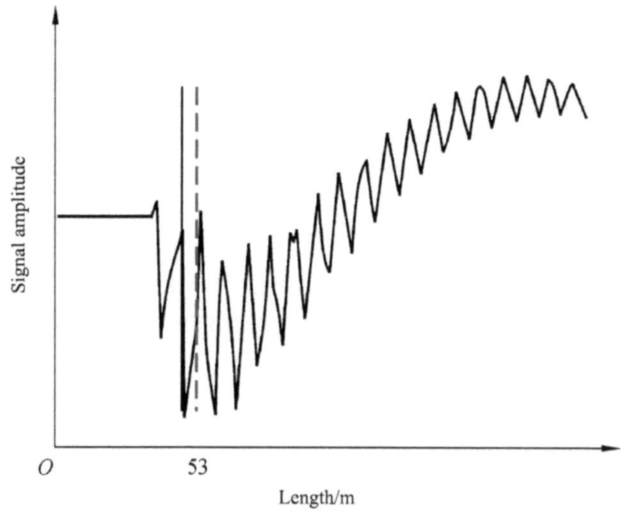

Fig. 2.12 Pulse Current Method Fault Waveform for Distance Measurement

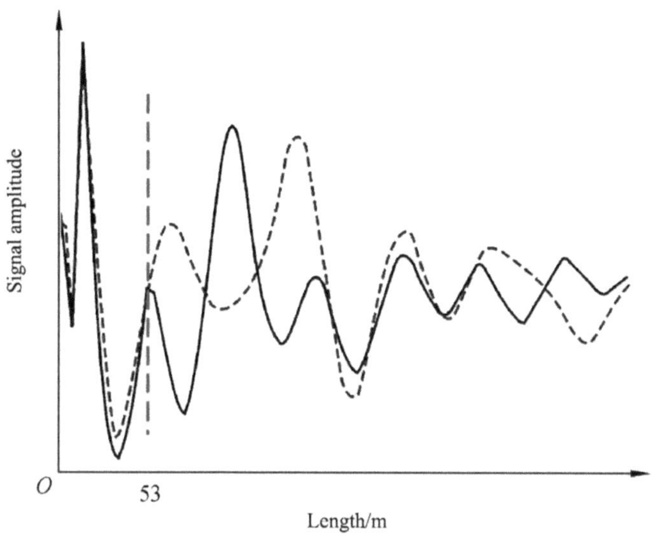

Fig. 2.13 Secondary Pulse Method Fault Waveform for Distance Measurement

Capacitive Current Detection Method

Under normal operating conditions, a power cables forms a corresponding capacitance between the conductor and the ground, exhibiting a symmetrical distribution. There is a certain correlation between the length of the cable and its capacitance, which follows a linear relationship. The so-called capacitive current detection method is based on this principle, effectively distinguishing between power and cable faults. This method is commonly used to detect faults in the conductors of cables. In this method is typically used to measure the capacitive current at both ends of the cable. By comparing the results from normal and faulty core wires, the location of the fault within the cable is accurately identified.

2.2 Traction Power Supply Equipment Fault Diagnosis Based on Particle Swarm Optimization Neural Network

The traction transformer is a pivotal element of the traction power supply system, tasked with the transformation, transmission, and distribution of electrical power. The reliable and stable functioning of traction transformers is crucial for the entire system's performance. Identifying hidden faults early in these transformers is vital to ensure operational safety. A fault in a traction transformer often manifests as a significant alteration in the oil gas content. While the traditional modified three-ratio method effectively extracts fault information, it does suffer from issues like coding loss. Neural networks, known for their robust feature extraction capabilities, can overcome these limitations. Thus, this section starts with analyzing the fault characteristics based on the gas content in the traction transformer oil, integrating the three-ratio method coding with a neural network trained using particle swarm optimization to effectively diagnose faults in traction transformers.

2.2.1 Traction Transformer Fault Diagnosis Methods

2.2.1.1 Fault Diagnosis Process

The entire diagnostic process is divided into three parts: the first part is the processing of the original data; the second part involves calculations using multiple sets of neural networks; the third part uses the output results from each neural network to determine the fault. Figure 2.14 shows the fault diagnosis process of the traction transformer.

1. Data Processing

   The more distinct the features of the neural networks' inputs, the better the classification performance. Therefore, simple processing of the input oil and gas data can be done, using the percentage of dissolved gases in oil for each transformer and the codes under the three-ratio method rules as inputs for the neural networks. In this chapter, CH_4, C_2H_6, C_2H_4, C_2H_2, and H_2 are selected as the feature gases, resulting in a total of 5 inputs. The percentage expression is: as follows:

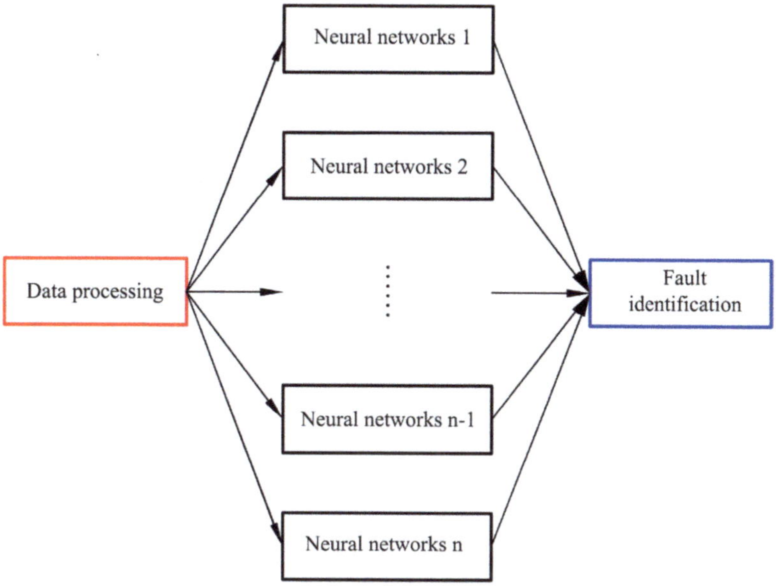

Fig. 2.14 Detailed Fault Diagnosis Process for Traction Transformers

$$P_{Vj} = \frac{V_j}{\sum_{i=1}^{5} V_i} \tag{2.14}$$

2. Multiple Neural Networks

   In 1985, Rumelhart and McClelland introduced a type of multilayer feedforward neural networks trained using the backpropagation algorithm (BP neural network). Mathematical theories have since proven that a three-layer neural network can approximate any nonlinear continuous function with arbitrary precision. This makes it well-suited for addressing complex internal mechanism issues in traction transformer fault diagnostics. The fundamental idea involves using the gradient descent method to adjust the neural network's weights and thresholds, minimizing the mean squared error between the network's actual output and the expected output value. Through learning, the network autonomously discerns a relationship between input and output data, adaptively storing the acquired knowledge as network weights. Traditional BP neural network training comprises forward propagation for output calculation and back propagation for adjusting the neural network's weights and thresholds. This section retains the forward propagation process of the BP neural network and introduces particle swarm optimization to improve the fitness function for training neural networks.

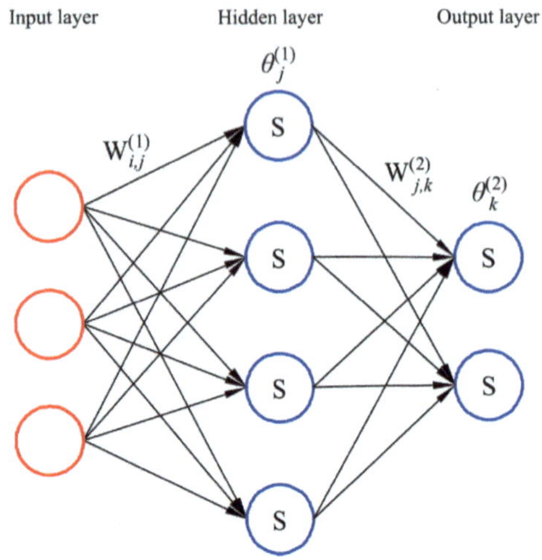

Fig. 2.15 BP Neural Network Model

The forward propagation process involves processing from the input layer through the hidden layer to the output layer, ultimately yielding the computation result.

As shown in Fig. 2.15, a typical BP neural network is divided into three layers: the input layer, hidden layer, and output layer. Each layer contains several neurons, with neurons in adjacent layers connected by connection weights and thresholds. The input to the hidden layer is given by:

$$O_j^{(1)} = \sum_{i=1}^{n_1} w_{i,j}^{(1)} \times x_i - \theta_j^{(1)} \tag{2.15}$$

where x_i represents the inputs from the input layer; $w_{ij}^{(1)}$ denotes the weight between the i neuron in the input layer and the j neuron in the hidden layer; θ_j is the threshold for the j neuron in the hidden layer; and $O_j^{(1)}$ is the input quantity for the j neuron in the hidden layer. The output of the hidden layer is then processed to the output layer.

$$y_j = S\left(O_j^{(1)}\right) \tag{2.16}$$

The S function is commonly used as the Sigmoid function. In this section, the Tan-Sigmoid function serves as the activation function in the hidden layer:

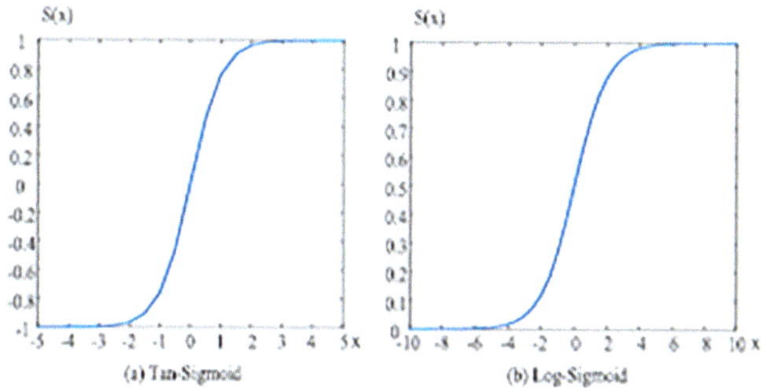

(a) Tan-Sigmoid

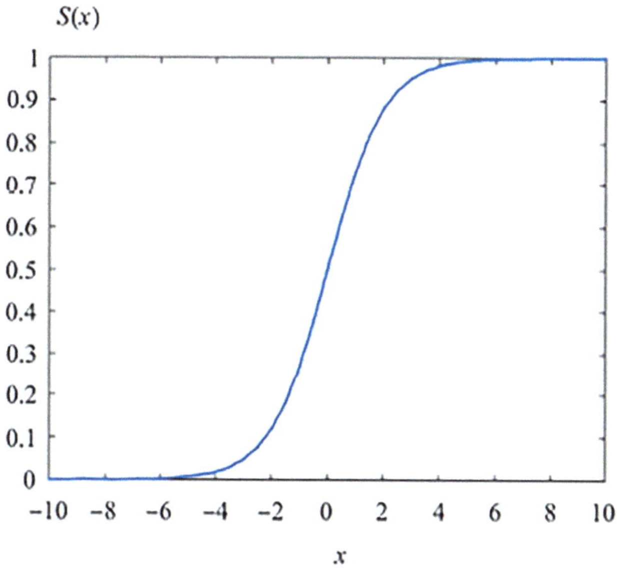

(b) Log-Sigmoid

Fig. 2.16 Sigmoid Function

$$S(x) = \frac{2}{1 + \mathrm{e}^{-2x}} - 1 \tag{2.17}$$

whose expression maps the output values to the range $(-1, 1)$. The function graph is shown in Fig. 2.16a.

The transfer of data from the hidden layer to the output layer is similar to the transfer from the input layer to the hidden layer. First, calculate the input to the output layer:

$$O_k^{(2)} = \sum_{j=1}^{n_2} w_{j,k}^{(2)} \times y_j - \theta_k^{(2)} \tag{2.18}$$

where y_j is the output from the previous layer; $w_{j,k}^{(2)}$ is the weight between the j neuron in the hidden layer and the k neuron in the output layer; θ_k is the threshold for the k neuron in the output layer; and $O_k^{(2)}$ is the input for the k neuron in the output layer. The output of the output layer, which is the final result, is as follows:

$$Y_k = S\left(O_k^{(2)}\right) \tag{2.19}$$

The output layer uses the Log-sigmoid function as its activation function,

$$S(x) = \frac{1}{1 + \mathrm{e}^{-x}} \tag{2.20}$$

The function expression maps the output values to the range (0, 1), as illustrated in Fig. 2.16b.

There are seven types of faults in traction transformers. If only one neural network is used for classification, at least 7 output neurons are required. However, the complexity of the target function can lead to the *sawtooth phenomenon*, affecting the training effectiveness. Therefore, this chapter employs seven neural networks, each corresponding to one of the seven fault types (1—low-temperature overheating, 2—high-temperature overheating, 3—partial discharge, 4—low-energy discharge, 5—high-energy discharge, 6—low-energy discharge with overheating, 7—high-energy discharge with overheating). Each neural network identifies whether it pertains to its specific type of fault, which not only reduces the complexity of the target function but also increases the negative sample size.

Based on the results of the original data processing, the input layer is set with 8 neurons. After comparing training results by adjusting the number of neurons, the hidden layer is determined to have 15 neurons, and the output layer has 2 neurons, known as the fault neuron and the normal neuron. The output results $\left[Y_1^{(i)}, Y_2^{(i)}\right]$ represent the two outputs of the i neural network, where $Y_1^{(i)}$ is the output of the fault neuron, and $Y_2^{(i)}$ is the output of the normal neuron. During the neural network training, the expected output values are set as [1, 0] and [0, 1], representing the fault state and normal condition, respectively.

3. Fault Identification.

   Initially, select all results where $Y_1^{(i)} > Y_2^{(i)}$.

If no such $Y_1^{(i)} > Y_2^{(i)}$ results exist, the system is considered to be operating under normal conditions. The confidence level P for this normal state is then calculated.

$$P = \frac{\sum_{i=1}^{7} Y_2^{(i)}}{\sum_{i=1}^{7} Y_1^{(i)} + \sum_{i=1}^{7} Y_2^{(i)}} \times 100\% \tag{2.21}$$

Conversely, if only one result meets, it is identified as a specific type of fault, and the fault confidence level P is computed accordingly.

$$P = \frac{Y_1^{(j)}}{Y_1^{(j)} + Y_2^{(j)}} \times 100\% \tag{2.22}$$

If multiple results, the result with the highest Y_1 value is chosen for fault identification, with the confidence level P still calculated based on eq. (2.22).

2.2.1.2 Training Neural Networks with Particle Swarm Optimization by Improving the Fitness Function

Training neural networks through Particle Swarm Optimization involves optimizing the network's weights and thresholds. The training procedure is depicted in Fig. 2.17.

Step 1: Determine the neuron count in each layer.

Step 2: Initialize a swarm of particles, where each particle represents potential network weights and thresholds; set the iteration number and learning factor.

Step 3: Assess the fitness of each particle.

Step 4: Calculate the distance of each particle from both the current and the global optimum, updating particle positions accordingly.

Step 5: End training if the maximum iteration number is reached or if the optimal configuration meets the predefined training criteria. If not, repeat from Step 3.

The essence of using PSO for optimizing neural networks lies in the particle-based search for the best weights and thresholds. The effectiveness of this method hinges on the fitness function, typically utilizing the mean squared error as a measure of network performance, described as follows:

$$Fitness = 1 - E_{\text{total}} \tag{2.23}$$

where *Fitness* denotes the measure of suitability, and E_{total} represents the total mean squared error:

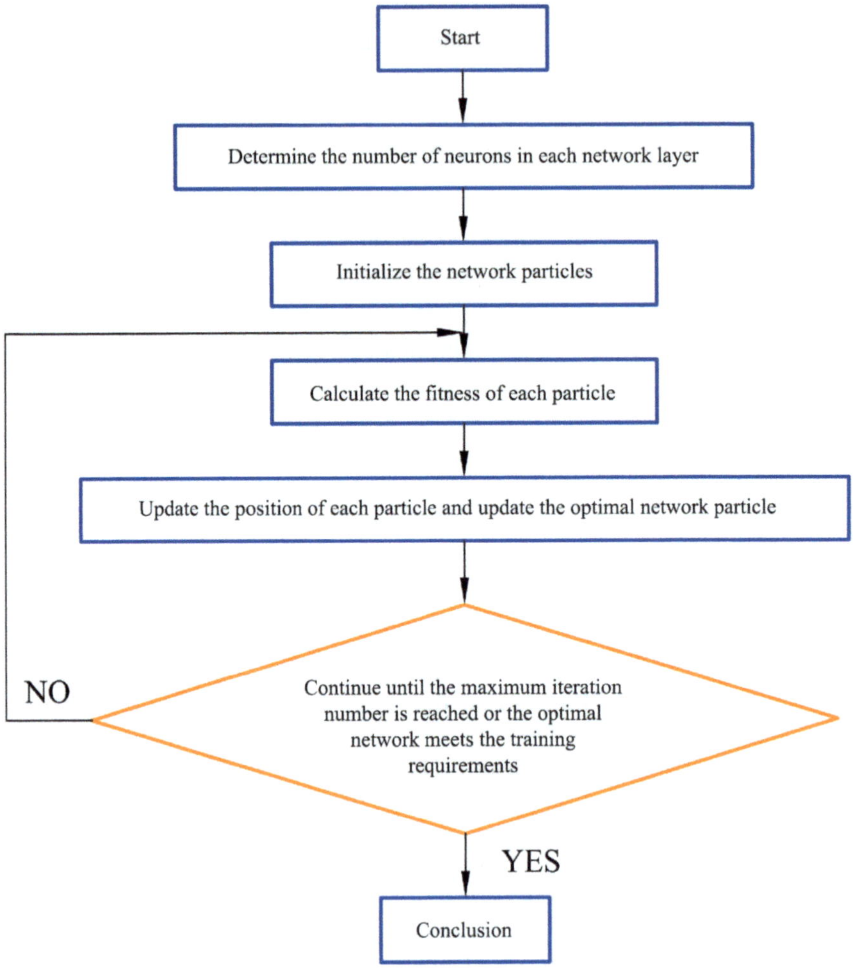

Fig. 2.17 Particle Swarm Optimization Neural Network Process Flow

$$E_{\text{total}} = \sum_{k=1}^{n_2} \frac{1}{2}\left(target_k - Y_k\right)^2 \tag{2.24}$$

where Y_k represents the output of each neuron in the output layer during the forward propagation process, and $target_k$ is the expected output for the sample data.

It is widely accepted that a smaller MSE during neural network training indicates better performance. However, when neural networks are used for classification tasks and training reaches a certain level of maturity, MSE alone may not fully capture the network's true performance. Classification accuracy emerges as a crucial metric for evaluating neural network quality. Consequently, incorporating classification accuracy into the fitness function provides a more accurate representation of network performance. The modified fitness function is defined as follows:

$$Fitness = 1 - E_{\text{total}} + \lambda \tag{2.25}$$

where, λ represents the classification accuracy.

In the early stages of network training, both the mean squared error and the classification accuracy show significant changes, jointly dominating the fitness function. In the later stages of training, the network's output mean squared error E_{total} changes very little, while the classification accuracy still shows relatively larger changes, better reflecting the network's performance, thereby dominating the fitness function.

2.2.2 *Case Study of Traction Transformer Fault Diagnosis*

Initially, a set of data is designated as training samples, which includes: 22 samples for low-temperature overheating faults, 16 for high-temperature overheating faults, 13 for partial discharge faults, 16 for low-energy discharge faults, 17 for high-energy discharge faults, 9 for combined low-energy discharge and overheating faults, and 6 for high-energy discharge with overheating faults, totaling 99 samples. These samples are used to train a neural network with a structure of 8-15-2 using particle swarm optimization and the improved fitness function. Taking the high-energy discharge neural network as a case study, Fig. 2.18 illustrates the optimization effect of training neural networks with the improved fitness function using particle swarm optimization. Figure 2.18a displays the trend in MSE changes; Fig. 2.18b shows the trend in fault diagnosis accuracy for the training samples; Fig. 2.18c demonstrates the fitness variation trend; Figs. 2.18d–j present the output results of the training samples. As depicted, MSE begins to converge from the 100th iteration, and the accuracy rate of the training samples reaches 100% by the 121st iteration. The fitness function also starts to converge at this point, as shown in Fig. 2.18c.

Table 2.18 shows a set of dissolved gases in oil data for a traction transformer.

According to the coding rules in Table 2.6, the three-ratio coding for this set of data is (0, 0, 1). The volume fraction is shown in Table 2.19.

The results obtained by inputting the three-ratio coding and the volume fraction of dissolved gases in oil into the seven-particle swarm-optimized neural networks are shown in Table 2.20.

According to the fault identification rules, the diagnostic result is a low-energy discharge with overheating fault, and its confidence level is:

$$P = \frac{0.9743}{0.9743 + 0.2352} \times 100\% \approx 80.55\% \tag{2.26}$$

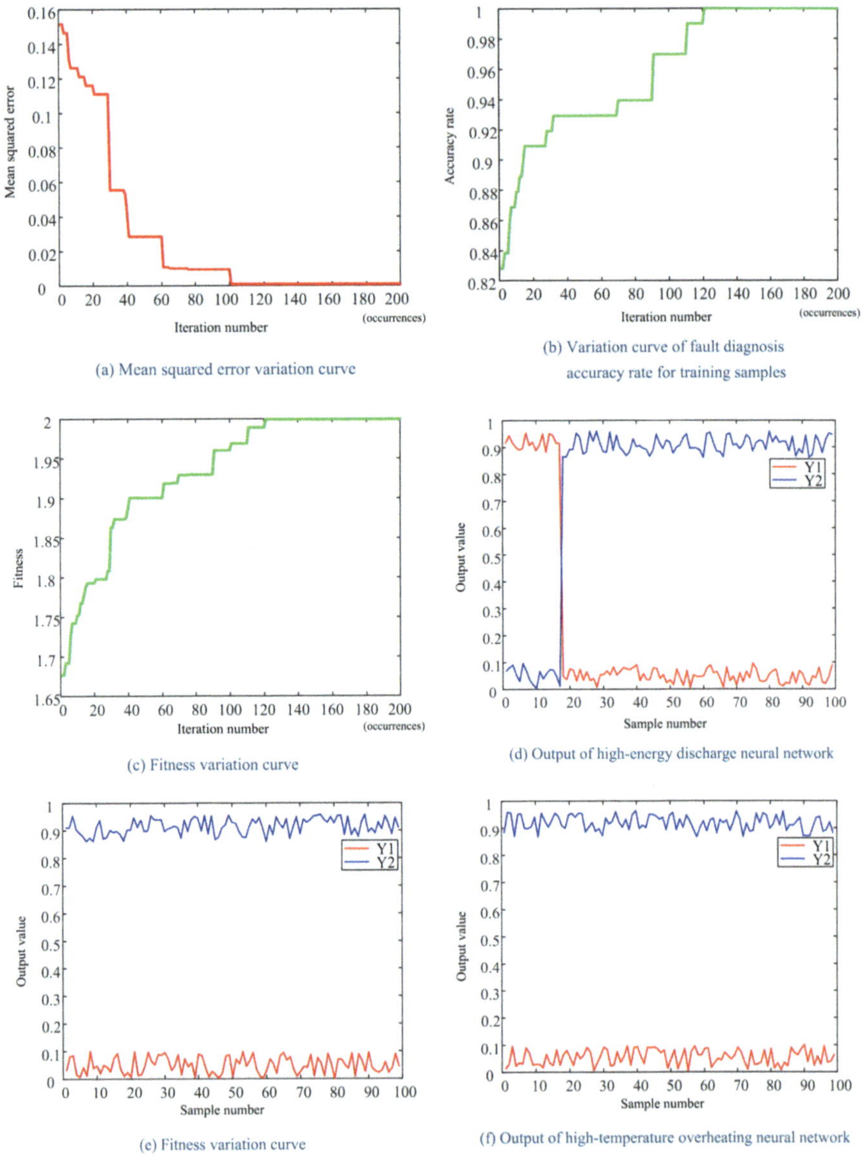

Fig. 2.18 Effects of Particle Swarm Optimization Training Neural Networks with Improved Fitness Function

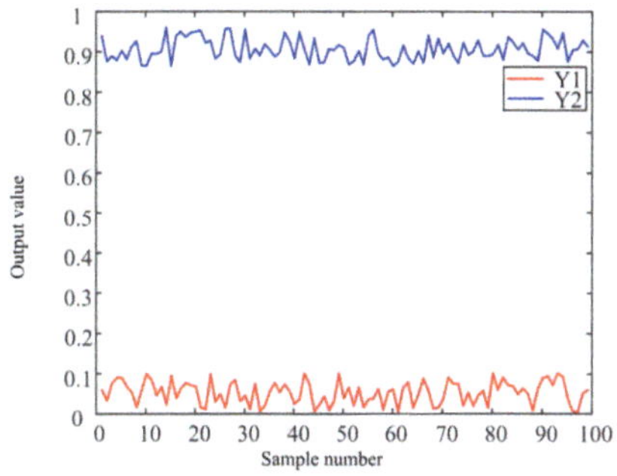

(g) Output of partial discharge neural network

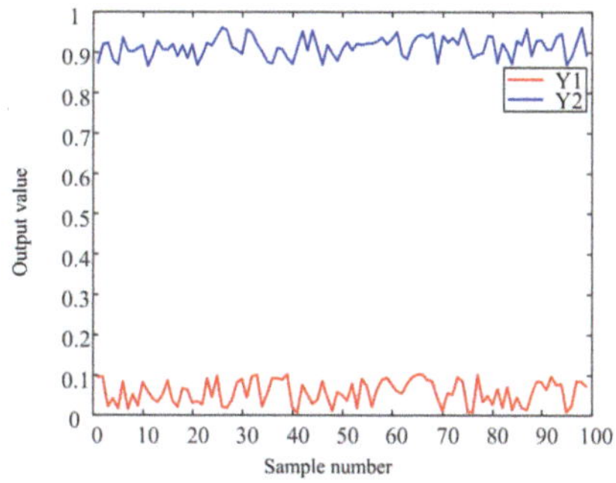

(h) Output of neural network for low-energy discharge

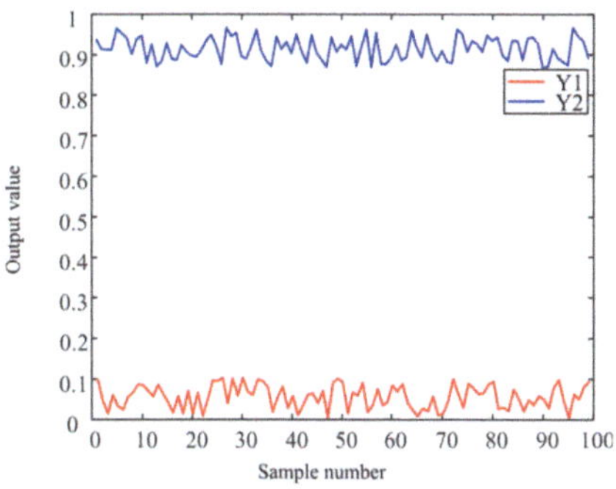

(i) Output of low-energy discharge with overheating neural network

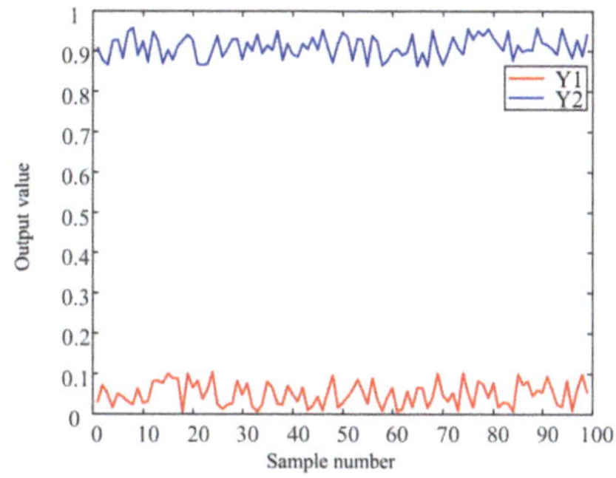

(j) Output of high-energy discharge with overheating neural network

Fig. 2.18 (continued)

Table 2.18 Dissolved gases in oil data for a traction transformer

H_2/(μL/L)	CH_4/(μL/L)	C_2H_6/(μL/L)	C_2H_4/(μL/L)	C_2H_2/(μL/L)
82.29	22.90	4.91	5.09	0

Table 2.19 Volume fraction of dissolved gases in oil for the traction transformer

H_2/(%)	CH_4/(%)	C_2H_6/(%)	C_2H_4/(%)	C_2H_2/(%)
71.44	19.88	4.26	4.42	0

This result aligns with the actual fault type of the traction transformer. This case study underscores that while the three-ratio method alone was insufficient in identifying the fault, the incorporation of neural networks effectively addressed the limitations of incomplete coding inherent in the three-ratio method.

Compare the results of fault diagnosis using the particle swarm optimization neural network model introduced in this section, the BP neural network, and standard particle-optimized neural networks. Take the high-energy discharge fault diagnosis neural networks as a case study, trained with the particle swarm

Table 2.20 Output results from the neural networks after inputting the three-ratio coding and volume fractions

Y	Low-temperature overheating neural network	High-temperature overheating neural network	Partial discharge neural network	Low-energy discharge neural network	High-energy discharge neural network	Neural network for low-energy discharge with overheating	Neural network for high-energy discharge with overheating
Y_1	0.1625	0.2139	0.1391	0.1326	0.1728	0.9743	0.1963
Y_2	0.8786	0.8246	0.8492	0.9989	0.7965	0.2352	0.7223

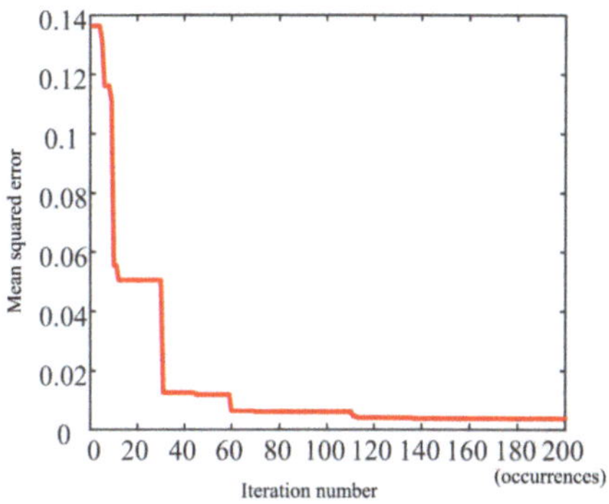

(a) mean squared error variation curve

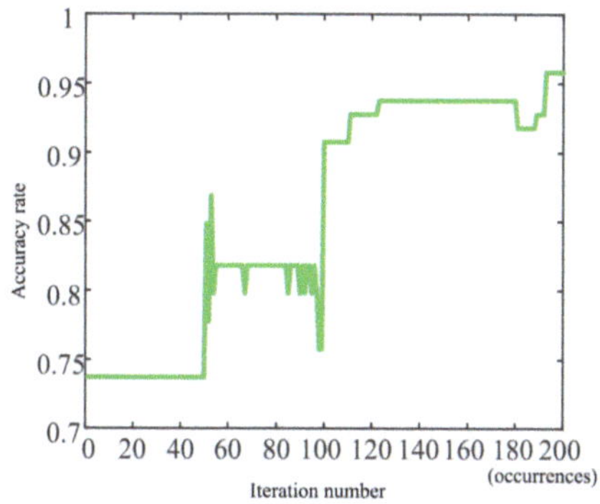

(b) training sample fault diagnosis accuracy variation curve

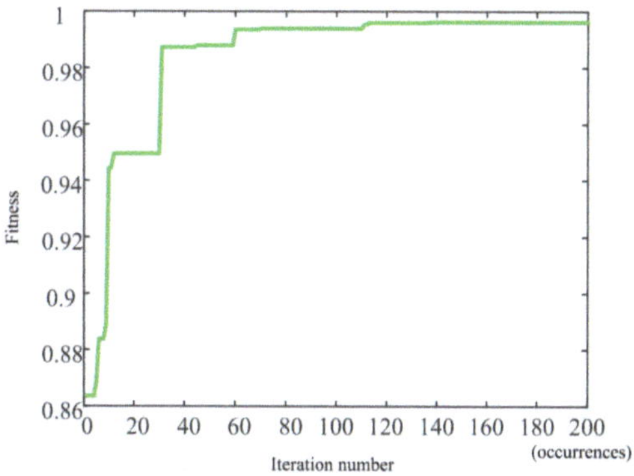

(c) fitness variation curve

Fig. 2.19 Effects of Training Neural Networks with Particle Swarm Optimization

optimization using the fitness function defined by eq. (2.23). Figure 2.19 illustrates the optimization effects, where Fig. 2.19a shows the trend of mean squared error variation; Fig. 2.19b displays the variation trend of fault diagnosis accuracy for training samples; and Fig. 2.19c shows the fitness variation trend. From Figs. 2.19a, c, it can be observed that the mean squared error begins to converge from the 107th iteration, and the fault diagnosis accuracy of the training samples ultimately reaches 96%.

As shown in Fig. 2.20, the BP neural network training effects are illustrated. From Fig. 2.20a, it can be seen that the BP neural network begins to converge starting from the 700th iteration. Compared to training with particle swarm optimization, its convergence speed is slower.

From the data, 22 low-temperature overheating fault data, 16 high-temperature overheating fault data, 13 partial discharge fault data, 16 low-energy discharge fault data, 17 high-energy discharge fault data, 9 low-energy discharge with overheating fault data, and 6 high-energy discharge with overheating fault data were selected, totaling 99 test data. The test results are shown in Table 2.21.

Fig. 2.20 BP Neural Network Training Effects

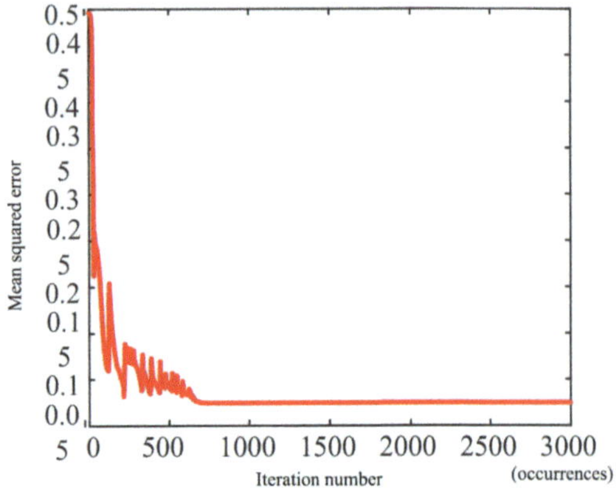

(a) mean squared error variation curve

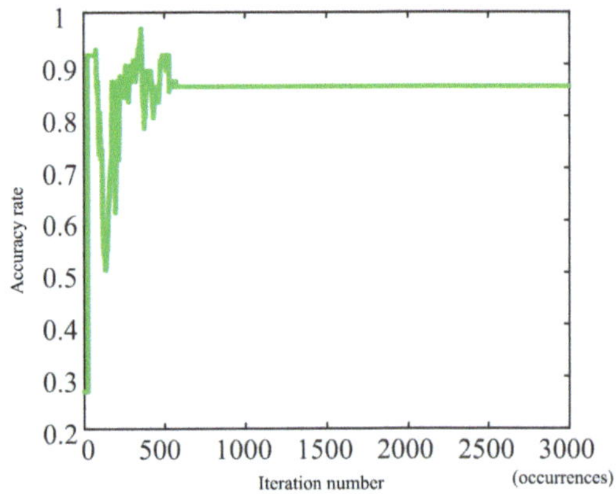

(b) training sample fault diagnosis accuracy variation curve

Table 2.21 Comparison of fault diagnosis results from three methods

Fault type	Accuracy rate (%)		
	This chapter's model	BP neural network	PSO neural network
Low-temperature overheating	89.90	71.72	81.82
High-temperature overheating	92.93	72.73	75.76
Partial discharge	87.88	67.68	83.84
Low-energy discharge	90.91	77.78	86.87
High-energy discharge	91.92	76.77	87.88
Low-energy discharge with overheating	92.93	74.75	78.79
High-energy discharge with overheating	88.89	79.80	80.81

In this section, the model's fault diagnosis accuracy is calculated solely based on its determination results, without considering its confidence level. The correct diagnosis count divided by the total number of test samples represents the accuracy rate.

The particle swarm optimization neural network model presented in this chapter markedly surpasses the other models in terms of fault diagnosis accuracy, notably outperforming the BP neural network, which exhibits the least effective results. This comparison leads to two important conclusions:

1. The mean squared error of the neural network output does not comprehensively reflect the network's true performance.
2. Incorporating fault diagnosis accuracy into the fitness function of PSO-optimized neural networks effectively enhances diagnostic precision without leading to overfitting.

2.3 Traction Power Supply System Fault Diagnosis Based on Fuzzy Petri Net

Traction power supply systems, operating under severe conditions, are prone to faults that can halt trains and disrupt subsequent operations, triggering a cascading effect. Rapid and precise fault diagnosis is vital for minimizing accident-related losses and mitigating impacts.

Fuzzy Petri nets, an advancement on traditional Petri nets, are increasingly used in the fault diagnosis of power grids and traction power supply systems. These models, defined by rigorous mathematical principles and intuitive inference processes, align well with the dynamics of fault occurrences and management in electrical grids. They demonstrate significant fault tolerance to data uncertainty. Fuzzy Petri nets, in particular, use the accuracy rate of protection actions, derived from statistical laws, to represent the genuine effectiveness of protection and circuit breaker operations. The diagnostic results are presented probabilistically, offering a confidence level that enhances the credibility of the inference and provides a realistic basis for explanations. However, the reliance on long-term statistical averages to handle uncertainty does not reflect the specific moment's availability level and fails to accurately represent the operational availability of protection and circuit breaker systems in traction power supply contexts.

Therefore, this section examines the effects of uncertainty in protection equipment and circuit breaker data on fault diagnostic outcomes. By studying the time-dependent availability patterns of these equipments, their Weibull distribution curves are derived. These curves represent the uncertainty in information sources for protection and circuit breakers, thus establishing the link between fault diagnostic outcomes and the operational duration of protection and circuit breaker equipment.

2.3.1 Fuzzy Petri Net Fault Diagnostic Methods

A fuzzy Petri net (FPN) enhances the fundamental Petri net structure. It assigns a real number between 0 and 1 to both places and transitions, indicating the confidence that a place is accurate and the likelihood of a transition occurring. A fuzzy Petri net is defined as follows:

$$S_{FPN} = (P, T, D, \boldsymbol{I}, \boldsymbol{O}, \alpha, \beta, Th, \boldsymbol{U}, \boldsymbol{W}) \tag{2.27}$$

where $P = (p_1, p_2, \cdots p_n)$ represents a finite set of places; $T = (t_1, t_2, \cdots t_m)$ denotes a finite set of transitions; $D = (d_1, d_2, \cdots d_n)$ signifies a finite set of propositions; $\boldsymbol{I}$: $P \rightarrow T$ is the input matrix, which maps places to transitions, with $I = \{\delta_{ij}\}$, δ_{ij} being a binary value, $\delta_{ij} \in \{0, 1\}$. $\delta_{ij} = 1$ indicates that P_i is an input to T_j; $\delta_{ij} = 0$ means it is not, where $i = 1, 2, \cdots n, j = 1, 2, \cdots m$. $\boldsymbol{O}$: $T \rightarrow P$ is the output matrix, $O = \{\gamma_{ij}\}$, γ_{ij} being a binary value, $\gamma_{ij} \in \{0, 1\}$. $\gamma_{ij} = 1$ indicates that P_i is an output of T_j; $\gamma_{ij} = 0$ means it is not, where $i = 1, 2, \cdots n$, $j = 1, 2, \cdots m$; α: $P \rightarrow [0, 1]$ indicates the confidence level of the proposition associated with a place, reflecting the credibility of the information from the place; β: $P \rightarrow D$ maps places to propositions. If $\alpha(p_i) = y_i$, where $y_i \in [0, 1]$, and $\beta(p_i) = d_i$, then the confidence level for the proposition d_i is y_i; Th: $Th \rightarrow [0, 1]$ sets a threshold t_i for transitions $t_i (t_i \in T)$, with $Th = \{\lambda_1, \lambda_2, \cdots, \lambda_m\}$; $\boldsymbol{U}$: Certainty Factor (CF) matrix, $U = diag\ (\mu_1, \mu_2, \cdots, \mu_m)$, μ_j represents the likelihood of transition t_j occurring, $\mu_j \in [0, 1]$; $\boldsymbol{W} = \boldsymbol{W}_I \cup \boldsymbol{W}_o$, represents the set of all input and output weights on flow relationships, with $\boldsymbol{W}_I = \{w_1, w_2, \cdots w_n\}$ and $\boldsymbol{W}_o = \{\mu_1, \mu_2, \cdots \mu_n\}$.

Using a fuzzy Petri net for fault diagnosis in power grids primarily involves mapping the action information flow between protection, circuit breakers, and component faults. The true degree of the resultant proposition is deduced from the initial proposition's confidence level, based on established rules.

In a fuzzy Petri net, if the sum $\sum \alpha(p_{Ij}) \cdot w_{Ij} > Th(t)$, $j = 1, 2 \cdots n$, then the transition t can be triggered, creating a new confidence level in the subsequent place P_{Oj}:

$$CF(t) = \begin{cases} \sum_j a(pI_j) \cdot w_{Ij}, & \sum_j a(pI_j) \cdot w_{Ij} > Th(t) \\ 0, & \sum_j a(pI_j) \cdot w_{Ij} < Th(t) \end{cases} \tag{2.28}$$

The piecewise function $CF\ (t)$ can be approximated by a continuous function $G(x)$:

$$G(x) = x \cdot f_a(x) \cdot \mu \tag{2.29}$$

where

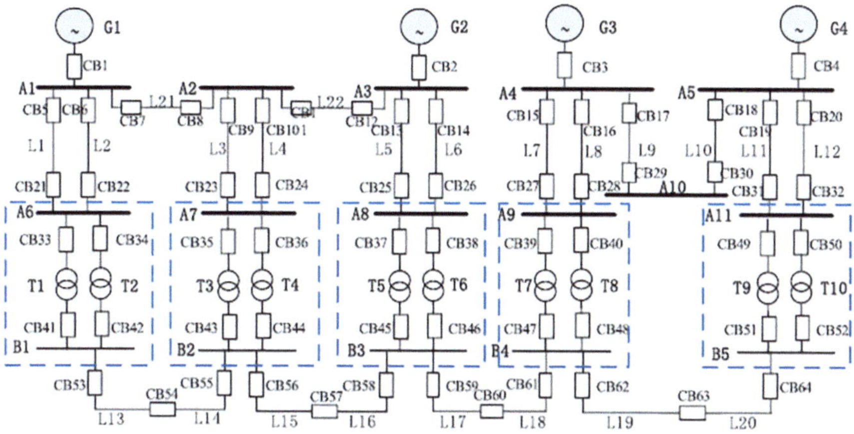

Fig. 2.21 Power Grid Topology of a Certain Section of the Zhejiang-Gan Traction Power Supply System

$$x = \sum_j a(pI_j) \cdot w_{Ij} \tag{2.30}$$

$$f_a(x) = \frac{1}{1 + e^{-b(x - Th(t))}} \tag{2.31}$$

where b is sufficiently large such that when $x > Th(t), f_a(x) \approx 1$, and when $x < Th(t)$, $f_a(x) \approx 0$.

When a place's transition meets the trigger condition, the transition is activated, and the confidence level in the next place becomes $G(x)$. When multiple transitions are triggered, the new confidence level at the place selects the highest result among the incoming transitions, ultimately determining the component fault's confidence level. Based on this confidence value, the likelihood of a component fault can be assessed.

In a segment of the traction power supply system within Jiangxi territory along the Zhejiang-Jiangxi line, as illustrated in Fig. 2.21, there are five traction substations: Pingxiang North, Xicun, Jieshui, Huangtugang, and Zhangshu, depicted by dashed boxes in the diagram. They are interconnected through four sectioning posts: Luxi, Binjiang, Shuixi, and Zhangjiashan, represented as CB54, CB57, CB60, and CB63 in the diagram. The system comprises 52 components and 64 circuit breakers, with the components being: busbars Al–A11, B1–B5, lines L1–L22, transformers T1–T10, power plants G1–G4, and the 64 circuit breakers as CB1–CB64.

When conducting fault diagnosis using a fuzzy Petri net, action information from circuit breakers and protection equipment collected via SCADA is utilized to determine the faulted component or equipment. As depicted in Fig. 2.21, when a fault occurs in line L1, the fuzzy Petri net model, considering line primary protection, backup protection, and the corresponding circuit breaker actions, is shown in Fig. 2.22.

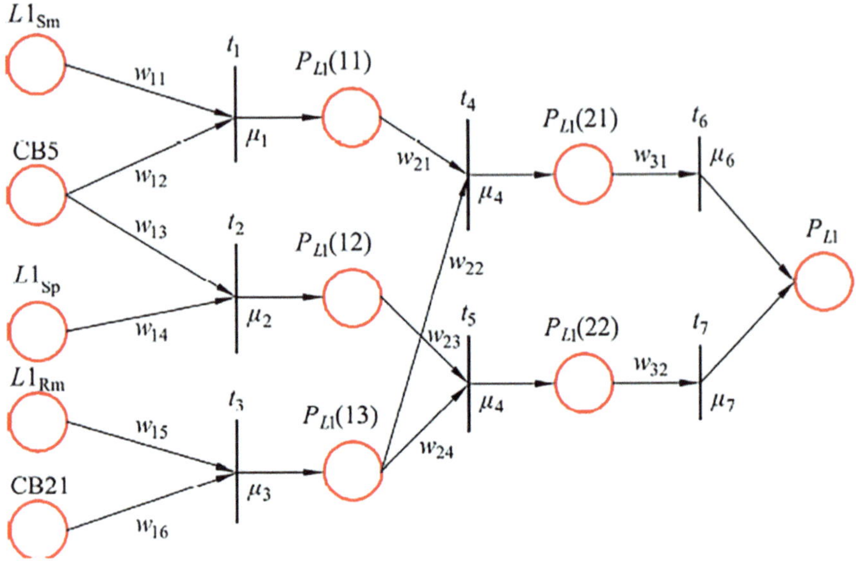

Fig. 2.22 FPN Model for Fault Diagnosis of Line L1

In the diagram, $\mathrm{Ll}_{\mathrm{Sm}}$ and $\mathrm{Ll}_{\mathrm{Rm}}$ denote the primary protection at the upper and lower ends of line L1, respectively, while $\mathrm{L1}_{\mathrm{Sp}}$ represents the backup protection at the upper end. When a fault occurs on line L1, the primary protections $\mathrm{Ll}_{\mathrm{Sm}}$ and $\mathrm{Ll}_{\mathrm{Rm}}$ initially act, tripping the corresponding circuit breakers CB5 and CB21 at both ends of the line. If the primary protection fails to act or their corresponding circuit breakers do not operate, the backup protection $\mathrm{L1}_{\mathrm{Sp}}$ is activated, disconnecting the circuit breakers at both ends of the line. In the figure, w, μ and t are defined as in the previous formula (2.27) for the fuzzy Petri net.

The common inference method used in fault diagnosis compares the value α at the given initial place with the transition's threshold value. When α exceeds this threshold, the transition is triggered, and the α value of the next place is obtained, continuing this process until the confidence value of the output place is determined. The fuzzy inference process, as illustrated in Fig. 2.22:

1. Initially, triggers t_1, t_2, t_3 within T_1 are activated.

$$\begin{cases} x_1 = \alpha(\mathrm{Ll}_{\mathrm{Sm}}) \cdot w_{11} + \alpha(\mathrm{CB5}) \cdot w_{12} \\ x_2 = \alpha\left(\mathrm{Ll}_{\mathrm{Sp}}\right) \cdot w_{14} + \alpha(\mathrm{CB5}) \cdot w_{13} \\ x_3 = \alpha(\mathrm{Ll}_{\mathrm{Rm}}) \cdot w_{15} + \alpha(\mathrm{CB21}) \cdot w_{16} \end{cases} \tag{2.32}$$

Using function approximation, the intermediate place's α value is obtained.

$$\begin{cases} \alpha(P_{L1}(11)) = x_1 \cdot \mu_1 / \left(1 + e^{-b(x_1 - Th(t_1))}\right) \\ \alpha(P_{L1}(12)) = x_2 \cdot \mu_2 / \left(1 + e^{-b(x_2 - Th(t_2))}\right) \\ \alpha(P_{L1}(13)) = x_3 \cdot \mu_3 / \left(1 + e^{-b(x_3 - Th(t_3))}\right) \end{cases} \tag{2.33}$$

2. Next, triggers t_4, t_5 within T_2 are activated,

$$\begin{cases} x_4 = \alpha(P_{L1}(11)) \cdot w_{21} + \alpha(P_{L1}(13)) \cdot w_{22} \\ x_5 = \alpha(P_{L1}(12)) \cdot w_{23} + \alpha(P_{L1}(13)) \cdot w_{24} \end{cases} \tag{2.34}$$

resulting in:

$$\begin{cases} \alpha(P_{L1}(21)) = x_4 \cdot \mu_4 / \left(1 + e^{-b(x_4 - Th(t_4))}\right) \\ \alpha(P_{L1}(22)) = x_5 \cdot \mu_5 / \left(1 + e^{-b(x_5 - Th(t_5))}\right) \end{cases} \tag{2.35}$$

3. Finally, triggerst_6, t_7 within T_3 are activated.

$$\begin{cases} x_6 = \alpha(P_{L1}(21)) \cdot \mu_6 / \left(1 + e^{-b(\alpha(P_{L1}(21)) - Th(t_6))}\right) \\ x_7 = \alpha(P_{L1}(22)) \cdot \mu_7 / \left(1 + e^{-b(\alpha(P_{L1}(22)) - Th((t_7))}\right) \end{cases} \tag{2.36}$$

4. Ultimately, the confidence value of the output place is determined.

$$\alpha(P_{L1}) = max(x_6, x_7) \tag{2.37}$$

The diagnostic process begins with the status information from the protection equipment and circuit breaker. The diagnostic outcome is linked to the availability of this status information. Currently, the confidence level of the status information in the initial place is typically calculated using statistical data to determine the average correct action rate:

$$R = \frac{\sum N_r}{N} \tag{2.38}$$

where N_r represents the number of occurrences of correct action, and N denotes the total number of operations.

Assuming a correct action rate of 0.9913 for primary protection, 0.8 for backup protection, and 0.9833 for the correct action rate of the circuit breaker, with a b value of 200, a weight w of 0.5 for all, a threshold of 0.3, and a CF μ of 0.95, the probability of failure for line L1, as calculated in eqs. (2.32)–(2.37) in Fig. 2.22,

$$\begin{cases} x_1 = 0.9913 \times 0.5 + 0.9833 \times 0.5 = 0.9873 \\ x_2 = 0.8000 \times 0.5 + 0.9833 \times 0.5 = 0.8916 \\ x_3 = 0.9913 \times 0.5 + 0.9833 \times 0.5 = 0.9873 \end{cases}$$

$$\begin{cases} \alpha(P_{\mathrm{L1}}(11)) = 0.9873 \times 0.95 / \left(1 + \mathrm{e}^{-200 \times (0.9873 - 0.3)}\right) = 0.9379 \\ \alpha(P_{\mathrm{L1}}(12)) = 0.8916 \times 0.95 / \left(1 + \mathrm{e}^{-200 \times (0.8916 - 0.3)}\right) = 0.8417 \\ \alpha(P_{\mathrm{L1}}(13)) = 0.9873 \times 0.95 / \left(1 + \mathrm{e}^{-200 \times (0.9873 - 0.3)}\right) = 0.9379 \end{cases}$$

$$\begin{cases} x_4 = 0.9379 \\ x_5 = 0.8925 \end{cases}$$

$$\begin{cases} \alpha(P_{\mathrm{LI}}(21)) = 0.8910 \\ \alpha(P_{\mathrm{LI}}(22)) = 0.8479 \end{cases}$$

$$\begin{cases} x_6 = 0.8465 \\ x_7 = 0.8055 \end{cases}$$

Thus, the final diagnostic result is

$$\alpha(P_{\mathrm{L}}) = max(0.8465, 0.8055) = 0.8465$$

That is, the probability of failure for line L1 is 0.8465.

From this analysis, the use of a fuzzy Petri net for fault diagnosis in traction power supply systems exhibits significant characteristics:

1. The credibility of the diagnostic results depends on the confidence level of the initial place. In the fault diagnosis of power systems, this confidence level is often represented by the correct action rate of the protection equipment and circuit breakers. Given that the correct action rate is closely related to equipment availability, the outcomes of a fuzzy Petri net diagnosis should correlate with the availability of the protection equipment and circuit breakers.
2. Currently, the confidence level of the initial places for protection and circuit breakers is calculated based on statistical data representing the correct action rate. This represents the long-term average value of availability, which masks the variability and uncertainty of the protection equipment and circuit breaker information, failing to reflect the actual situation of their availability in real-time. This significantly reduces the confidence level of the fault diagnostic result.
3. The weight assigned to protection and circuit breaker information in fault diagnosis reasoning lacks a solid basis for determination and is highly subjective.

It is evident that the availability of protection equipment and circuit breakers significantly impacts the results of fault diagnosis. Higher availability of equipment not only enhances the confidence level of the information it provides but also assigns greater weight in the inference process, thus leading to more accurate and realistic diagnostic results.

2.3.2 *Place Confidence Model Based on Weibull Distribution*

The availability of protection equipment and circuit breakers is intrinsically linked to their operational duration. As time progresses and equipment advances through different stages of their service lives, their failure rates, and consequently their availability, vary. This variability renders the equipment's failure rate a function of time. The Weibull distribution, originally proposed by Swedish scientist Weibull while investigating material strength and other phenomena, offers numerous application benefits and has become a staple in availability theory. It stands as one of the most prevalent models for analyzing equipment lifespan and availability in recent years. The Weibull distribution boasts a well-defined physical basis and broad practical applications, characterized by high compatibility and robust data fitting capabilities. Crucially, it can effectively encapsulate various failure processes and characteristics across different failure periods of a product. Other distributions, such as the exponential and normal distributions, are considered special cases of the Weibull distribution, underscoring its suitability for modeling the availability of protection equipment and circuit breakers.

2.3.2.1 Equipment Availability Weibull Distribution

The commonly utilized two-parameter Weibull distribution exhibits the following features:

$$F(t) = 1 - \exp\left[-\left(\frac{t}{\eta}\right)^m\right] \tag{2.39}$$

Distribution density function:

$$f(t) = \frac{mt^{m-1}}{\eta^m}\exp\left[-\left(\frac{t}{\eta}\right)^m\right] \tag{2.40}$$

Availability function:

$$R(t) = 1 - F(t) = \exp\left[-\left(\frac{t}{\eta}\right)^m\right] \tag{2.41}$$

Failure rate function:

$$\lambda(t) = \frac{mt^{m-1}}{\eta^m} \tag{2.42}$$

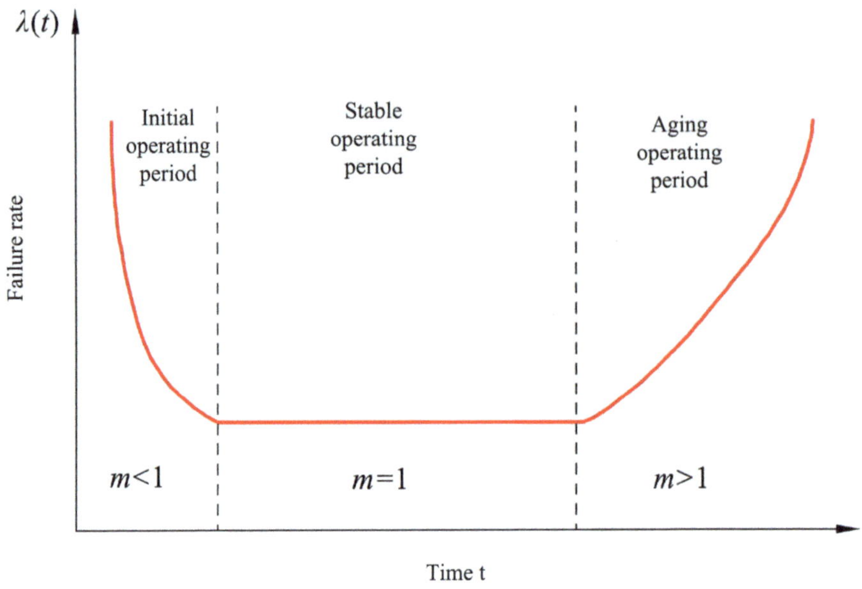

Fig. 2.23 Relationship Between the Bathtub Curve of Equipment Failure Rate and the Shape Parameter

where t denotes time; m is the shape parameter, which determines the trend of the failure rate; and η is the scale parameter, influencing the minimal failure rate.

From eq. (2.42), the following observations can be made: When $m < 1$, the failure rate function $\lambda(t)$ decreases, implying a diminishing failure rate with t, which is typically suitable for modeling early failures. When $m = 1$, the failure rate $\lambda(t)$ remains constant, indicative of the equipment operating within a period of random failures, and is appropriate for modeling random failures. When $m > 1$, the failure rate function $\lambda(t)$ increases, signifying that the equipment is in a wear-out failure period, primarily suited for modeling wear or aging failures. These three shapes correspond precisely to the three stages of the bathtub curve depicted in Fig. 2.23, illustrating the extensive applicability of the Weibull distribution in representing various stages of equipment availability.

2.3.2.2 Weibull Distribution Parameter Estimation

It is evident that only the shape parameter m and the scale parameter η need to be determined to define the Weibull distribution. Due to the non-linear nature of these parameters, a transformation into linear parameters is necessary to apply the least squares method effectively. This transformation is achieved by taking the logarithm of both side of the eq. (2.39):

$$\ln[-\ln(1-F(t)]=m\ln t-m\ln\eta \tag{2.43}$$

If we set $y = \ln[-\ln(1-F(t)]$, $x = \ln t$, eq. (2.43) can then be expressed in a linear form:

$$y=mx-m\ln\eta \tag{2.44}$$

For any given sample, the equation becomes:

$$\ln\left[-\ln\left(1-\widehat{F}_{\mathrm{i}}\right)\right]=m\ln t_{(\mathrm{i})}-m\ln\eta \tag{2.45}$$

where i is the index of the failure record; $\widehat{F}_{\mathrm{i}}$ is the non-parametric estimated value of the distribution function $F(t)$. A common method to determine $\widehat{F}_{\mathrm{i}}$ is the Bernard median rank estimation:

$$\widehat{F}_{\mathrm{i}}=\frac{i-0.3}{n+0.4} \tag{2.46}$$

where i is the failure record index and n is the number of fault data samples. When the sample data is large, $\widehat{F}_{\mathrm{i}}$ obtained by various methods are typically very close, making the Bernard median rank estimation a reliable choice for estimating the fault distribution function of protection equipment and circuit breakers.

After estimating the distribution function using observed sample data $\left\{\ln t_i, \ln\left[\ln\frac{1}{1-F(t)}\right]\right\}$, the least squares method is used to determine the shape parameter m and the scale parameter η, leading to the formulation of the equipment availability function.

The least squares method estimates the parameters m and η with the objective of minimizing the deviation, which is typically the residual between the actual values and the estimated values. When using x and y as independent variables, the deviations obtained generally differ; thus, parameters are estimated using two methods: minimizing the vertical residual sum of squares $(y_i-\widehat{y}_i)^2$ and minimizing the horizontal residual sum of squares $(x_{\mathrm{i}}-\widehat{x}_{\mathrm{i}})^2$.

1. Parameter Estimation with Minimum Vertical Residual Sum of Squares

   Using y as the variable, the vertical residual sum of squares is defined as:

$$R_{yx}=\sum_{i=1}^{r}e_i^2=\sum_{i=1}^{r}(y_i-\widehat{y}_i)^2 \tag{2.47}$$

The estimates for the shape parameter and scale parameter are given by

$$\begin{cases} m_{yx} = \dfrac{\sum_{i=1}^{r} [(x_i - \overline{x})(y_i - \overline{y})]}{\sum_{i=1}^{r} (x_i - \overline{x})^2} \\ \eta_{yx} = \exp\left[- \left(\dfrac{\overline{y}}{m_{yx}} - \overline{x} \right) \right] \end{cases} \tag{2.48}$$

where $\widehat{y}_i = mx_i - m \ln \eta$, $\overline{x}$ is the mean of x, and $\overline{y}$ is the mean of y.

2. Parameter Estimation with Minimum Horizontal Residual Sum of Squares
   Using x as the variable, the horizontal residual sum of squares is defined as:

$$R_{xy} = \sum_{i=1}^{r} e_i^2 = \sum_{i=1}^{r} \left(x_i - \hat{x}_i \right)^2 \tag{2.49}$$

The estimates for the shape parameter and scale parameter are calculated by

$$\begin{cases} m_{xy} = \dfrac{\sum\limits_{i=1}^{r} (y_i - \overline{y})^2}{\sum\limits_{i=1}^{r} [(x_i - \overline{x})(y_i - \overline{y})]} \\ \eta_{xy} = \exp\left[\left(\overline{x} - \dfrac{\overline{y}}{m_{xy}} \right) \right] \end{cases} \tag{2.50}$$

where, $\widehat{x} = \frac{y}{m} + \ln \eta$

From eqs. (2.47) and (2.49), we obtain

$$\frac{R_{yx}}{R_{xy}} = m^2 \tag{2.51}$$

It is clear that when the coefficient $m > 1$, the parameter estimation derived from the model minimizing the horizontal residual sum of squares performs better; otherwise, the model minimizing the vertical residual sum of squares is more effective.

Data pertaining to a certain type of protection equipment's operational fault records is displayed in Table 2.22, with corresponding calculations detailed in Table 2.23.

Utilizing both the horizontal and vertical residual methods, the estimated parameters obtained are summarized in Table 2.24.

The results indicate that the horizontal residual sum of squares method yields a smaller residual sum of squares and a higher goodness of fit, suggesting a more precise fit compared to the vertical residual sum of squares method. This finding aligns with the conclusions drawn from eq. (2.51). Consequently, the parameter estimation results obtained using the horizontal residual sum of squares method are utilized, and the cumulative distribution function and availability function for this group of relay protection equipment are established:

Table 2.22 Fault statistics of a type of protection equipment

	Fault condition			
Basic information	Serial number	Fault timing	Operating time (h)	Status serial number
	1	2003-01-01	25,560	F_1
Commissioning date	2	2003-09-20	31,848	F_2
2000-02-01	3	2003-03-07	36,648	F_3
Termination date	4	2005-01-01	43,104	F_4
2008-02-01	5	2005-06-07	46,872	F_5
Period normal	6	2005-10-01	49,656	F_6
	7	2006-01-01	51,864	F_7
A total of 21 records	8	2006-06-06	55,608	F_8
	9	2007-01-01	60,264	F_9

$$\begin{aligned} F_{re}(t) &= 1 - \exp\left[-\left(\frac{t}{82740.890}\right)^{3.0932}\right] \\ R_{re}(t) &= \exp\left[-\left(\frac{t}{82740.890}\right)^{3.0932}\right] \end{aligned} \tag{2.52}$$

The availability function curve of the protection equipment is displayed in Fig. 2.24.

In the context of using a Petri net for fault diagnosis, another critical input is derived from the action information of circuit breakers. Given the demanding working conditions of the traction power supply system, circuit breaker actions are frequent. With an increasing number of operations, the wear on contacts and mechanical components of the circuit breakers intensifies, consequently diminishing their availability.

The fault statistics data for the vacuum circuit breaker of this model are shown in Table 2.25.

Unlike protection equipment, the lifespan of a circuit breaker is mainly assessed by the number of operations. For this model of circuit breaker, the statistical duration is $T = 50{,}000$ operations, with the number of failures during the experiment being $r = 5$.

The data calculated using the horizontal residual and vertical residual formulas are shown in Tables 2.26 and 2.27.

The analysis indicates that the parameter estimation using the method of minimizing horizontal residual sum of squares yields smaller residuals and a higher goodness of fit, demonstrating a better fitting effect. Therefore, this result is selected as the estimated parameters for the availability distribution of this model of circuit breaker, leading to the formulation of the following cumulative distribution function and availability function:

$$\begin{aligned} F_{CB}(t) &= 1 - \exp\left[-\left(\frac{t}{37403.3894}\right)^{1.6453}\right] \\ R_{CB}(t) &= \exp\left[-\left(\frac{t}{37403.3894}\right)^{1.6453}\right] \end{aligned} \tag{2.53}$$

The availability curve of the circuit breaker is illustrated in Fig. 2.25.

Table 2.23 Weibull distribution comparison table of calculation results for the protection equipment

i	t_i	F_i	$x_i = \ln t_i$	$y_i = \ln[-\ln(1-F_i)]$	$x_i - \overline{x}$	$y_i - \overline{y}$	$(x_i - \overline{x})(y_i - \overline{y})$	$(x_i - \overline{x})^2$	$(y_i - \overline{y})^2$
1	25,560	0.0230	10.1488	−3.7595	−0.5247	−1.7488	0.9175	0.2753	3.0583
2	31,848	0.0559	10.3687	−2.8552	−0.3047	−0.8445	0.2573	0.0928	0.7132
3	36,648	0.0888	10.5091	−2.3750	−0.1643	−0.3644	0.0599	0.0270	0.1328
4	43,104	0.1217	10.6714	−2.0419	−0.0021	−0.0312	0.0001	0.0000	0.0010
5	46,872	0.1546	10.7552	−1.7841	0.0817	0.2266	0.0185	0.0067	0.0513
6	49,656	0.1875	10.8129	−1.5720	0.1394	0.4387	0.0612	0.0194	0.1925
7	51,864	0.2204	10.8564	−1.3904	0.1829	0.6202	0.1135	0.0335	0.3847
8	55,608	0.2533	10.9261	−1.2307	0.2526	0.7799	0.1970	0.0638	0.6083
9	60,264	0.2862	11.0124	−1.0873	0.3390	0.9234	0.3130	0.1149	0.8527

Table 2.24 Weibull distribution parameter estimation and comparison for the protection equipment

Method used	Result comparison			
	m	η	Residual sum of squares (RSS)	Goodness of fit
Vertical residual sum of squares method	3.0594	83,336.843	0.0655	0.9982
Horizontal residual sum of squares method	3.0932	82,740.890	0.0069	0.9999

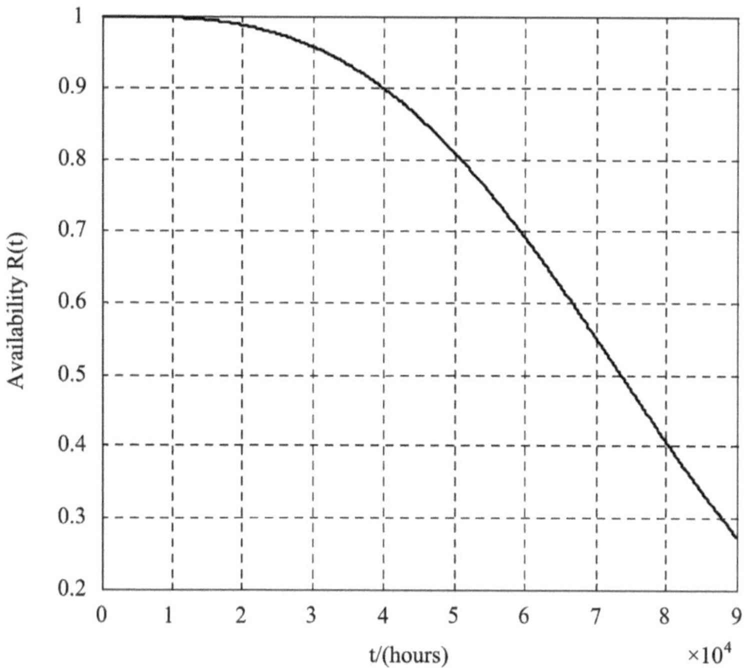

Fig. 2.24 Availability Curve of the Protection Equipment

For the mentioned protection equipment and the circuit breaker, parameter estimation resulting in $m > 1$ suggests that the equipment is in an aging operational phase, with its availability diminishing over time. Consequently, for protection equipment and circuit breakers with varying service durations, the credibility of the information they provide will alter as their operational time advances. When using signals from protection equipment and circuit breakers as sources of information for fault diagnosis, it is crucial to consider the impact of the equipment's availability on the credibility of the information source.

Table 2.25 Fault statistics of a vacuum circuit breaker

Serial number	Number of failures/次	Fault cause	Status serial number
1	10,010	Deviation of insulating rod from the guide frame	F_1
2	18,015	Transmission shaft failure	F_2
3	26,000	Exceeding technical range	F_3
4	36,483	Permanent magnet mechanism coil burnout during closure	F_4
5	42,000	Vacuum arc-extinguishing chamber pressure resistance failure	F_5
6	50,000	No failure occurred, test stopped	–

2.3.3 *Case Study of Traction Power Supply System Fault Diagnosis Based on Fuzzy Petri Net*

2.3.3.1 Fuzzy Petri Net Diagnosis in Traction Power Supply Systems

When using a fuzzy Petri net for fault diagnosis, the confidence level α of the initial place directly affects the credibility of the final diagnosis result. α represents the degree of trust in the protection equipment and circuit breaker information, which is related to the availability of the protection equipment and circuit breaker. The more reliable the equipment is, the higher the credibility of the information it provides, and thus, this information is given more weight in the reasoning process. Based on this understanding, this article proposes a modified approach to represent the information confidence level in the fuzzy Petri net fault diagnosis process. It utilizes the equipment's availability function to depict the credibility of the information provided. This adjustment ensures that the fault diagnostic result correlates with the service time of the equipment, enabling the Petri net's diagnostic outcomes to reflect time characteristics more accurately and align with the actual operational conditions of the equipment. Additionally, the weight of information in the inference process is adjusted based on the degree of availability-information from equipment with high availability carries more weight in the inference process, whereas information from equipment with lower availability has less influence. This adjustment enhances the credibility of the inference results and can minimize the impact of equipment misoperations and failures to operate.

The Petri net utilized for fault diagnosis in this article is defined as $S_{FPN} = (P, T, D, \boldsymbol{I}, \boldsymbol{O}, \alpha, \beta, Th, \boldsymbol{U}, \boldsymbol{W})$.

This structure follows the format of eq. (2.27) but differs in that the initial confidence level α of the place is not determined solely by the method specified in eq. (2.38), and the weights w are not arbitrarily assigned.

The initial confidence level of a protection place:

Table 2.26 Weibull distribution comparison table of calculation results for the circuit breaker

i	t_i	F_i	$x_i = \ln t_i$	$y_i = \ln[-\ln(1 - F_i)]$	$x_i - \overline{x}$	$y_i - \overline{y}$	$(x_i - \overline{x})(y_i - \overline{y})$	$(x_i - \overline{x})^2$	$(y_i - \overline{y})^2$
1	10,010	0.1094	9.2113	−2.1556	−0.8539	−1.3961	1.1921	0.7291	1.9490
2	18,015	0.2656	9.7990	−1.1753	−0.2663	−0.4157	0.1107	0.0709	0.1728
3	26,000	0.4219	10.1659	−0.6015	0.1006	0.1580	0.0159	0.0101	0.025
4	36,483	0.5781	10.5046	−0.1473	0.4394	0.6123	0.2690	0.193	0.3749
5	42,000	0.7344	10.6454	0.2819	0.5802	1.0415	0.6043	0.3366	1.0847

Table 2.27 Weibull distribution parameter estimation and comparison for the circuit breaker

Method used	Result comparison			
	m	η	Residual sum of squares	Goodness of fit R^2
Vertical residual sum of squares method	1.6360	37,305.7331	0.0203	0.9931
Horizontal residual sum of squares method	1.6453	374,033,894	0.0075	0.9998

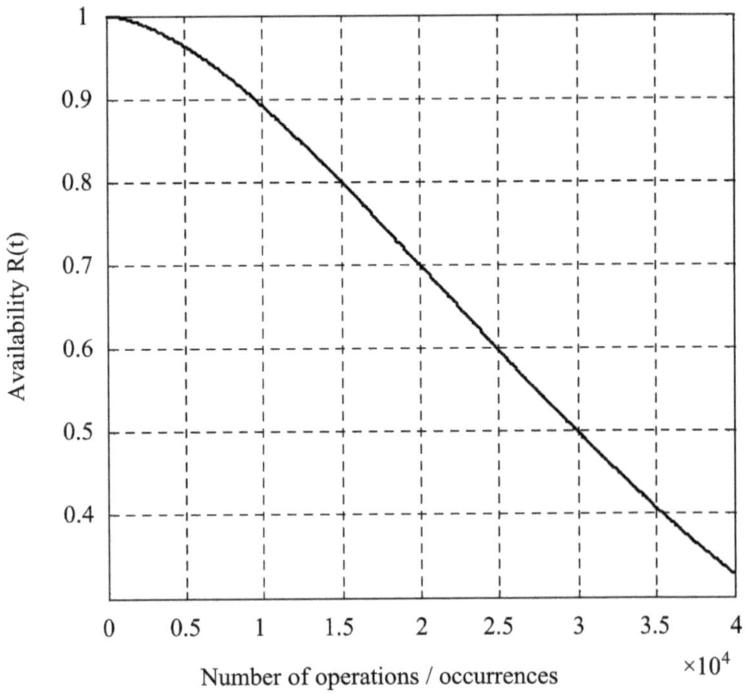

Fig. 2.25 Availability Curve of a Model Vacuum Circuit Breaker

$$\alpha(P_{re}) = R_{re}(t) \tag{2.54}$$

The initial confidence level of a circuit breaker place:

$$\alpha(P_{CB}) = R_{CB}(t) \tag{2.55}$$

Weights of Each Place in Inference:

$$w(P(i)) = \frac{R_i}{\sum_j R_j} \tag{2.56}$$

where R_i is the availability corresponding to that place; $\sum R_j$ represents the sum of reliabilities of all places involved in this inference. For initial places, R_i can be directly provided by the equipment's availability function. For intermediate places, it can be represented by an equivalent availability, which is determined by the availability of the preceding level places involved in the inference, as follows:

$$R_i(P(kl)) = \prod_m R_r(P((k-1)m)) \tag{2.57}$$

$R_r(P((k-1)m))$ represents the availability of the preceding level place involved in the inference of place $P(kl)$.

It follows that the confidence level of the initial place is a function of time and alters as equipment availability changes. Furthermore, the weight of the information in the inference process is dynamically adjusted in response to shifts in equipment availability.

Taking the case study of line L1's fuzzy Petri net fault diagnosis, as shown in Fig. 2.22, the availability function of the protection equipment is articulated by eq. (2.52), and the availability of the circuit breaker by eq. (2.53). Figures 2.26 and 2.27 depict how the weight of each piece of information in the fuzzy inference changes with the service time of the equipment.

From Fig. 2.26, it is observed that as the operating time of the protection equipment increases, the availability of the protection equipment declines. Consequently, the weights w_{11}, w_{14}, and w_{15}, which are associated with the protection information in fault diagnosis, gradually diminish. This trend indicates that the influence of information provided by the protection equipment in fault inference progressively decreases, while the weight of the circuit breaker information incrementally increases. Similarly, as the number of operations of the circuit breakers increases, their availability gradually decreases, and their weight in fuzzy inference also reduces. Thus, this method allows for real-time dynamic adjustment of information weight in fault inference based on the degree of equipment availability, making the diagnostic results more aligned with reality.

Since the intermediate places $P_{L1}(11)$, $P_{L1}(12)$, and $P_{L1}(13)$ are all related to the availability of the protection equipment, changes in the protection equipment's availability similarly affect them, resulting in the subsequent intermediate places' weights w_{21}, w_{22}, w_{23}, w_{31}, w_{32} being less impacted by the availability of protection equipment and circuit breakers, as shown in Figs. 2.27 and 2.28.

The relationship between the fault confidence level of diagnostic results and time, as shown in Fig. 2.28, reveals that the confidence level of diagnostic results changes over time. As the protection equipment and circuit breakers operate over a certain period and their availability decreases to a certain level, the confidence level of

Fig. 2.26 Variation Curve of the Input Layer's Weights w_1 Over Time. (**a**) Input Place Weight w_1 as a Function of Protection Operating Time (Circuit Breaker Number of Operations $t_2 = 1$). (**b**) Input Place Weight w_1 as a Function of Circuit Breaker Operating Time (Protection Operating Time $t_1 = 1$)

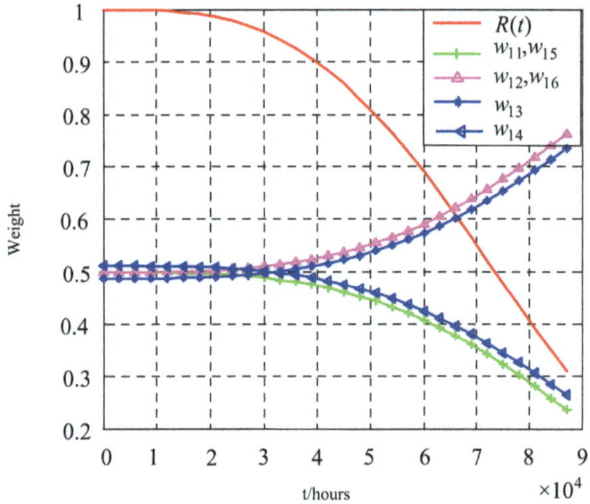

(a) Input Place Weight w_1 as a Function of Protection Operating Time (Circuit Breaker Number of Operations t_2=1):

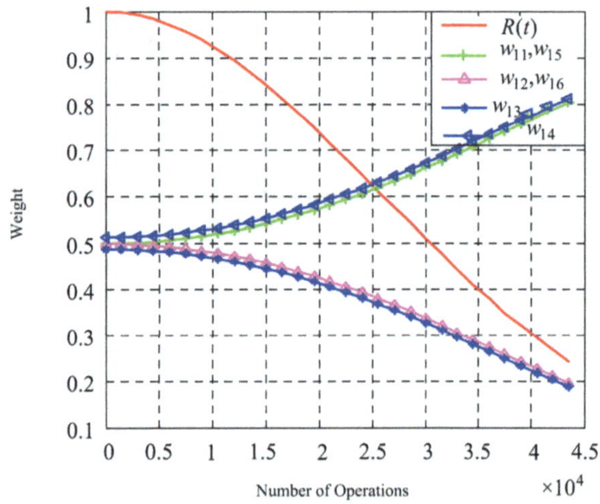

(b) Input Place Weight w_1 as a Function of Circuit Breaker Operating Time (Protection Operating Time t_1=1):

diagnostic results will decrease to zero. This inference concludes that at this point, the equipment's action is considered a misoperation.

Due to the use of dynamic weights, when the availability of equipment decreases, its weight in the inference process is reduced, correspondingly increasing the weight of other information. As a result, in the later stages of operating time, the credibility of the diagnostic results is improved, as shown in Fig. 2.28.

Fig. 2.27 Variation Curve of the Intermediate Layer's Weights w_2 Over Time. (**a**) Variation curve of the intermediate layer's weight w_2 with protection operating time (circuit breaker number of operations $t_2 = 1$). (**b**) Variation curve of the intermediate layer's weight w_2 with circuit breaker operating time (protection operating time $t_1 = 1$)

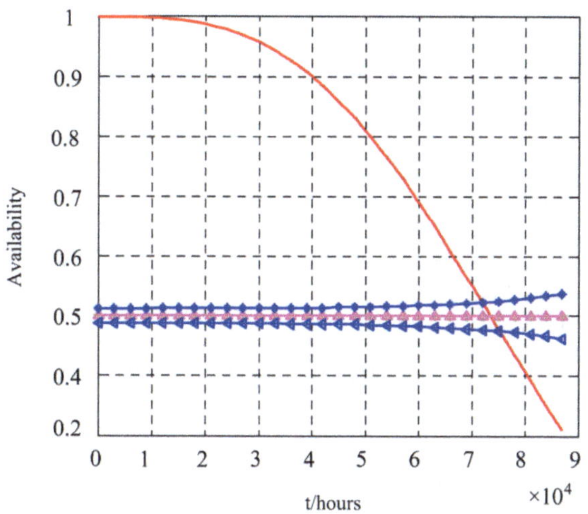

(a) Variation curve of the intermediate layer's weight w_2 with protection operating time (circuit breaker number of operations t_2=1)

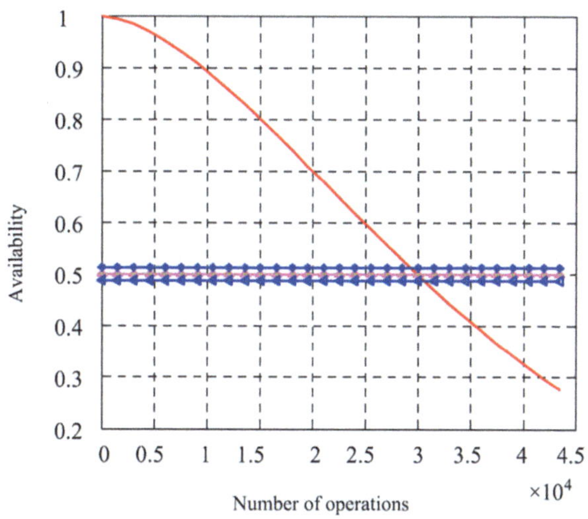

(b) Variation curve of the intermediate layer's weight w_2 with circuit breaker operating time (protection operating time t_1=1)

The analysis above demonstrates that the confidence level of diagnostic results is influenced by the source equipment, namely the availability of the protection equipment and circuit breakers. Therefore, the confidence level of diagnostic results is actually related to both the operating time of the protection equipment and the number of operations of the circuit breakers; it is a function of both. Figure 2.29 shows the relationship between the fault confidence level of diagnostic results and the operating time of the protection equipment and the number of operations of the

Fig. 2.28 Variation Curve of the Intermediate Layer's Weights w_3 Over Time. (**a**) Variation curve of the intermediate layer's weight w_3 with protection operating time (circuit breaker number of operations $t_2 = 1$). (**b**) Variation curve of the intermediate layer's weight w_3 with circuit breaker operating time (protection operating time $t_1 = 1$)

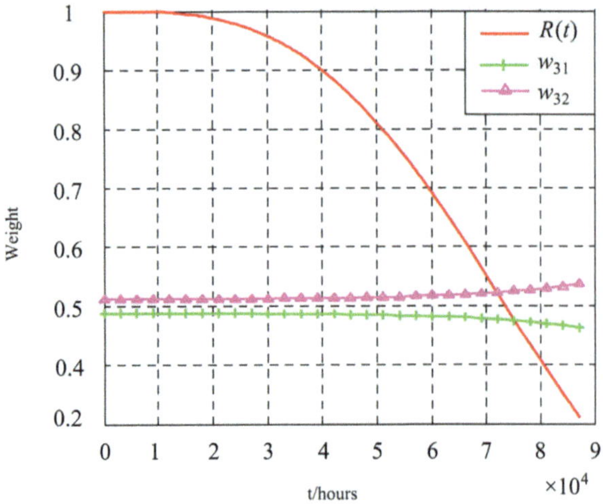

(a) Variation curve of the intermediate layer's weight w_3 with protection operating time (circuit breaker number of operations t_2=1)

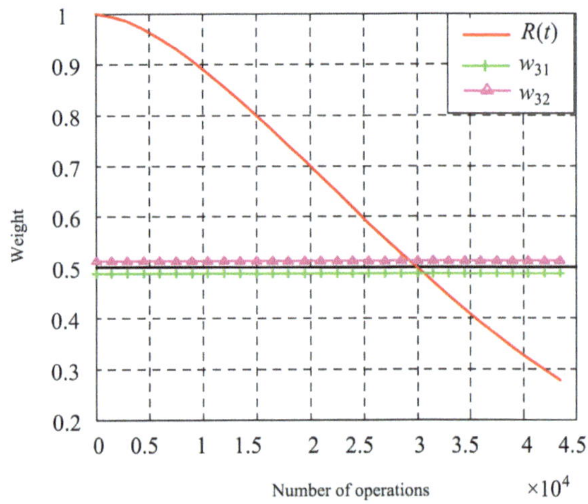

(b) Variation curve of the intermediate layer's weight w_3 with circuit breaker operating time (protection operating time t_1=1)

circuit breaker. From the figure, it can be seen that as the operating time of the protection equipment and the circuit breaker increases, the availability of their actions decreases, and consequently, the confidence level of the fault diagnosis results based on the information they provide also decreases (Fig. 2.30).

Fig. 2.29 Relationship Between the Confidence Level of Diagnostic Results and Time. (**a**) Relationship between the confidence level of diagnostic results and the operating time of the protection equipment. (**b**) Relationship between the confidence level of diagnostic results and the number of operations of the circuit breakers

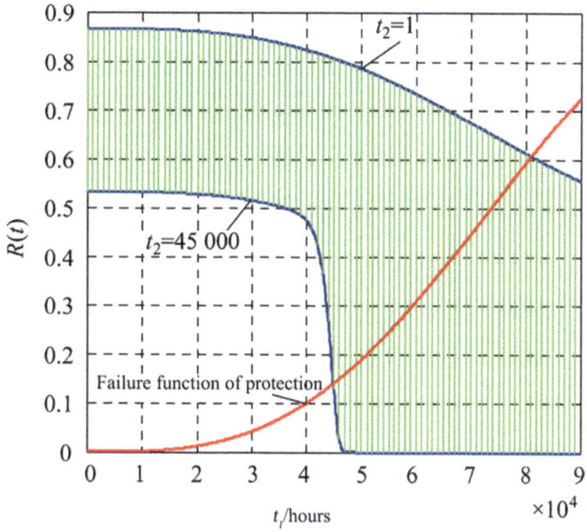

(a) Relationship between the confidence level of diagnostic results and the operating time of the protection equipment.

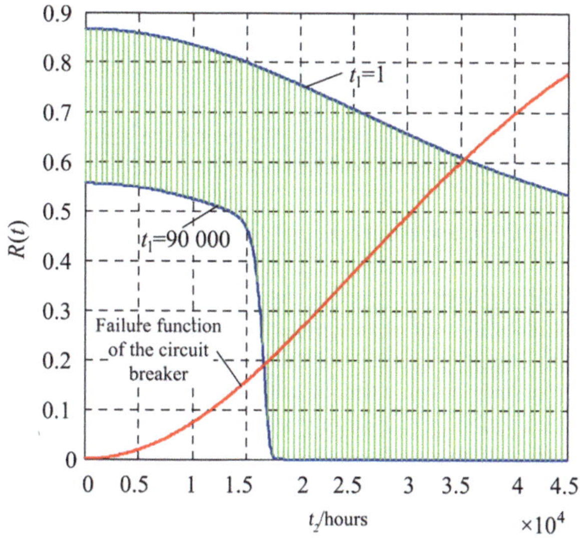

(b) Relationship between the confidence level of diagnostic results and the number of operations of the circuit breakers.

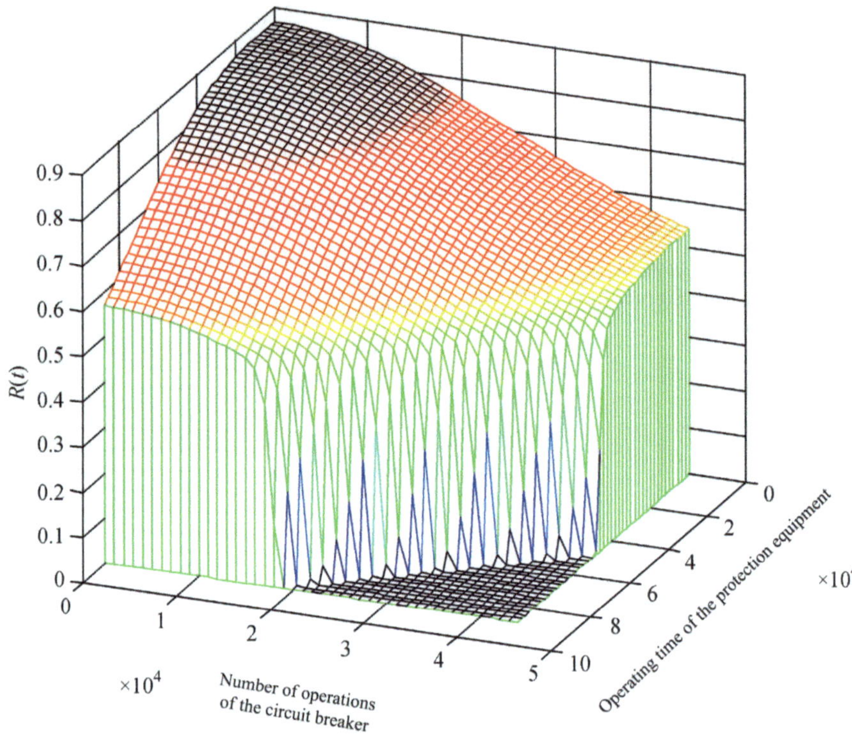

Fig. 2.30 Relationship Between the Confidence Level of Diagnostic Results and the Operating Time of the Protection Equipment and Circuit Breakers

2.3.3.2 Case Analysis

Utilizing the fuzzy Petri net fault diagnosis model that incorporates equipment availability, and leveraging the availability functions of the protection equipment and circuit breakers established earlier as the basis for determining the initial place confidence level in the fuzzy Petri net, a diagnostic model for various components in the traction power supply system depicted in Fig. 2.21 has been developed. For typical components such as bus B1, traction transformer T_1, and line L13, the established FPN diagnostic models are illustrated in Figs. 2.31, 2.32, and 2.33 respectively.

This section of the traction power supply system has been operational since January 2007, with most equipment continuously in use to date. Notably, the protection equipment for line L13 was replaced in June 2008, the protection equipment for bus A6 was replaced following a fault occurrence in July 2010, and the circuit breaker CB42 was replaced in January 2013. Tables 2.28 and 2.29 provide the operating time information for the equipment used in the FPN models depicted in Figs. 2.31, 2.32, and 2.33. It is observed that the availability of the same equipment varies at different times and generally decreases as operating time increases.

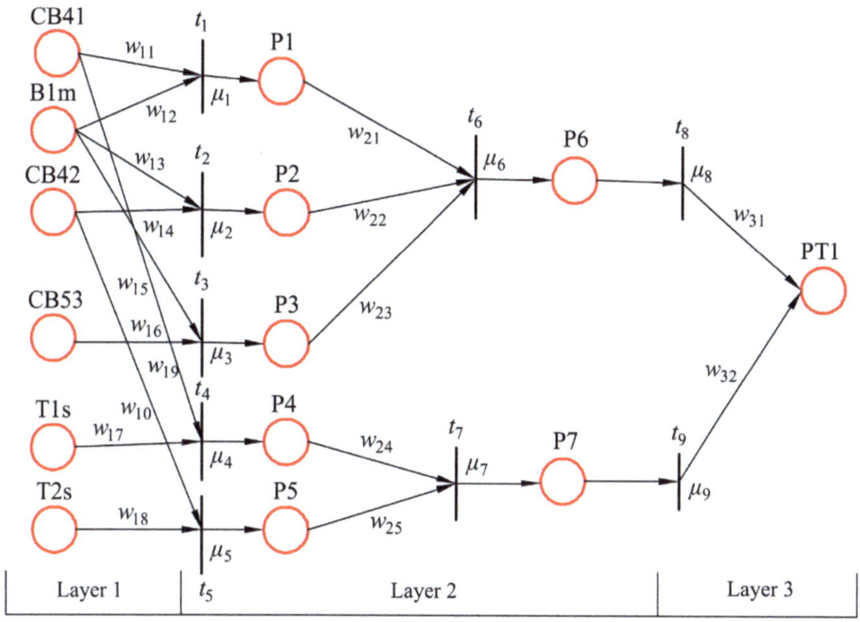

Fig. 2.31 FPN Model of Bus B1

Using the availability listed in the table and according to eq. (2.56), the weight coefficients of each part in the FPN for bus fault diagnosis in Fig. 2.31 are shown in Table 2.30. The changes in the weight coefficients for each part in Figs. 2.32 and 2.33 are shown in Tables 2.31 and 2.32, respectively.

The data from Tables 2.30, 2.31, and 2.32 demonstrate that using the described method, the weights of information from each place during inference are no longer set arbitrarily but dynamically changed according to the availability relationships of the equipment.

Using the weights calculated from Tables 2.30, 2.31, and 2.32, and combining the availability data of protection equipment and circuit breakers from Tables 2.28 and 2.29, the method is applied to diagnose various typical fault conditions. The diagnosis is evaluated using the average correct action rate method, and the results are presented in Table 2.33.

It is evident from the FPN that the correct diagnostic result for faulty components can be obtained. This method integrates the time characteristics of the operational availability of protection equipment and circuit breakers, and it correlates the source availability of the FPN fault diagnosis with the operating time of the equipment. Therefore, the credibility of the diagnostic result is influenced by the service time of the equipment, making it more aligned with actual conditions and enhancing the persuasiveness of the diagnostic result.

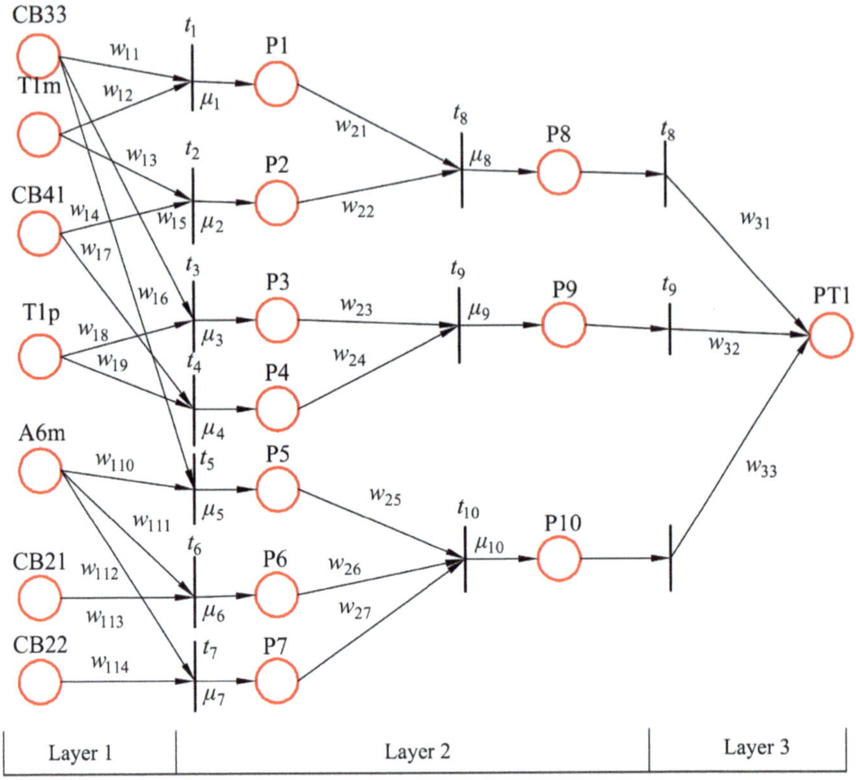

Fig. 2.32 FPN Model of Transformer T_1

2.4 Bayesian Network-Based Traction Power Supply System Fault Diagnosis

China's vast territory features significant climate variations across different regions. The impact of various weather conditions on the traction power supply system and its relay protection equipment varies, allowing weather conditions to be simplified into three statuses: normal weather, bad weather, and severe weather. The influence of uncertainties in relay protection equipment and circuit breakers under these different weather conditions on fault diagnosis is considered. The uncertainties in relay protection equipment and circuit breakers primarily include their inherent design, manufacturing, and aging defects, as well as inaccuracies in measurement data from current transformers and other components under different weather conditions. These uncertainties are manifested through relay protection action information and circuit breaker trip information. Relay protection action information and circuit breaker trip information can be divided into historical statistical information and online information. Historical statistical information refers to relay protection

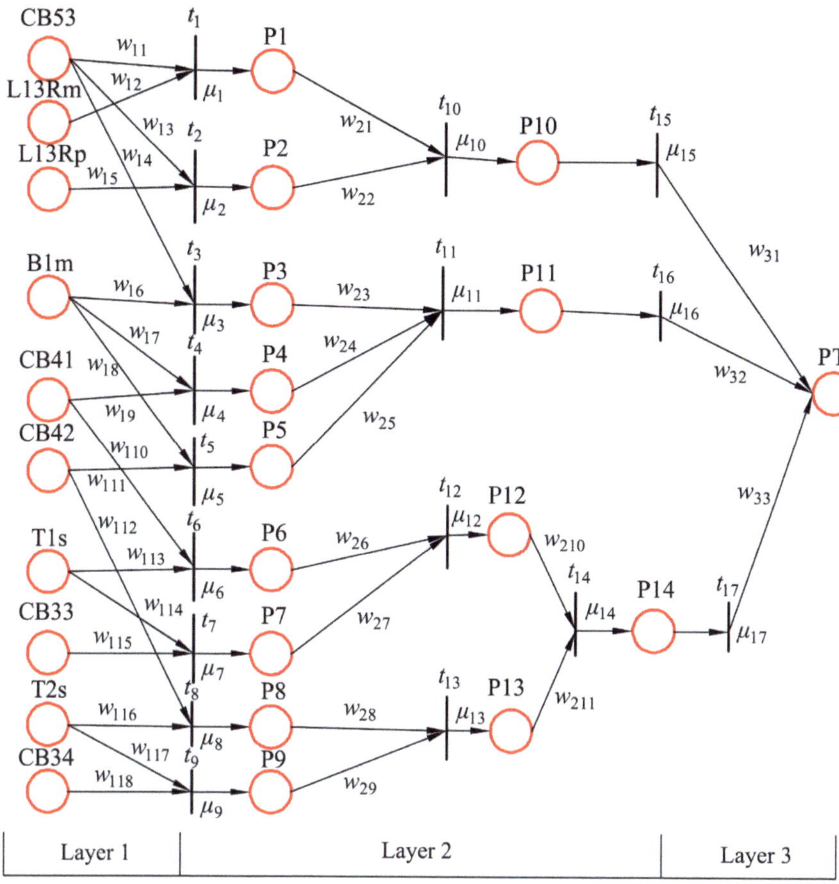

Fig. 2.33 FPN Model of Catenary L13

action information and circuit breaker trip information from past events, which is used for parameter estimation; online information refers to real-time acquired relay protection action information and circuit breaker tripping information, used for inference calculation. The uncertainty of historical statistical information is characterized by erroneous data from relay protection equipment and circuit breaker failures to operate and misoperations being recorded as normal data. Since equipment failures to operate and misoperations are greatly influenced by weather conditions, it is necessary to consider the uncertainty brought by failure rates under different weather conditions. Additionally, for newly operational traction power supply systems lacking sufficient historical statistical information, it is essential to establish simulation programs to generate enough historical statistical information.

Therefore, this section proposes a traction power supply system fault diagnosis model based on a Bayesian network that considers information uncertainty. The Bayesian network inherently possesses the capability to handle uncertain

Table 2.28 Operating time information for protection equipment in fault inference (as of December 2013)

Protection name	January 2009		June 2011		December 2013	
	Operating time/h	Availability	Operating time/h	Availability	Operating time/h	Availability
Protection of bus B1	17,280	0.9922	38,160	0.9128	59,760	0.6938
Protection of bus A6	17,280	0.9922	7920	0.9993	20,880	0.9860
Protection of line L13	5040	0.9998	25,920	0.9728	47,520	0.8354
Protection of trans-former T_1	17,280	0.9922	38,160	0.9128	59,760	0.6938
Protection of trans-former T_2	17,280	0.9922	38,160	0.9128	59,760	0.6938

Table 2.29 Operating time information for circuit breakers in fault inference (as of December 2013)

Circuit breaker name	January 2009		June 2011		December 2013	
	Number of operations	Availability	Number of operations	Availability	Number of operations	Availability
CB21	360	0.9995	780	0.9983	1143	0.9968
CB22	336	0.9996	673	0.9987	1276	0.9962
CB33	424	0.9994	840	0.9981	1366	0.9957
CB34	472	0.9992	957	0.9976	1407	0.9955
CB41	640	0.9988	1356	0.9957	1867	0.9928
CB42	680	0.9986	1421	0.9954	2018	0.9918
CB53	1360	0.9957	3243	0.9823	217	0.9998

information and reduces the uncertainty in statistical data used for parameter estimation, significantly enhancing the effectiveness of the fault diagnostic results for the traction power supply system.

2.4.1 Bayesian Network Model

Initially, the relay protection configuration and the topology of the traction power supply system form the basis for constructing a Bayesian network structure for each component within the system. Then, the misoperation rates and failure-to-operate rates of relay protection equipment and circuit breakers under various weather conditions are utilized to refine the Bayesian network parameter estimation, thereby reducing the uncertainty of historical information. Subsequently, fault diagnosis of

Table 2.30 Weights in the FPN model for bus B1 fault

Weight ID number	Weight value			Weight ID number	Weight value			Weight ID number	Weight value		
	2009.1	2011.6	2013.12		2009.1	2011.6	2013.12		2009.1	2011.6	2013.12
w_{11}	0.5017	0.5217	0.5886	w_{17}	0.4983	0.4783	0.4114	w_{23}	0.3327	0.3303	0.3350
w_{12}	0.4983	0.4783	0.4114	w_{18}	0.4984	0.4783	0.4116	w_{24}	0.5000	0.5001	0.5002
w_{13}	0.4984	0.4783	0.4116	w_{19}	0.5017	0.5217	0.5886	w_{25}	0.5000	0.4999	0.4998
w_{14}	0.5016	0.5217	0.5884	w_{110}	0.5016	0.5217	0.5884	w_{31}	0.4970	0.4727	0.4096
w_{15}	0.4991	0.4817	0.4097	w_{21}	0.3337	0.3349	0.3327	w_{32}	0.5030	0.5273	0.5904
w_{16}	0.5009	0.5183	0.5903	w_{22}	0.3336	0.3348	0.3323				

Table 2.31 Weights in the FPN model for transformer T_1 fault

Weight ID number	Weight value			Weight ID number	Weight value			Weight ID number	Weight value		
	2009.1	2011.6	2013.1		2009.1	2011.6	2013.1		2009.1	2011.6	2013.12
w_{11}	0.5018	0.5223	0.5893	w_{19}	0.4983	0.4783	0.4114	w_{23}	0.5002	0.5006	0.5007
w_{12}	0.4982	0.4777	0.4107	w_{110}	0.4982	0.5003	0.4975	w_{24}	0.4998	0.4994	0.4993
w_{13}	0.4983	0.4783	0.4114	w_{111}	0.4982	0.5003	0.4973	w_{25}	0.3333	0.3332	0.3332
w_{14}	0.5017	0.5217	0.5886	w_{112}	0.4981	0.5002	0.4974	w_{26}	0.3333	0.3333	0.3335
w_{15}	0.5018	0.5223	0.5893	w_{113}	0.5018	0.4997	0.5027	w_{27}	0.3334	0.3334	0.3333
w_{16}	0.5018	0.4997	0.5025	w_{114}	0.5019	0.4998	0.5026	w_{31}	0.3342	0.3126	0.2505
w_{17}	0.5017	0.5217	0.5886	w_{21}	0.5002	0.5006	0.5007	w_{32}	0.3342	0.3126	0.2505
w_{18}	0.4982	0.4777	0.4107	w_{22}	0.4998	0.4994	0.4993	w_{33}	0.3317	0.3748	0.4989

Table 2.32 Weights in the FPN model for catenary L13 fault

Weight ID number	Weight value			Weight ID number	Weight value			Weight ID number	Weight value		
	2009.1	2011.6	2013.12		2009.1	2011.6	2013.12		2009.1	2011.6	2013.12
w_{11}	0.4990	0.5024	0.5448	w_{112}	0.5016	0.5078	0.5884	w_{25}	0.3336	0.3348	0.3323
w_{12}	0.5010	0.4976	0.4552	w_{113}	0.4983	0.4783	0.4114	w_{26}	0.4998	0.4994	0.4993
w_{13}	0.4990	0.5024	0.5448	w_{114}	0.4982	0.4777	0.4107	w_{27}	0.5002	0.5006	0.5007
w_{14}	0.5009	0.5183	0.5903	w_{115}	0.5018	0.5223	0.5893	w_{28}	0.4998	0.4994	0.499 1
w_{15}	0.5010	0.4976	0.4552	w_{116}	0.4984	0.4922	0.4116	w_{29}	0.5002	0.5006	0.5009
w_{16}	0.4991	0.4817	0.4097	w_{117}	0.4982	0.4917	0.4107	w_{210}	0.5001	0.4724	0.5003
w_{17}	0.4983	0.4783	0.4114	w_{118}	0.5018	0.5083	0.5893	w_{211}	0.4999	0.5276	0.4997
w_{18}	0.4984	0.4783	0.4116	w_{21}	0.5000	0.5000	0.5000	w_{31}	0.3387	0.3774	0.5569
w_{19}	0.5017	0.5217	0.5886	w_{22}	0.5000	0.5000	0.5000	w_{32}	0.3315	0.3061	0.2625
w_{110}	0.5017	0.5217	0.5886	w_{23}	0.3327	0.3303	0.3350	w_{33}	0.3298	0.3165	0.1806
w_{111}	0.5016	0.5217	0.5884	w_{24}	0.3337	0.3349	0.3327	–	–	–	–

Table 2.33 Case study diagnostic results

Fault signal	Diagnostic result/credibility		
	2009.1	2011.6	2013.12
Blm action; CB41, CB42, CB53, CB54 action	Bus B1 fault, credibility 0.8535	Bus B1 fault, credibility 0.8196	Bus B1 fault, credibility 0.7469
T1m, A6m action; CB33, CB41, CB21 action	Transformer T_1 fault, credibility 0.8538	Transformer T_1 fault, credibility 0.8564	Transformer T_1 fault, credibility 0.8498
L13Rm action; CB53, CB54 action	Line L13 fault, credibility 0.8555	Line L13 fault, credibility 0.8381	Line L13 fault, credibility 0.7930

the traction transformer is conducted through the inference capabilities of the Bayesian network.

2.4.1.1 Bayesian Network Model

A Bayesian network is a DAG, where each node represents a random variable, and the edges between nodes illustrate the direct dependency relationships between these variables. A directed edge from node *X* to node *Y* indicates that *X* is the parent node of *Y*, with *Y* being the subnode of *X*. Nodes that have no parent are known as root nodes. Each node in a Bayesian network is associated with a probability distribution: root nodes have a marginal distribution attached, while non-root nodes have a conditional probability distribution that depends on their parent nodes.

Bayesian networks can be approached from both quantitative and qualitative perspectives: Quantitative Perspective: Bayesian networks describe the dependency relationships of subnodes on their parent nodes through conditional probability distributions. Qualitative Perspective: Bayesian networks utilize directed acyclic graphs to depict the independent or dependent relationships between nodes.

2.4.1.2 Bayesian Network Structure

The structure of a Bayesian network reveals the qualitative relationships between nodes. Methods for constructing such structures include data analysis and manual construction, often with the consultation of experts. Due to the clear causal relationship between the actions of the relay protection equipment and circuit breakers in the traction power supply system, this chapter builds a Bayesian network structure based on the fundamental principle that a component fault triggers the action of protection equipment, which in turn controls the tripping of circuit breakers. For instance, a Bayesian network is established for a traction transformer in a traction power supply system, as shown in Fig. 2.34.

The protection configuration of traction transformer T includes primary protection Tm and backup protection Tp, both controlling two circuit breakers, CB1 and

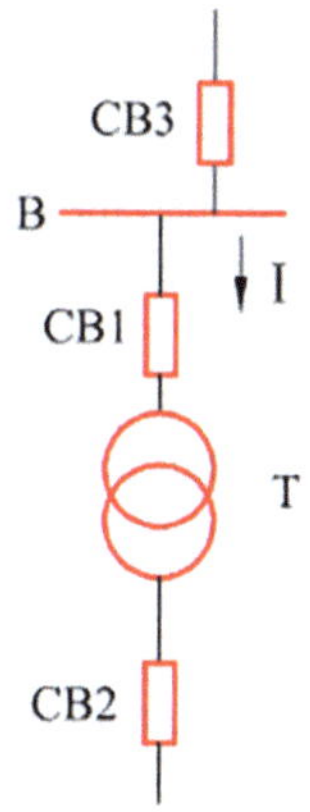

Fig. 2.34 Connection Diagram of a Certain Traction Transformer

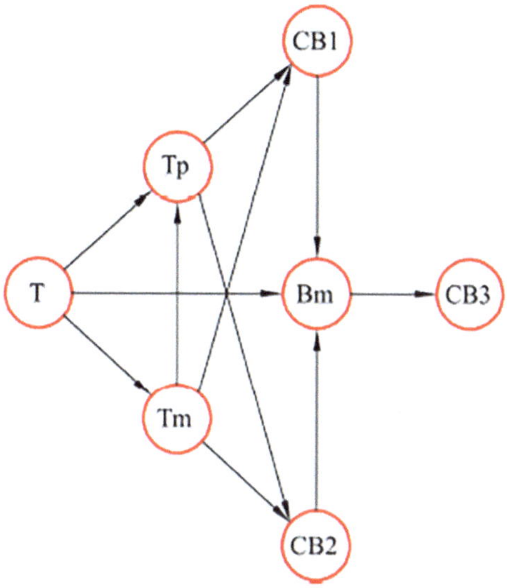

Fig. 2.35 Bayesian Network of Traction Transformer T

CB2. Additionally, the bus B protection configuration includes primary protection Bm controlling CB1 and CB3 and serves as the remote backup protection for traction transformer T. According to the relay protection configuration, when a fault occurs in traction transformer T, its primary protection Tm should act to control the tripping of circuit breakers CB1 and CB2. If the primary protection Tm fails to operate, its backup protection Tp will act to control the tripping of CB1 and CB2. If both the backup protection Tp and circuit breaker CB1 fail to operate, the primary protection of bus B will act as the remote backup protection for transformer T, controlling the tripping of circuit breaker CB3 to isolate the faulty component. Based on this series of protection actions, a Bayesian network as shown in Fig. 2.35 can be established.

2.4.1.3 Consideration of Historical Information Uncertainty in Bayesian Network Parameter Learning

The parameters of a Bayesian network, which are the probability distributions attached to each node, reveal the quantitative relationships between nodes. Bayesian network parameter learning, also known in statistics as parameter estimation, includes two fundamental methods: maximum likelihood estimation and Bayesian estimation. Due to the lack of prior knowledge about the nodes in the Bayesian networks discussed in this paper, the maximum likelihood estimation method is used.

Assume there exists a traction power supply system and a Bayesian network constructed from it, consisting of n nodes $X = \{X_1, X_2, \cdots, X_n\}$. As depicted in Fig. 2.35, nodes in this model represent components, protection equipment, and circuit breakers. Therefore, each node X_i can take on only two values, 0 and 1, and has q_i combinations of parent nodes. If X_i is a root node, then $q_i = 1$, and the network parameters can be expressed as:

$$\theta_{ijk} = \mathrm{P}(X_i = k \mid \pi(X_i) = j) \tag{2.58}$$

where $i = 1, 2, \cdots, n$, representing the node sequence number; $j = 1, 2, \cdots, q_i$, representing the parent node combination sequence number; $k = 0, 1$, representing the node status. By the normativity of probability distribution, we have:

$$\sum_{k=0}^{1} \theta_{ijk} = \sum_{k=0}^{1} \mathrm{P}(X_i = k \mid \pi(X_i) = j) = 1, \forall i, j \tag{2.59}$$

Given a set of complete data $D = (D_1, D_2, \cdots, D_m)$ for the above Bayesian network, the log-likelihood function of θ is defined as:

$$l(\theta \mid D) = \log\left(\prod_{l=1}^{m} \mathrm{P}(D_l \mid \theta)\right) = \sum_{l=1}^{m} \log(P(D_l \mid \theta)) \tag{2.60}$$

Define the feature function $\chi(i, j, k : D_l)$ of the samples D_l:

$$\chi(i, j, k : D_l) = \begin{cases} 1, & X_i = k \text{ and } \pi(X_i) = j \\ 0, & X_i \neq k \text{ or } \pi(X_i) \neq j \end{cases} \tag{2.61}$$

Thus, we have:

$$\log(P(D_l \mid \theta)) = \sum_{i=1}^{n} \sum_{j=1}^{q_i} \sum_{k=0}^{1} \chi(i,j,k : D_l) \log \theta_{ijk} \tag{2.62}$$

Definition

$$m_{ijk} = \sum_{l=1}^{m} \chi(i,j,k : D_l) \tag{2.63}$$

where m_{ijk} is the sample size in the data that satisfies $X_i = k$ and $\pi(X_i) = j$.

$$\begin{aligned} l(\theta \mid D) &= \sum_{l=1}^{m} \log(\mathrm{P}(D_l \mid \theta)) \\ &= \sum_{l=1}^{m} \sum_{i=1}^{n} \sum_{j=1}^{q_i} \sum_{k=0}^{1} \chi(i,j,k : D_l) \log \theta_{ijk} \\ &= \sum_{i=1}^{n} \sum_{j=1}^{q_i} \sum_{k=0}^{1} \sum_{l=1}^{m} \chi(i,j,k : D_l) \log \theta_{ijk} \\ &= \sum_{i=1}^{n} \sum_{j=1}^{q_i} \sum_{k=0}^{1} m_{ijk} \log \theta_{ijk}. \end{aligned} \tag{2.64}$$

where $\{m_{ijk} \mid i = 1, \cdots, n; j = 1, \cdots, q_i; k = 0, 1\}$ is the sufficient statistic. For any fixed i and j, since $\sum_{k=0}^{1} \theta_{ijk} = 1$, when θ_{ijk} takes the following values:

$$\theta_{ijk}^{*} = \begin{cases} \dfrac{m_{ijk}}{\sum_{k=0}^{1} m_{ijk}}, & \sum_{k=0}^{1} m_{ijk} > 0 \\ \dfrac{1}{2}, & \sum_{k=0}^{1} m_{ijk} \leqslant 0 \end{cases} \tag{2.65}$$

The expression $\sum_{k=0}^{1} m_{ijk} \log \theta_{ijk}$ reaches its maximum value, thereby maximizing the likelihood function $l(\theta \mid D)$. The maximum likelihood estimate of θ_{ijk}^{*} for θ_{ijk}, given by eq. (2.65), can be expressed based on its physical meaning as

$$\theta_{ijk}^{*} = \frac{\textit{number of samples in D satisfying } X_i = k \text{ and } \pi(X_i) = j}{\text{number of samples in } D \text{ satisfying } \pi(X_i) = j}$$

As previously mentioned, there is a significant amount of uncertainty in the historical information. Therefore, correction coefficients λ_0 and λ_1 are introduced to adjust the Bayesian network parameter maximum likelihood estimation as follows:

$$\theta_{ijk}^{*} = \begin{cases} \dfrac{m_{ijk}\lambda_k}{\sum\limits_{k=0}^{1} m_{ijk}\lambda_k}, & \sum\limits_{k=0}^{1} m_{ijk} > 0 \\ \dfrac{1}{2}, & \sum\limits_{k=0}^{1} m_{ijk} \leqslant 0 \end{cases}. \tag{2.66}$$

where

$$\lambda_0 = 1 - \lambda_{ref-x-i} \tag{2.67}$$

$$\lambda_1 = 1 - \lambda_{inc-x-i} \tag{2.68}$$

$\lambda_{ref-x-i}$ and $\lambda_{inc-x-i}$ are the failure-to-operate rate and misoperation rate of equipment x under weather condition i, respectively, calculated using the method described below. Therefore, the network parameter values under three different weather conditions can be obtained, forming a three-mode Bayesian network.

2.4.1.4 Bayesian Network Inference

Based on the direction of inference, Bayesian network inference can be categorized into predictive inference and diagnostic inference. A Bayesian network structure is established according to the topological structure of the traction power supply system and the relay protection configuration. Faulty components are diagnosed from the effects to the causes based on relay protection action information and circuit breaker trip information, classifying it as diagnostic inference. Depending on the computation methods, Bayesian network inference can also be divided into exact inference and approximate inference. Exact inference is suitable for relatively simple Bayesian networks, while approximate inference can handle the large computational load brought by complex Bayesian networks with many nodes. Given that the Bayesian network constructed here is not complex, the exact inference method combined with the variable elimination algorithm is employed. Let Y be the set of all variables in the Bayesian network. F represents a decomposition of the probability distribution P(Y) represented by the Bayesian network. Upon observing evidence E in the factors, set each evidence variable to its observed value, resulting in a set of functions F', termed evidence instantiation. Here, F' is a decomposition of P(Z,E), where $Z = Y/E$. Let X be a subset of Z; eliminating all variables in Z that are not in X from F' results in another set of functions, denoted as F″. Since F″ is a decomposition of P(X,E), multiplying all factors in F″ yields P(X,E). According to the definition of conditional probability, we can further obtain:

$$
\begin{aligned}
\mathrm{P}(x_i = x_{ij} \mid E) = & \frac{P(x_i = x_{ij}, \mathrm{E})}{P(E)} \\
& = \frac{P(x_i = x_{ij}, \mathrm{E})}{\sum_{k=1}^{m} P(x_i = x_{ik}, \mathrm{E})} \\
& = \frac{P(x_i = x_{ij}) \prod_{s=1, s \neq i}^{n} P(x_s \mid \pi(x_s))}{\sum_{k=1}^{m} \left[P(x_i = x_{ik}) \prod_{s=1, s \neq i}^{n} P(x_s \mid \pi(x_s)) \right]}
\end{aligned}
\tag{2.69}
$$

where x_i represents a suspected component node, x_{ij} is its node status, and E is the evidence information, which in this chapter refers to online relay protection action information and circuit breaker trip information.

2.4.2 *Equipment Failure Rate Under Different Weather Conditions*

The uncertainty in the historical statistical information of relay protection and circuit breakers arises from their failures to operate and misoperations. Under varying weather conditions, the failure-to-operate and misoperation rates of relay protection equipment and circuit breakers differ, generally showing that more severe weather conditions lead to higher equipment failure rates. Weather conditions are categorized into normal weather, bad weather, and severe weather. The failure rates of equipment under these different weather conditions are calculated, and these rates are integrated into Bayesian network parameter estimation to reduce information uncertainty. Relay protection equipment and circuit breakers exhibit statuses of action and non-action (for circuit breakers, this corresponds to tripping and not tripping), so their failure rates can be divided into misoperation rates and failure-to-operate rates.

2.4.2.1 Calculation of Equipment Failure-to-Operate Rate Under Different Weather Conditions

Let λ_{ref-x} be the average failure-to-operate rate of equipment x, and $\lambda_{ref-x-i}$ be the failure-to-operate rate of equipment x under weather condition i. Let N represent the total number of failures to operate per year, and M represents the total number of correct operations per year. Additionally, let N_i be the number of failures to operate under weather condition i, and M_i be the number of correct operations under weather condition i.

$$
\lambda_{ref-x} = N/(M+N) \tag{2.70}
$$

$$\lambda_{ref-x-i} = N_i/(M_i + N_i) \tag{2.71}$$

Therefore, from eq. (2.71):

$$N = \sum_{i=1}^{3} N_i = \sum_{i=1}^{3} \lambda_{ref-x-i}(M_i + N_i) \tag{2.72}$$

Further, from eqs. (2.70) and (2.72):

$$\lambda_{ref-x} = \frac{\sum_{i=1}^{3} \lambda_{ref-x-i}(M_i + N_i)}{M + N} = \sum_{i=1}^{3} \lambda_{ref-x-i} P_i \tag{2.73}$$

where

$$P_i = \frac{M_i + N_i}{M + N} \tag{2.74}$$

The eq. (2.73) defines the failure-to-operate rate correction model under the three weather status model.

The goal of this model is to calculate the failure-to-operate rates of equipment under different weather conditions, which are typically determined statistically by obtaining the average failure-to-operate rate. Define F_i as the percentage of the number of failures to operate under weather condition i relative to the total number of failure-to-operate occurrences per year.

$$F_i = N_i/N \tag{2.75}$$

From eqs. (2.70), (2.73), and (2.75), we can obtain

$$\lambda_{ref-x-i} = \lambda_{rref-x} F_i / P_i \tag{2.76}$$

2.4.2.2 Calculation of Equipment Misoperation Rate Under Different Weather Conditions

Define λ_{inc-x} as the average misoperation rate of equipment x, and $\lambda_{inc-x-i}$ as the misoperation rate of equipment x under weather condition i. Let S be the total number of misoperations per year, T be the total statistical time, S_i be the number of misoperations under weather condition i per year, and T_i be the duration of weather condition i.

$$\lambda_{inc-x} = S/T \tag{2.77}$$

$$\lambda_{inc-x-i} = S_i/T_i \tag{2.78}$$

From eqs. (2.77) and (2.78), the following can be derived:

$$\lambda_{inc-x} = \frac{\sum_{i=1}^{3} \lambda_{inc-x-i} T_i}{T} = \sum_{i=1}^{3} \lambda_{inc-x-i} P_{ti} \tag{2.79}$$

where

$$P_{ti} = T_i/T \tag{2.80}$$

The eq. (2.79) establishes the three-state model for misoperation rate adjustment.

Similar to the calculation of the failure-to-operate rate, the average misoperation rate is typically the only available metric. Therefore, define C_i as the percentage of the equipment's misoperation occurrences under weather condition i relative to the total number of misoperation occurrences per year.

$$C_i = S_i/S \tag{2.81}$$

From eqs. (2.77), (2.79), and (2.81), we can obtain

$$\lambda_{inc-x-i} = \lambda_{inc-x} C_i / P_{ti} \tag{2.82}$$

2.4.3 Traction Power Supply Fault Simulation Model Based on Monte Carlo Simulation

Given the challenge of gathering extensive data necessary for Bayesian network parameter estimation, it becomes essential to establish a fault simulation model for the traction power supply system to generate the required data. The fundamental logic of this model development involves a component experiencing a fault, which triggers a series of relay protection equipment actions and circuit breaker trips to isolate the fault source. The causal relationships in this process align with the Bayesian network structure, thus the DAG of the Bayesian network serves as the foundation for the simulation. Relay protection equipment and circuit breakers do not always operate under ideal conditions; failures to operate and misoperations also occur. This process has significant randomness, which Monte Carlo simulation can capture by modeling the systematic features of protection equipment and circuit breaker failure rates through mathematical methods, specifically through random sampling of the entire random process. By conducting multiple simulation

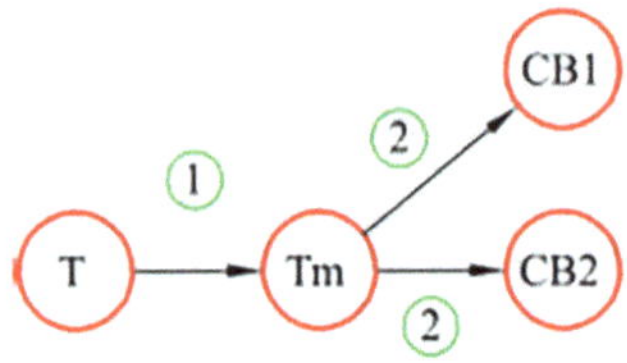

Fig. 2.36 Traction Transformer T Fault Simulation Scenario One

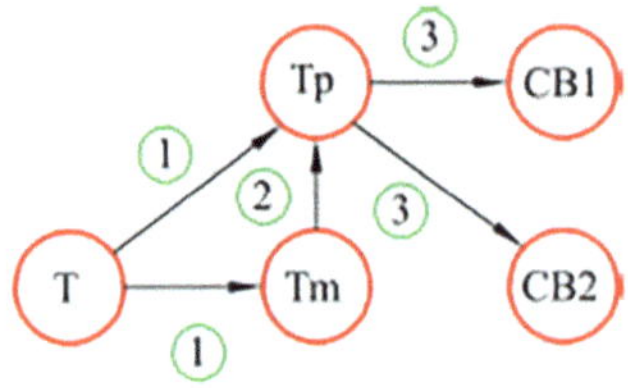

Fig. 2.37 Traction Transformer T Fault Simulation Scenario Two

experiments, the results can be approximated as reflecting actual conditions. This section, in combination with the directed acyclic graph based on relay protection configuration, establishes a traction power supply fault simulation model based on Monte Carlo simulation.

Taking traction transformer T from Fig. 2.34 as a case study, assume that a fault occurs in traction transformer T. The action status of its primary protection Tm is then generated based on the failure-to-operate rate of that protection equipment through random sampling, producing a status value of 1 or 0, which respectively represents the protection equipment acting or not acting. If the primary protection acts, then based on the failure-to-operate rate of the circuit breakers, the corresponding circuit breaker status values of 1 or 0 are similarly generated through random sampling, representing the respective circuit breakers tripping or not tripping. If all function normally, the process is illustrated in Fig. 2.36.

If the primary protection Tm fails to operate, then based on the failure rate of the backup protection Tp, its status is randomly sampled to generate a 1 or 0. If the backup protection acts, then based on the failure-to-operate rate of the circuit breakers, the corresponding circuit breaker status values of 1 or 0 are generated, as shown in Fig. 2.37.

If both the primary protection Tm and the backup protection Tp fail to operate, or if the circuit breaker CB2 fails to operate, then based on the failure rate of the remote backup protection Bm, its status value is randomly generated. If the remote backup protection acts, then based on the circuit breaker CB3's failure-to-operate rate, the corresponding circuit breaker status value is randomly generated, as depicted in Fig. 2.38.

In the simulation scenarios, nodes or other components of the Bayesian network not directly involved are also generated through random sampling methods. Component nodes utilize each component's relative fault probability as their random sampling probability, while other nodes use the corresponding misoperation rate as their random sampling probability. Simulation data for each node is detailed in Tables 2.34, 2.35, 2.36, 2.37, 2.38, 2.39, and 2.40.

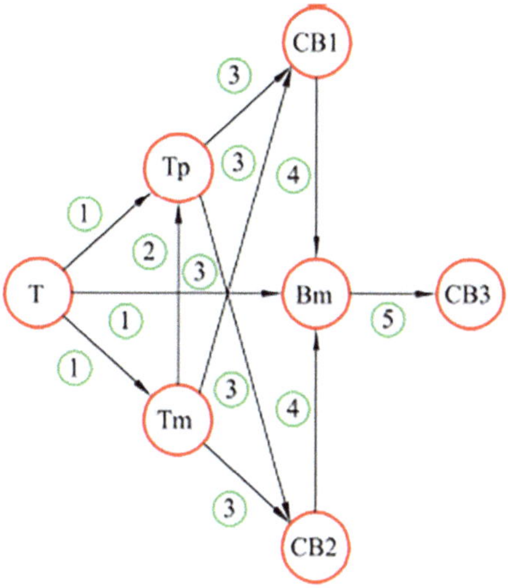

Fig. 2.38 Traction Transformer T Fault Simulation Scenario Three

Table 2.34 Simulation data for node T

T	
0	1
964,513	13,720

Table 2.35 Simulation data for node Tm

	Tm	
T	0	1
0	960,823	3690
1	49	13,671

Table 2.36 Simulation data for node Tp

		Tp	
T	Tm	0	1
0	0	956,990	3833
0	1	11	3679
1	0	1	48
1	1	13,622	49

Table 2.37 Simulation data for node CB1

		CB1	
Tm	Tp	0	1
0	0	931,660	25,331
0	1	50	3831
1	0	192	13,441
1	1	42	3686

Table 2.38 Simulation data for node CB2

Tm	Tp	CB2	
		0	1
0	0	933,610	23,381
0	1	45	3836
1	0	202	13,431
1	1	53	3697

Table 2.39 Simulation data for node Bm

T	CB1	CB2	Bm	
			0	1
0	0	0	933,335	23,655
0	0	1	3754	79
0	1	0	11	0
0	1	1	3583	96
1	0	0	0	1
1	0	1	48	0
1	1	0	13,598	24
1	1	1	48	1

Table 2.40 Simulation data for node CB3

Bm	CB3	
	0	1
0	911,948	42,455
1	1381	22,449

Based on the above, the construction steps for the fault diagnosis model of the traction power supply system proposed in this section are as follows: First, establish a Bayesian network structure based on the topological structure of the traction power supply system and the relay protection configuration; then use a Monte Carlo-based method to simulate the actions of relay protection and circuit breakers following component faults, and generate data; calculate the equipment failure rates under various weather conditions, and combine this with relay protection and circuit breaker data for Bayesian network parameter learning; finally, input the real-time action information of relay protection and circuit breakers as evidence into the Bayesian networks for each component under the corresponding mode, to infer the probability of failure of each component in this event, as shown in Fig. 2.39.

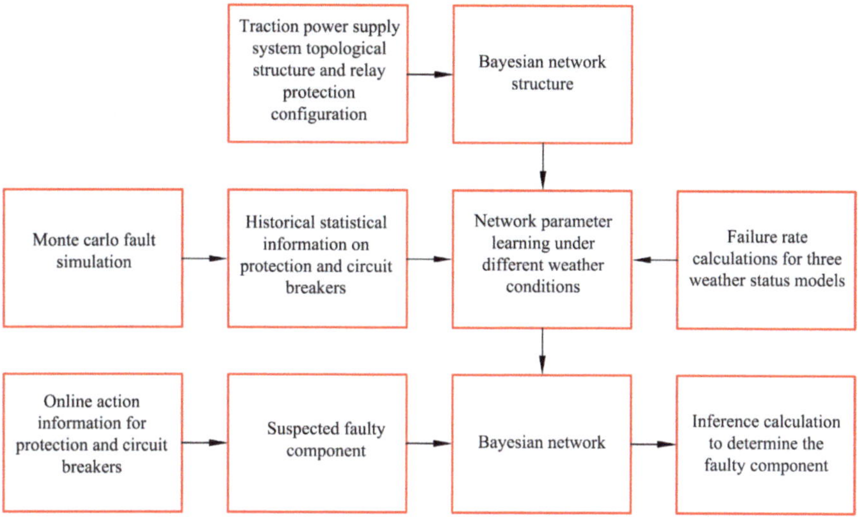

Fig. 2.39 Traction Power Supply System Fault Diagnosis Process

2.4.4 Traction Power Supply System Fault Diagnosis Based on Bayesian Network Case Study

2.4.4.1 Topological Structure of the Traction Power Supply System in the Case

The topology of this traction power supply system, shown in Fig. 2.40, includes 14 components, 25 protections, and 22 circuit breakers. The red parts represent operating equipment and lines, green for backup equipment, and dotted lines for indoor equipment. A and B represent busbars, T for traction transformers, AT for auto-transformer. FS denotes the catenary line, with a and b indicating the line's up and down direction. Protection configuration is as follows: A1 main busbar protection Alm controls circuit breakers CB101 and CB102, serving as remote backup protection for traction transformers T3 and T4; T4 traction main transformer protection T4m and close backup protection T4p both control circuit breakers CB102 and CB204, with T4p also acting as backup protection for B1 busbar; T3 traction main transformer protection T3m and close backup protection T3p both control circuit breakers CB102 and CB202; T3p also acts as backup protection for B2 busbar; B1 main busbar protection Blm controls circuit breakers CB204, CB213, and CB214, and serves as remote backup protection for FS2 up and down line; line FS2a primary protection FS2am and close backup protection FS2ap control circuit breakers CB213, CB2131, and CB2132; line FS2b primary protection FS2bm and close backup protection FS2bp. Controls circuit breakers CB214, CB2141, and CB2142; FS2ap and FS2bp act as remote backup protection for 7AT and 5AT auto-transformers; 5AT auto-transformer main transformer protection 5ATm and

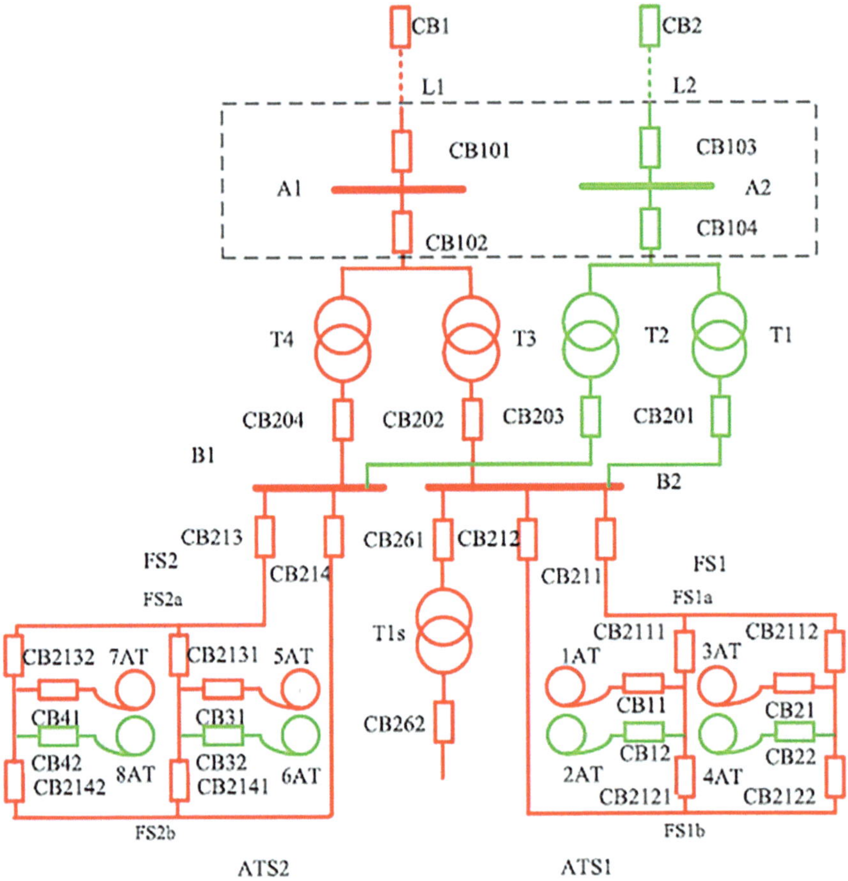

Fig. 2.40 A Certain Traction Power Supply System Topology

backup protection 5ATp both control circuit breaker CB31; 7AT auto-transformer main transformer protection 7ATm and backup protection 7ATp both control circuit breaker CB41; B2 main busbar protection B2m controls circuit breakers CB202, CB261, CB212, and CB211, and serves as remote backup protection for FS1 up and down line and Tls transformer; line FS1a primary protection FS1am and close backup protection FS1ap control circuit breakers CB211, CB2111, and CB2112; line FS1b primary protection FS1bm and close backup protection FS1bp control circuit breakers CB212, CB2121, and CB2122; FS1ap and FS1bp act as remote backup protection for 3AT and 1AT auto-transformers; 3AT auto-transformer main transformer protection 3ATm and backup protection 3ATp both control circuit breaker CB21; 1AT auto-transformer main protection 1ATm and backup protection 1ATp both control circuit breaker CB11; Tls main transformer protection Tlsm and backup protection Tlsp control circuit breakers CB261 and CB262.

2.4.4.2 Establishing the Bayesian Network Structure for Each Component of the Traction Power Supply System

Combine the methods previously described and the provided topology of the traction power supply system to establish the Bayesian networks for each component. Figure 2.41 shows the Bayesian networks for the traction transformer and the station service transformer. Figure 2.42 depicts the Bayesian networks for auto-transformers. Figure 2.43 shows the Bayesian networks for busbars. Figure 2.44 illustrates the Bayesian network for the up and down line, specifically for the FS2 down line.

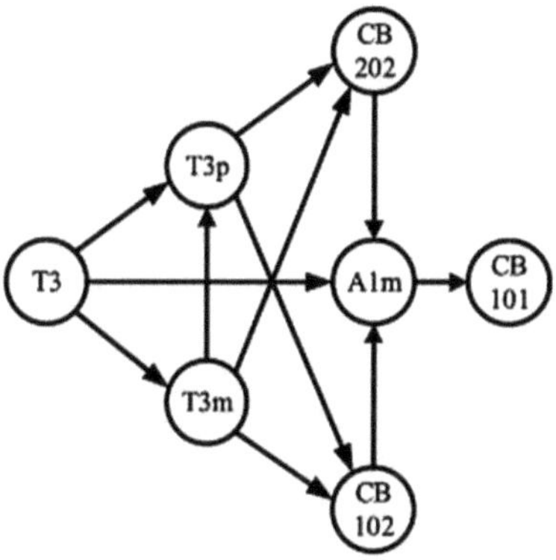

(a) Bayesian network for the T3 traction transformer

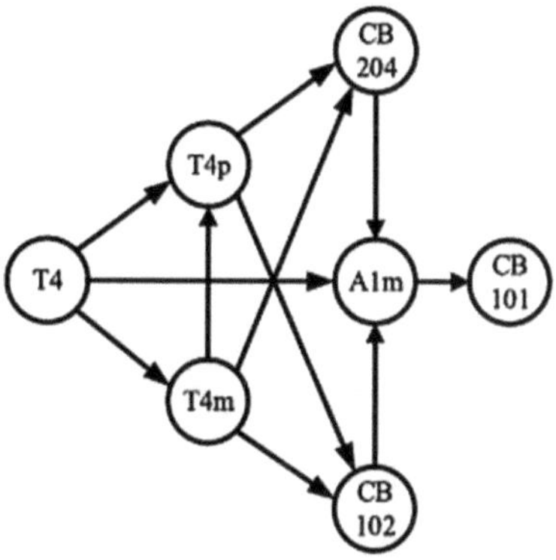

(b) Bayesian network for the T4 traction transformer.

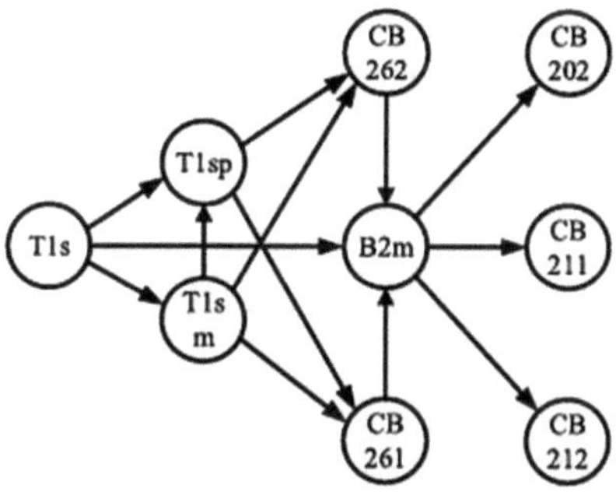

(c) Bayesian network for the T4 traction transformer.

Fig. 2.41 Bayesian Networks for the Traction Transformer and the Station Service Transformer. (**a**) Bayesian network for the T3 traction transformer. (**b**) Bayesian network for the T4 traction transformer. (**c**) Bayesian network for the T4 traction transformer

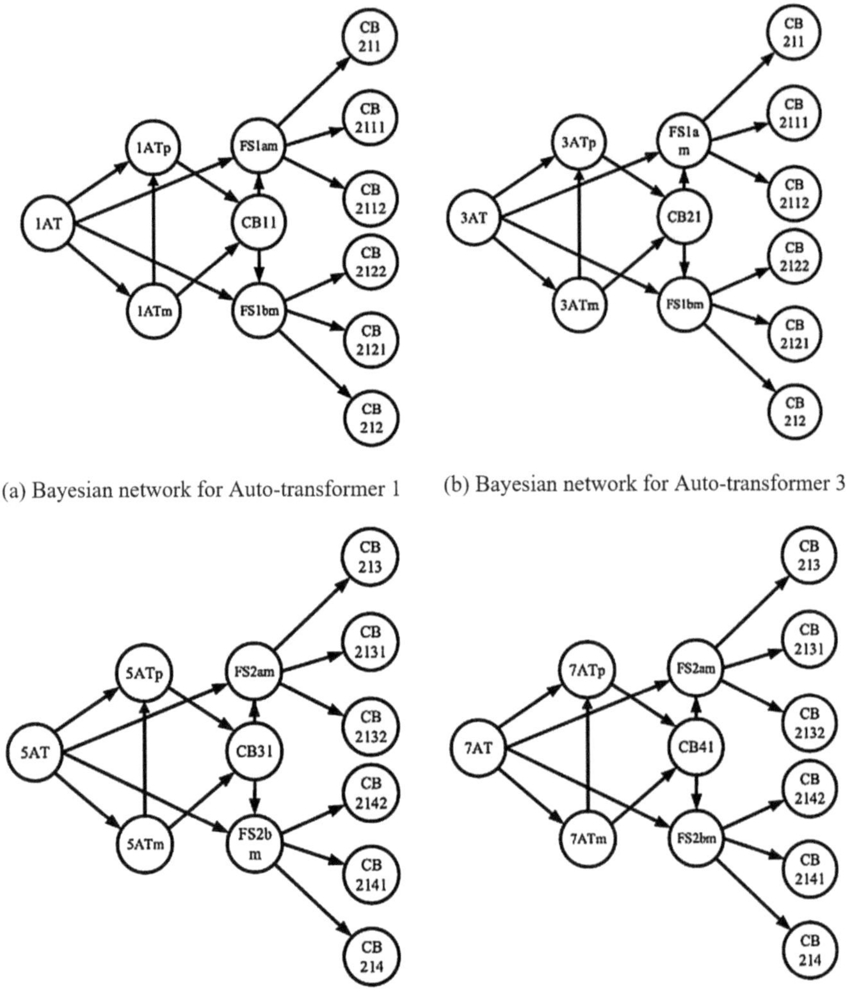

(a) Bayesian network for Auto-transformer 1 (b) Bayesian network for Auto-transformer 3

(c) Bayesian network for Auto-transformer 5 (d) Bayesian network for Auto-transformer 7

Fig. 2.42 Bayesian networks for auto-transformers. (**a**) Bayesian network for Auto-transformer 1. (**b**) Bayesian network for Auto-transformer 3. (**c**) Bayesian network for Auto-transformer 5. (**d**) Bayesian network for Auto-transformer 7

2.4.4.3 Bayesian Network Parameter Estimation

Table 2.41 lists the component fault probabilities under various weather conditions calculated from data collected on-site. From these data, the relative fault probabilities for each component are determined.

Under normal weather conditions, the relative fault probability of the T3 traction transformer is calculated along $P_{T3} = \frac{0.0118}{(0.3904/2)*4+0.0185*3+0.0118*7} \approx 0.01284$

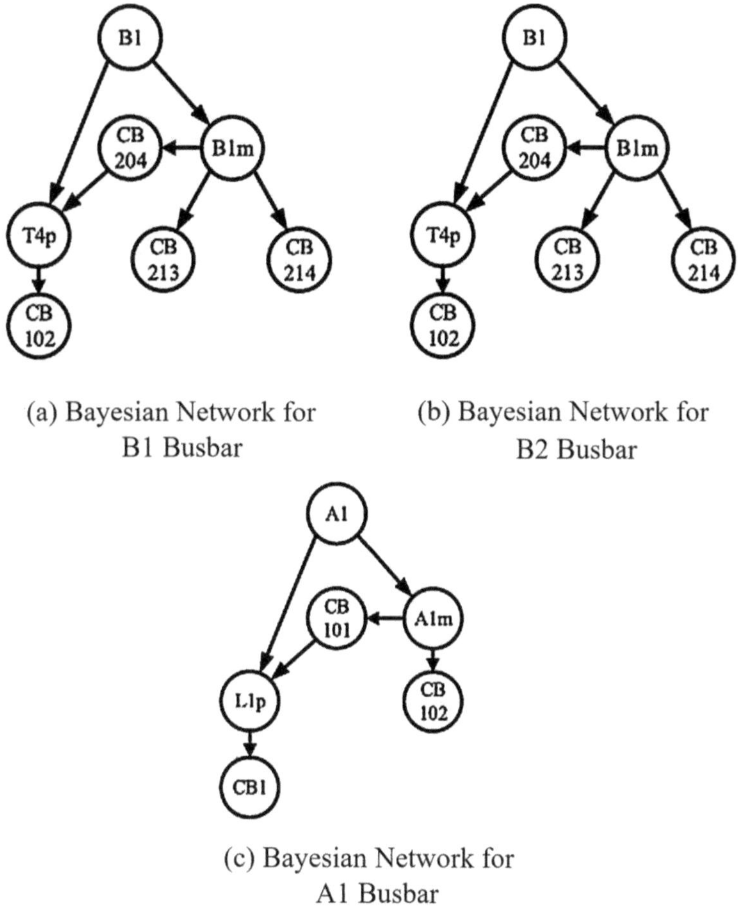

Fig. 2.43 Bayesian Networks for Busbars. (**a**) Bayesian Network for B1 Busbar. (**b**) Bayesian Network for B2 Busbar. (**c**) Bayesian Network for A1 Busbar

Similarly, the relative fault probability for other components is obtained as follows:

$$P_{T1s} = P_{T4} = P_{1AT} = P_{3AT} = P_{5AT} = P_{7AT} \approx 0.01284$$

$$P_{A1} = P_{B1} = P_{B2} \approx 0.02013$$

$$P_{FS1a} = P_{FS1b} = P_{FS2a} = P_{FS2b} \approx 0.21243$$

The average fault rate of protection and circuit breakers for each component is shown in Table 2.42. Under various weather conditions, time and fault data statistics are presented in Table 2.43. Using the Monte Carlo algorithm, simulated data are

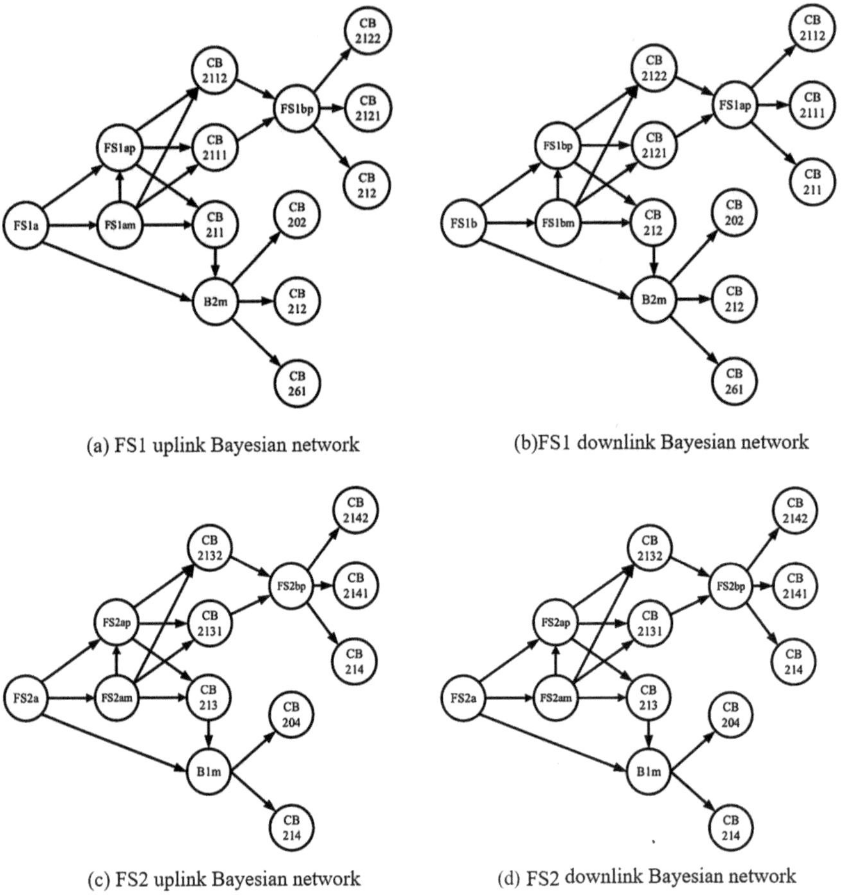

Fig. 2.44 Bayesian Network for the Up and Down Line. (**a**) FS1 uplink Bayesian network. (**b**) FS1 downlink Bayesian network. (**c**) FS2 uplink Bayesian network. (**d**) FS2 downlink Bayesian network

Table 2.41 Component fault probability

	Weather condition	Line/100 km	Busbar	Transformer
Failure rate	Normal weather (1)	0.3904	0.0185	0.0118
	Bad weather (2)	0.5127	0.0251	0.0174
	Severe weather (3)	0.6932	0.0372	0.0256

Table 2.42 Failure rates for protection and circuit breakers

Failure rate	Main line protection	Main busbar protection	Main transformer protection	Circuit breaker
Failure-to-operate rate	0.0019	0.0013	0.0027	0.0145
Misoperation rate	0.0085	0.0022	0.0039	0.0268

Table 2.43 Time and fault data statistics

Weather condition	Time proportion	Percentage of total statistics for failure-to-operate and correct actions	Percentage of total refusal to operate occurrences	Percentage of total misoperation occurrences
Normal weather (1)	90%	86%	50%	50%
Bad weather (2)	9.5%	13.25%	45%	45%
Severe weather (3)	0.5%	0.75%	5%	5%

generated based on the relative fault probabilities of each component, along with the failure rates of protection and circuit breakers.

Based on the data from Tables 2.42 and 2.43, the failure rate of the catenary line under normal weather conditions is calculated

$$\lambda_{ref-l-1} = \frac{\lambda_{ref-l} F_{r-l-1}}{P_1} = \frac{0.0019 \times 50\%}{86\%} \approx 0.0011$$

$$\lambda_{inc-l-1} = \frac{\lambda_{inc-l} C_{i-l-1}}{P_1} = \frac{0.0085 \times 50\%}{90\%} \approx 0.00472$$

Then the correction coefficient is determined.

$$\lambda_0 = 1 - \lambda_{ref-l-1} = 0.9989$$

$$\lambda_1 = 1 - \lambda_{inc-l-1} = 0.99528$$

Similarly, the failure rates and correction coefficients for various types of equipment under different weather conditions are calculated, as shown in Table 2.44.

Using the correction coefficients and simulated data calculated from Table 2.44, Bayesian network parameters are estimated using Eq. (2.66).

2.4.4.4 Bayesian Network Inference Calculation

As shown in Table 2.45, the relay protection and circuit breaker operation information for a normal weather condition fault event is presented.

First, determine the suspected faulty components as line FS1a and busbar B2, then select the normal weather mode, and finally calculate the posterior probability using Eq. (2.69).

Table 2.44 Failure rates and correction coefficients for different components under various weather conditions

Component	Failure rate and correction coefficient	Normal weather (1)	Bad weather (2)	Severe weather (3)
Line	$\lambda_{ref-l-i}$	0.0011	0.006453	0.0127
	$\lambda_{inc-l-i}$	0.00472	0.0425	0.085
	λ_0	0.9989	0.993547	0.9873
	λ_1	0.99528	0.9575	0.915
Busbar	$\lambda_{ref-b-i}$	0.000756	0.004415	0.00867
	$\lambda_{inc-b-i}$	0.00122	0.011	0.022
	λ_0	0.999244	0.995585	0.99133
	λ_1	0.99878	0.989	0.978
Transformer	$\lambda_{ref-t-i}$	0.00157	0.00917	0.018
	$\lambda_{inc-t-i}$	0.00217	0.0195	0.039
	λ_0	0.99843	0.99083	0.982
	λ_1	0.99783	0.9805	0.961
Circuit breaker	$\lambda_{ref-c-i}$	0.00843	0.04925	0.09667
	$\lambda_{ine-c-i}$	0.01489	0.134	0.268
	λ_0	0.99157	0.95075	0.90333
	λ_1	0.98511	0.866	0.732

$$P(FS1a = 1D) = 0.0000732$$

$$P(B2 = 1D) = 0.8058$$

It is determined that busbar B2 is the faulty component. Event analysis is as follows: The B2 main busbar protection and its controlled circuit breakers all trip, the FS1a line close backup protection action controls three circuit breakers, only CB211 acts, and CB211 is also controlled by B2 main busbar protection, thus, it is reasonable to determine B2 busbar as the faulty component. The FS1a line close backup protection is determined to be a misoperation.

2.5 Summary

This chapter provides an overview of fault diagnosis for high-speed railway traction power supply equipment and systems, introducing fault diagnostic methods and processes for traction transformers, switching equipment, power cables, and other high-speed rail traction power supply equipment. Utilizing various data collected by advanced sensors, and leveraging various intelligent reasoning algorithms such as particle swarm optimization neural networks, fuzzy Petri nets, Bayesian networks, and other intelligent algorithms, along with multi-source data, information, resources, etc., the following models are established: Fuzzy Petri net fault diagnosis model considering equipment availability, confidence model based on Weibull

Table 2.45 Relay protection and circuit breaker operation information

Component	FSlam	FSlap	B2m	CB211	CB101	CB212	CB2111	CB2112	CB202
Status	0	1	1	1	0	1	0	0	1

distribution, Bayesian network model, and traction power supply fault simulation model based on Monte Carlo simulation. These models assess the availability and health condition of high-speed railway traction power supply systems and equipment. Finally, this chapter explores practical case studies of fault diagnosis for high-speed railway traction power supply systems and equipment.

Chapter 3
Failure Prediction of High-Speed Railway Traction Power Supply System

In high-speed railway traction power supply systems, effective intelligent fault diagnosis measures assist operation and maintenance staff in swiftly addressing faults, promptly restoring power supply, or minimizing the power outage area. After accumulating a substantial amount of historical data, monitoring systems can discern the operational and transitional trends of equipment. By integrating current structural characteristics, parameters, and environmental conditions, it is possible to predict the future status of equipment, foreseeing the likelihood of imminent faults. This provides a reliable basis for early fault detection and prevention of fault-related losses. Predictive science is an emerging discipline where prognostics involves diagnosing parts or system functional status predictively, based on current or historical performance. Guided by relevant theories and methods, it analyzes and infers the future development status and trends of the study object. Prognostics technologies are currently widely applied in industries such as industrial, commercial, financial, and meteorological fields. In recent years, researchers at home and abroad have utilized methods such as artificial neural networks, support vector machines, Bayesian networks, and hidden Markov models in the field of power system prognostics. With the continuous development of high-tech, traction power supply systems and large industrial equipment are becoming increasingly complex, with higher levels of automation and integration. This complexity in the causes and signs of faults poses greater challenges for prognostics, necessitating the integration of various prognostics techniques and the application of new prognostics theories and classification methods. This promotes the co-development of prognostics, state management technology, and fault diagnosis.

Z. He et al., *Prognostics and Health Management of High-Speed Railway Traction Power Substations*, https://doi.org/10.1007/978-981-95-4538-4_3

3.1 Overview of Prognostics for High-Speed Railway Traction Power Supply System

Prognostics in high-speed railway traction power supply systems or equipment involve starting from the existing status or fault data, identifying patterns, and using these patterns to predict future statuses or statuses that are unobservable. Traditional prognostics primarily utilize statistical analysis models (such as regression analysis, time series analysis) or intelligent learning models (such as neural networks, support vector machines) to derive future time series and trends, without considering the influence of many related factors. Big data analysis techniques can explore the relationships between equipment state parameters and numerous factors related to traction power supply system operation and environmental meteorology. Based on these relationships, multi-parameter prognostics models are optimized and corrected to enhance the prognostics results with self-correction and adaptability, thus improving the accuracy of prognostics.

Prognostics of traction power equipment is a crucial aspect of high-speed railway traction power supply system prognostics. It primarily involves analyzing the evolution patterns of traction power supply equipment failures and the relationships between equipment fault characteristic parameters and faults. By integrating multi-parameter prognostics models and fault diagnosis models, it is possible to achieve real-time prognostics of the probability of equipment faults, fault types, and fault locations. Current research mainly utilizes algorithms such as Bayesian networks and Apriori to mine the relationships between fault characteristic parameters and employs methods like Markov models and time series similarity fault matching to achieve prognostics at different time scales.

3.2 Prognostics of Traction Transformer Based on Improved Grey Prognostics

Once a fault occurs in a traction transformer, it can likely lead to a power supply interruption, thereby affecting the entire high-speed railway system. Therefore, it is necessary to adopt effective fault diagnosis measures to assist personnel in timely accident handling. Traction transformer fault diagnosis can promptly detect existing faults, while traction transformer prognostics can foresee the possibility of fault occurrence, providing a basis for devising maintenance plans. This is crucial for enhancing the availability and economic efficiency of the traction power supply system. This section predicts changes in the dissolved gas content in traction transformer oil using an improved grey prognostics method, combined with the fault diagnosis model presented in this section, to achieve traction transformer prognostics.

3.2.1 Prognostics Methods for Traction Transformers Based on Improved Grey Prognostics

The traction transformer prognostics method introduced in this section uses equally spaced dissolved gas data in transformer oil for prognostics. It then utilizes neural networks trained in Sect. 2.2 to compute the output results, finally determining the confidence level of the traction transformer fault. The main steps are shown in Fig. 3.1.

1. Oil and Gas Data Prognostics

   Through the improved grey prognostics model, prognostics are made on oil and gas data at equal time intervals. By predicting the error values, the predicted oil and gas data are corrected, resulting in more accurate prognostics data.
2. Data Processing

   Data processing primarily prepares the subsequent input into neural networks. The processing method is the same as introduced in Sect. 3.2.
3. Fault Diagnosis Model

   The traction transformer fault diagnosis model proposed in Sect. 2.2 is utilized to determine the type of fault.

3.2.2 Grey Prognostics Method

3.2.2.1 Principle of Grey Prognostics

Grey prognostics are a forecasting method based on grey system theory, which constructs grey models through the generating processing of the original series. This generation sequence exhibits strong regularity, commonly utilizing two methods: accumulation generation and reduction generation. Grey prognostics explores, discovers, and grasps the laws of system changes to establish corresponding differential equation models, thereby predicting the future evolution trends and status of the system.

The GM(1, 1) model is a single sequence model that requires equal intervals in the original time series. Initially, the original time series must be modeled. Let $X^{(0)}$ be the original series:

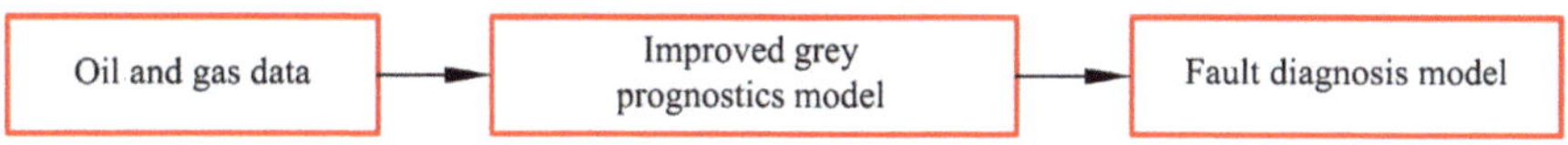

Fig. 3.1 Traction Transformer Prognostics Process

$$X^{(0)} = \left\{x^{(0)}(1), x^{(0)}(2), \cdots, x^{(0)}(\mathrm{n})\right\} \tag{3.1}$$

The data in the original series are accumulated in chronological order, and the resulting new series is called the accumulated series, as shown below:

$$x^{(1)}(k) = \sum\nolimits_{m=1}^{k} x^{(0)}(m) = x^{(1)}(k-1) + x^{(0)}(k), k = 1, 2, \cdots, n \tag{3.2}$$

$$X^{(1)} = \left\{x^{(1)}(1), x^{(1)}(2), \cdots, x^{(1)}(n)\right\} \tag{3.3}$$

Equation (3.2) describes the accumulation generation formula, where $X^{(1)}$ represents the new series from the first accumulation generation of $X^{(0)}$. Similarly, the k-th accumulated series $x^{(r)}(k)$ and $x^{(r-1)}(k)$ satisfy the following relationship:

$$x^{(r)}(k) = \sum_{m=1}^{k} x^{(r-1)}(m), k = 1, 2, \cdots, n \tag{3.4}$$

Let $Z^{(1)}$ be the adjacent mean generation series of $X^{(1)}$:

$$Z^{(1)}(k) = 0.5x^{(1)}(k) + 0.5x^{(1)}(k-1) \tag{3.5}$$

$$Z^{(1)} = \left\{z^{(1)}(2), z^{(1)}(3), \cdots, z^{(1)}(n)\right\} \tag{3.6}$$

The corresponding differential equation for the GM(1, 1) model is defined:

$$\frac{\mathrm{d}x^{(1)}}{\mathrm{d}t} + \alpha x^{(1)} = \mu \tag{3.7}$$

where α is called the development grey coefficient and μ is the endogenous control grey coefficient. This equation satisfies the initial condition when $t = t_0$, $x^{(1)} = x^{(1)}(t_0)$.

$$x^{(1)}(t) = \left[x^{(1)}(t_0) - \frac{\mu}{\alpha}\right] \times \mathrm{e}^{-\alpha(t-t_0)} + \frac{\mu}{\alpha} \tag{3.8}$$

The solution is provided for equidistant sampled discrete values, where $t_0 = 1$. The equation is: then:

$$x^{(1)}(k+1) = \left[x^{(1)}(1) - \frac{\mu}{\alpha}\right] \times \mathrm{e}^{-\alpha(k)} + \frac{\mu}{\alpha} \tag{3.9}$$

Substitute $x^{(1)}(2)$, $x^{(1)}(3)$, $\cdots$, $x^{(1)}(n)$ into Eq. (3.7), using differences instead of differentials because of equidistant sampling, with $\Delta t = (t + 1) - t = 1$:

$$\frac{\Delta x^{(1)}}{\Delta t} = x^{(1)}(t) - x^{(1)}(t-1) = x^{(0)}(t) \tag{3.10}$$

From Eqs. (3.7) and (3.10), we get:

$$\begin{cases} x^{(0)}(2) + \alpha x^{(1)}(2) = \mu \\ x^{(0)}(3) + \alpha x^{(0)}(3) = \mu \\ \cdots \\ x^{(0)}(n) + \alpha x^{(1)}(n) = \mu \end{cases} \tag{3.11}$$

Rearranging terms, we obtain:

$$\begin{cases} x^{(0)}(2) = \left[-x^{(1)}(2), 1\right] \begin{bmatrix} \alpha \\ \mu \end{bmatrix} \\ x^{(0)}(3) = \left[-x^{(1)}(3), 1\right] \begin{bmatrix} \alpha \\ \mu \end{bmatrix} \\ \cdots \\ x^{(0)}(n) = \left[-x^{(1)}(n), 1\right] \begin{bmatrix} \alpha \\ \mu \end{bmatrix} \end{cases} \tag{3.12}$$

Since $\frac{\Delta x^{(1)}}{\Delta t}$ involves values of $x^{(1)}$ at two times, use the adjacent mean generation series $z^{(1)}(k)$ in place of $x^{(1)}(k)$. From Eq. (3.12), we get:

$$\begin{bmatrix} \boldsymbol{x^{(0)}(2)} \\ \boldsymbol{x^{(0)}(3)} \\ \vdots \\ \boldsymbol{x^{(0)}(n)} \end{bmatrix} = \begin{bmatrix} -z^{(1)}(\boldsymbol{2}),\boldsymbol{1} \\ -z^{(1)}(\boldsymbol{3}),\boldsymbol{1} \\ \vdots \\ -z^{(1)}(\boldsymbol{n}),\boldsymbol{1} \end{bmatrix} \begin{bmatrix} \boldsymbol{\alpha} \\ \boldsymbol{\mu} \end{bmatrix} \tag{3.13}$$

Let

$$\boldsymbol{y} = \left(\boldsymbol{x^{(0)}(2), x^{(0)}(3), \cdots, x^{(0)}(n)}\right)^{\boldsymbol{T}}$$

$$\boldsymbol{B} = \begin{bmatrix} -z^{(1)}(\boldsymbol{2}), \boldsymbol{1} \\ -z^{(1)}(\boldsymbol{3}), \boldsymbol{1} \\ \vdots \\ -z^{(1)}(\boldsymbol{n}), \boldsymbol{1} \end{bmatrix}$$

$$\boldsymbol{U} = \begin{bmatrix} \boldsymbol{\alpha} \\ \boldsymbol{\mu} \end{bmatrix}$$

Equation (3.13) is expressed,

$$y = BU \tag{3.14}$$

Equation (3.14) is estimated using the least squares estimation method, yielding results:

$$U = \begin{bmatrix} \alpha \\ \mu \end{bmatrix} = \left(B^T B\right)^{-1} B^T y \tag{3.15}$$

Substitute the calculation results of Eq. (3.15) into Eq. (3.9) to calculate $x^{(1)}(n + 1)$, and then use Eq. (3.2) to calculate $x^{(0)}(n + 1)$, which is the grey prognostics value.

3.2.2.2 Improved Grey Prognostics Model

Although the traditional grey prognostics model exhibits a certain predictive capacity, there is still a significant discrepancy between its forecasts and actual values. This section introduces an improved grey prognostics model that uses grey prognostics of forecast errors to correct the predicted results. The process is illustrated in Fig. 3.2.

At first, equidistant sample data is processed through grey prognostics to generate the fitted sequence, which is then compared with the actual sequence to derive the first-order error series. Subsequently, grey prognostics are applied to this first-order error series to generate a first-order error prognostics series. This predicted error series is used to correct the initial fitted sequence, resulting in a first correction sequence. The first correction sequence is then compared with the actual sequence to determine the second-order error series. Grey prognostics of the second-order error series provide a second-order error prognostics series, which is used to adjust the first correction sequence, yielding a second correction sequence. This method is iteratively applied to derive higher-order error prognostics. Finally, by adjusting for each order's predictive error, the ultimate forecast values are obtained.

Suppose a sample sequence $X^{(0)}$ contains $4n$ data points. Using the previously introduced GM(1, 1) model, we use the first $2n$ actual data points to fit the $2n$ + 1th value, denoted as $x^{*(0)}(1)$. Using the first $(2n + 1)$ actual data points, we fit the

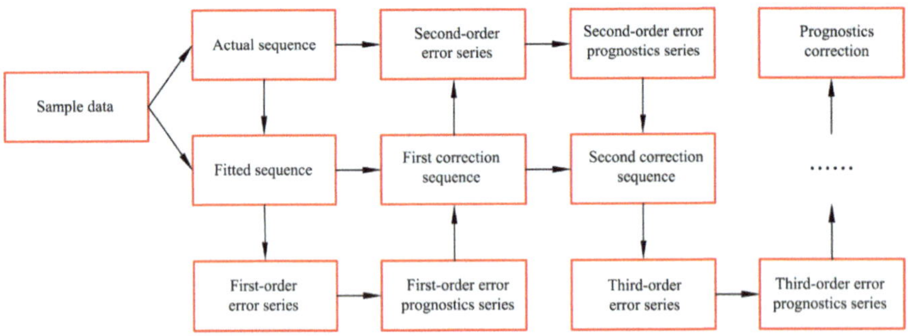

Fig. 3.2 Improved Grey Prognostics Process

$2n + 2$th value, $x^{*(0)}(2)$, and so on. This method results in the fitted sequence $X^{*(0)} = \{x^{*(0)}(1), x^{*(0)}(2), \cdots, x^{*(0)}(2n+1)\}$. Thus, the first-order error series $\delta^{(1)}$ is:

$$\delta^{(1)}(i) = \frac{x^{(0)}(i) - x^{(0)}(i)}{x^{(0)}(i)} \times 100\%, i = 1, 2, \cdots, 2n \tag{3.16}$$

where $x^{*(0)}(i)$ is the fitted value and $x^{(0)}(i)$ is the actual sample value.

Next, we compute the first-order prognostics error $\delta^{(1)}(2n+1)$. Since the $\delta^{(1)}$ series includes both positive and negative values, it is first transformed into a non-negative series. Given that $\delta^{(1)}(i) \geqslant -1$, we can convert the $\delta^{(1)}$ series into a $\gamma^{(1)}$ series as follows:

$$\gamma^{(1)}(i) = \delta^{(1)}(i) + 1 \tag{3.17}$$

Using the GM(1, 1) model and the $\gamma^{(1)}$ series, we predict $\gamma^{(1)}(2n+1)$, and then transform back to predict the error value:

$$\delta^{(1)}(2n+1) = \gamma^{(1)}(2n+1) - 1 \tag{3.18}$$

Using the actual error data from the first n periods to predict the n + 1th error, and using the actual error data from the first (n + 1) periods to predict the n + 2th error, and so on, we obtain the first-order error prediction series $\delta^{*(1)} = \{\delta^{*(1)}(1), \delta^{*(1)}(2), \cdots, \delta^{*(1)}(n)\}$, according to the following formula. Using this first-order error prognostics series, we correct the latter n items of the fitted sequence to get the first-order corrected series.

$$x^{*(1)}(i) = \frac{x^{*(0)}(n+i)}{1 + \delta^{*(1)}(i)}, i = 1, 2, \cdots, n \tag{3.19}$$

Comparing the first-order corrected series with the sample sequence yields the second-order error series $\delta^{(2)}$:

$$\delta^{(2)}(i) = \frac{x^{*(1)}(i) - x^{(0)}(i)}{x^{(0)}(i)} \times 100\%, i = 1, 2, \cdots, n \tag{3.20}$$

Using the same method, calculate the second-order predictive error $\delta^{(2)}(n+1)$, and so on up to higher-order prognostics errors based on the length of the sample data. Finally, correct the predicted values using the following formula to calculate the final predicted values:

$$x^{*(0)}(2n+1) = \frac{x^{*(0)}(2n+1)}{\left(1+\delta^{(1)}(2n+1)\right) \times \left(1+\delta^{(2)}(n+1)\right) \times \cdots} \tag{3.21}$$

3.2.3 Case Study of Traction Transformer Prognostics

3.2.3.1 Case Study 1

The historical data of a traction transformer before discovering a fault is shown in Table 3.1.

1. Improved Grey Prognostics of Oil Gas Content

   Due to the data having an approximate time interval of 15 days, we can use the improved grey prognostics model to prognosticate the content of gases such as H_2, CH_4, C_2H_6, C_2H_4, and C_2H_2 for the next time interval.

   First, calculate the traditional grey prediction value for each gas. Taking H_2 as an example, to predict the H_2 content on May 14, 2013, generate the H_2 Accumulated Series according to eq. (3.2).

$$X^{(1)} = \{6.3,17.5,46.2,79.5,131.3\} \tag{3.22}$$

   Generate the mean series according to eq. (3.5).

$$Z^{(1)} = \{11.9,31.85,62.85,105.4\} \tag{3.23}$$

   Therefore,

Table 3.1 Historical DGA data of a traction transformer

Sampling time	H_2/(μL/L)	CH_4/(μL/L)	C_2H_6/(μL/L)	C_2H_4/(μL/L)	C_2H_2/(μL/L)
2013-03-01	6.3	4.9	5.8	3.4	0
2013-03-15	11.2	23.3	6.3	15.4	0
2013-04-02	28.7	42.3	39.4	61.2	0.5
2013-04-17	33.3	63.7	41.6	145.4	0.4
2013-05-01	51.8	44.6	56.9	143.5	0.6
2013-05-14	46.4	138.2	41.5	164.3	0.5
2013-05-30	48.1	112.6	98.2	233.2	0.8
2013-06-15	52.0	176.3	103.1	202.5	1.2
2013-07-03	51.4	161.5	95.2	297.4	1.1
2013-07-16	67.2	199.1	123.9	335.8	1.4
2013-07-31	71.2	224.9	152.6	371.4	1.4

Table 3.2 DGA fitted sequence

Time	H_2/(μL/L)	CH_4/(μL/L)	C_2H_6/(μL/L)	C_2H_4/(μL/L)	C_2H_2/(μL/L)
2013-05-14	76.05	65.97	94.73	304.48	1.06
2013-05-30	72.84	118.82	69.79	299.97	0.85
2013-06-15	70.53	183.64	119.19	384.43	1.11
2013-07-03	71.65	244.34	151.80	371.98	1.55
2013-07-16	70.77	289.73	166.87	457.85	1.85
2013-07-31	80.45	337.90	194.56	539.94	2.23

Table 3.3 First-order error series

Time	δ (H_2)/%	δ (CH_4)/%	δ (C_2H_6)/%	δ (C_2H_4)/%	δ (C_2H_2)/%
2013-05-14	63.90	−52.27	128.27	85.32	111.03
2013-05-30	51.43	5.52	−28.93	28.63	5.90
2013-06-15	35.63	4.17	15.61	89.84	−7.09
2013-07-03	39.40	51.29	59.46	25.08	41.32
2013-07-16	5.31	45.52	34.68	36.36	32.47

$$B=\begin{bmatrix} -11.9 & 1 \\ -31.85 & 1 \\ -62.85 & 1 \\ -105.4 & 1 \end{bmatrix} \tag{3.24}$$

$$y=(11.2,28.7,33.3,51.8)^T \tag{3.25}$$

Calculate the development grey coefficient and endogenous control grey coefficient according to eq. (3.15).

$$\begin{bmatrix} \alpha \\ \mu \end{bmatrix} = \left(B^T B\right)^{-1} B^T y = \begin{bmatrix} -0.3966 \\ 10.228 \end{bmatrix} \tag{3.26}$$

By substituting the development grey coefficient and the endogenous control grey coefficient into Eq. (3.12), the calculations yield:

$$X^{(1)}(6)=207.3545 \tag{3.27}$$

Using Eq. (3.2) to calculate the final results rounded to two decimal places.

$$X^{(0)}(6)=76.05 \tag{3.28}$$

Similarly, the rest of the results can be calculated, as shown in Table 3.2.

Using Eq. (3.16) to calculate the error for H_2:

Table 3.4 First-order error grey prognostics values

Time	δ (H_2)/%	δ (CH_4)/%	δ (C_2H_6)/%	δ (C_2H_4)/%	δ (C_2H_2)/%
2013-07-31	3.11	44.83	36.53	36.35	31.9

Table 3.5 Comparison of final prognostics and actual DGA values

2013-07-31	H_2/(μL/L)	CH_4/(μL/L)	C_2H_6/(μL/L)	C_2H_4/(μL/L)	C_2H_2/(μL/L)
Actual values	71.2	224.9	152.6	371.4	1.4
Grey prognostics values	80.45	337.90	194.56	539.94	2.23
Improved grey prognostics values	78.02	233.31	142.51	396	1.69
Error	9.58%	3.74%	−6.61%	6.62%	20.71%

$$\delta(6) = \frac{76.05 - 46.4}{46.4} \times 100\% \approx 63.91\% \tag{3.29}$$

Similarly, the first-order error series for all gases can be obtained, the results of which are shown in Table 3.3.

Using Eq. (3.17) to transform the error series into the $\gamma^{(1)}$ series:

$$\gamma^{(1)}(H_2) = \{1.6390, 1.5143, 1.3563, 1.394, 1.0531\} \tag{3.30}$$

Further grey prognostics are then performed to obtain results.

$$\gamma^{(1)}(6) = 1.0311 \tag{3.31}$$

Using Eq. (3.18) to get:

$$\delta^{(1)}(6) = 0.0311 \tag{3.32}$$

Additional results are shown in Table 3.4.

Finally, using Eq. (3.21) to calculate the final prognostics results:

$$X^{**(0)}(11) = \frac{80.45}{1 + 0.0311} = 78.02 \tag{3.33}$$

Similarly, final prognostics results for other gases and their comparison with actual values are shown in Table 3.5.

From Table 3.5, it is evident that the prognostics error for C_2H_2 is relatively large, while the prognostics errors for other data are within an acceptable range. The reason for the significant error in C_2H_2 is that its own numerical values are small, and small changes in value cause significant fluctuations.

2. Fault Confidence Calculation

Table 3.6 Data processing

H_2/%	CH_4/%	C_2H_6/%	C_2H_4/%	C_2H_2/%	Three-ratio coding
9.16	27.40	16.74	46.50	0.20	0 2 1

Based on the modified three-ratio method encoding rules shown in Table 2.6, we process the predicted dissolved gases in oil data, obtaining the hydrogen gas percentage:

$$P_{V1} = \frac{78.02}{78.02 + 233.31 + 142.51 + 396 + 1.69} \times 100\% \approx 9.16\% \tag{3.34}$$

Similarly calculating the volume percentage of other gases is shown in Table 3.6.

Input the processed data into seven-particle swarm-optimized neural networks, with results displayed in Table 3.14.

Using Eq. (3.22) to calculate the fault confidence level.

$$P = \frac{0.8942}{0.8942 + 0.2548} \times 100\% \approx 77.82\% \tag{3.35}$$

In early August, during power outage maintenance, a bushing was found blackened, and this model's prognostics were consistent with the inspection results (Table 3.7).

3.2.3.2 Case Study 2

The historical data of a traction transformer before discovering a fault is shown in Table 3.8.

1. Improved Grey Prognostics of Oil Gas Content

   Due to the data having an approximate time interval of 15 days, we can use the Improved grey prognostics model to prognosticate the content of gases such as H_2, CH_4, C_2H_6, C_2H_4, and C_2H_2 for the next time interval.
2. Initially, calculate the Traditional Grey Prediction Value for each gas. Taking C_2H_6 as an example, to predict the H_2 content on December 24, 2015, generate the H_2 Accumulated Series according to eq. (3.2).

$$X^{(1)} = \{15.2,37,59.6,83.7,113.6\} \tag{3.36}$$

Generate the Mean Series according to eq. (3.5):

Table 3.7 Neural networks calculation results

Y	Low-temperature overheating neural networks	High-temperature overheating neural networks	Partial discharge neural networks	Low-energy discharge neural networks	High-energy discharge neural networks	Low-energy discharge overheating neural networks	High-energy discharge overheating neural networks
Y_1	0.8942	0.5329	0.1843	0.2472	0.2270	0.1341	0.1934
Y_2	0.2548	0.4672	0.9558	0.8874	0.7965	0.8629	0.8539

Table 3.8 Historical DGA data of a traction transformer

Sampling time	H_2/(μL/L)	CH_4/(μL/L)	C_2H_6/(μL/L)	C_2H_4/(μL/L)	C_2H_2/(μL/L)
2015-10-10	21.7	34.5	15.2	14.4	0
2015-10-25	23.1	33.7	21.8	17.3	0
2015-11-09	22.5	38.3	22.6	26.8	0
2015-11-26	29.6	40.2	24.1	38.4	0.1
2015-12-11	31.2	41.8	29.9	48.3	0.1
2015-12-24	36.3	46.2	34.1	61.4	0.1
2016-01-08	34.1	45.3	38.4	79.2	0.2
2016-01-24	41.8	57.6	44.6	78.5	0.2
2016-02-06	43.4	64.1	42.2	95.4	0.1
2016-02-23	42.6	62.2	53.9	112.8	0.3
2016-03-08	46.4	68.9	56.6	132.7	0.2

Table 3.9 DGA fitted sequence

Time	H_2/(μL/L)	CH_4/(μL/L)	C_2H_6/(μL/L)	C_2H_4/(μL/L)	C_2H_2/(μL/L)
2015-12-24	35.30	45.32	31.82	66.44	0.48
2016-01-08	40.47	49.18	37.42	83.22	0.30
2016-01-24	40.26	49.71	42.72	106.32	0.51
2016-02-06	45.20	57.86	49.24	119.13	0.55
2016-02-23	48.64	65.84	51.37	137.95	0.34
2016-03-08	50.08	69.53	58.61	161.92	0.54

$$Z^{(1)} = \{26.1,48.3,71.65,98.65\} \tag{3.37}$$

Therefore,

$$B = \begin{bmatrix} -26.11 & \\ -48.3 & 1 \\ -71.65 & 1 \\ -98.65 & 1 \end{bmatrix} \tag{3.38}$$

$$y = (21.8,22.6,24.1,29.9)^T \tag{3.39}$$

Calculate the Development Grey Coefficient and Endogenous Control Grey Coefficient according to eq. (3.15):

$$\begin{bmatrix} \alpha \\ \mu \end{bmatrix} = \left(B^T B\right)^{-1} B^T y = \begin{bmatrix} -0.109 & 0 \\ 17.925 & \end{bmatrix} \tag{3.40}$$

By substituting the development grey coefficient and the endogenous control grey coefficient into Eq. (3.12), the calculations yield:

Table 3.10 First-order error series

Time	δ (H_2)/%	δ (CH_4)/%	δ (C_2H_6)/%	δ (C_2H_4)/%	δ (C_2H_2)/%
2015-12-24	−2.75	−1.9	−6.67	8.21	375.79
2016-01-08	18.67	8.57	−2.54	5.07	48.43
2016-01-24	−3.7	−13.7	−4.21	35.44	155.05
2016-02-06	4.15	−9.73	16.67	24.88	452.65
2016-02-23	14.17	5.85	−4.7	22.29	12.08

Table 3.11 First-order error grey prognostics values

Time	δ (H_2)/%	δ (CH_4)/%	δ (C_2H_6)/%	δ (C_2H_4)/%	δ (C_2H_2)/%
2012-03-08	6.83	−3.38	4.82	31.97	202.36

$$X^{(1)}(6) = 145.4239 \tag{3.41}$$

Using Eq. (3.2) to calculate the final results rounded to two decimal places.

$$X^{(0)}(6) = 31.82 \tag{3.42}$$

Similarly, the rest of the results can be calculated, as shown in Table 3.9.

Using Eq. (3.16) to calculate the C_2H_6 error:

$$\delta(6) = \frac{31.82 - 34.1}{34.1} \times 100\% \approx -6.67\% \tag{3.43}$$

Similarly, all gas first-order error series are obtained, as shown in Table 3.10.

Using Eq. (3.17) to transform the error series into the $\gamma^{(1)}$ series:

$$\gamma^{(1)}(C_2H_6) = \{0.9333, 0.9746, 0.9579, 1.1667, 0.9530\} \tag{3.44}$$

Further grey prognostics are then performed to obtain results.

$$\gamma^{(1)}(6) = 1.0482 \tag{3.45}$$

Using Eq. (3.18) to get:

$$\delta^{(1)}(6) = 0.0482 \tag{3.46}$$

Results as shown in Table 3.11.

Finally, using Eq. (3.21) to calculate the final prognostics results:

Table 3.12 Comparison of final prognostics and actual DGA values

2016-03-08	H_2/(μL/L)	CH_4/(μL/L)	C_2H_6/(μL/L)	C_2H_4/(μL/L)	C_2H_2/(μL/L)
Actual values	46.4	68.9	56.6	132.7	0.2
Grey prognostics values	50.08	69.53	58.61	161.92	0.54
Improved grey prognostics values	46.88	71.96	55.92	122.70	0.18
Error	1.03%	4.44%	−1.2%	−7.54%	−10%

Table 3.13 Data processing

H_2/%	CH_4/%	C_2H_6/%	C_2H_4/%	C_2H_2/%	Three-ratio coding
15.75	24.18	18.79	41.22	0.06	0 2 1

$$X^{**(0)}(11) = \frac{58.61}{1 + 0.0482} = 55.92 \tag{3.47}$$

Similarly, calculate the final prognostics results for other gases and their comparison with actual values, as shown in Table 3.12.

From Table 3.12, it is evident that the prognostics errors are all within a ±10% range, indicating small errors.

3. Fault Confidence Calculation

Based on Table 2.6, code the predicted data using the three-ratio method and calculate the gas content percentages. The C_2H_6 gas percentage is calculated:

$$P_{V3} = \frac{55.92}{46.88 + 71.96 + 55.92 + 122.7 + 0.18} \times 100\% \approx 18.79\% \tag{3.48}$$

Similarly calculating the volume percentage of other gases is shown in Table 3.13.

Input the processed data into seven-particle swarm-optimized neural networks, with results displayed in Table 3.14.

Using Eq. (3.22) to calculate the fault confidence level.

$$P = \frac{0.7951}{0.7951 + 0.2392} \times 100\% \approx 76.87\% \tag{3.49}$$

In mid-March, during a power outage maintenance, burn marks were found inside the bushing, consistent with the prognostics results.

Table 3.14 Neural networks calculation results

Y	Low-temperature overheating neural networks	High-temperature overheating neural networks	Partial discharge neural networks	Low-energy discharge neural networks	High-energy discharge neural networks	Low-energy discharge overheating neural networks	High-energy discharge overheating neural networks
Y_1	0.7951	0.3945	0.1291	0.1422	0.1539	0.1903	0.1218
Y_2	0.2392	0.5820	0.9373	0.8329	0.8543	0.9167	0.9553

3.3 Prognostics of Traction Power Supply Equipment Based on Continuous-Time Markov Degradation Process

Scholars both domestically and internationally have conducted studies on equipment failure prognostics from three aspects: data classification, prognostics standards, and prognostics algorithms. Prognostics commonly employs intelligent algorithms such as support vector machines and neural networks, or methods like Monte Carlo simulations. These generally require extensive samples, have slow convergence speeds, and involve substantial computation, which needs further refinement for practical field applications. Additionally, these algorithms typically focus on individual units within a system, with limited research on the impact of individual differences on assessment prognostics, making it difficult to achieve prognostics for large-scale, similar equipment within a system. The lack of data on the equipment's current status impedes the development of comprehensive system-level maintenance strategies.

This section proposes a prognostics method based on a continuous-time Markov degradation process for traction power supply equipment of considerable quantity and identical type. This method aims to provide quantitative indicators for the assessment and prognostics of traction power supply equipment.

3.3.1 Markov Degradation Process Modeling for Traction Power Supply Equipment

3.3.1.1 State Division and Modeling of Equipment Degradation

According to IEEE standard guidelines, the operating status of transmission and transformation equipment is typically divided into four categories: normal condition, attentive condition, abnormal condition, and serious condition, as illustrated in Fig. 3.3. The equipment starts in a normal condition and degrades progressively through each operating status. When the equipment reaches a serious condition, it signifies an urgent need for maintenance and repair, thus generally treated as a fault status.

As illustrated in Fig. 3.3, under general conditions, the performance of traction power supply equipment begins to decline from the onset of its operation. The severity of the decline increases with service time, progressing to a fault state that necessitates taking the equipment out of operation. This performance degradation can be described through a continuous and gradual process. During the service life of traction power supply equipment, factors such as operating conditions, geographic and climatic environments, and daily maintenance and repair strategies inevitably affect its degradation process. Equipment located outdoors is particularly susceptible to geographic and climatic conditions. However, despite the variety of factors

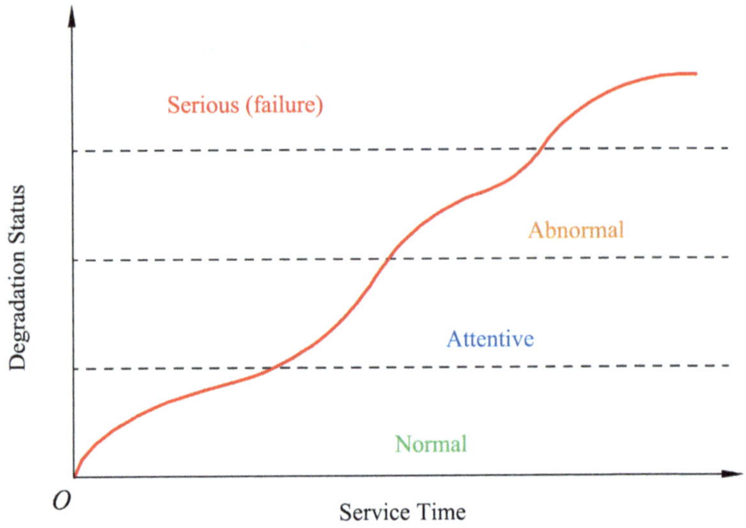

Fig. 3.3 Schematic Diagram of Traction Power Supply Equipment Performance Degradation Curve

impacting the traction power supply equipment, the degradation process ultimately results in a fault state. For instance, equipment failures caused by extreme weather conditions lead to outcomes consistent with those caused by performance degradation advancing to a fault. Consequently, this paper treats weather and other external influencing factors as conditions that accelerate the degradation process. Regardless of whether faults occur early or late in the equipment's service life, it is assumed that the equipment will undergo a gradual degradation process, albeit at different rates, but leading to the same outcome. The consideration of weather and other external influencing factors can be disregarded when modeling the degradation process using historical statistical data over extended time scales.

In the actual traction power supply system, fault maintenance records, including equipment failures that cause system shutdowns and equipment defect information, are maintained. Additionally, every recorded instance of maintenance indicates the involvement of maintenance and repair activities, thus reflecting the impact of operation and maintenance strategies on the equipment degradation process. Therefore, to facilitate integration with practical applications, this section treats the recorded maintenance data as fault states in the model.

Corresponding to IEEE standard guidelines, this section, combining historical fault records data of traction power supply equipment, has developed a four-state Markov degradation model, as shown in Fig. 3.4.

Since the data recorded in the field are mostly related to equipment failures or shutdowns, several points about this model are as follows:

1. The predicted traction power supply equipment is of the same type, meaning they share the same voltage level, type, specifications, and other parameters.

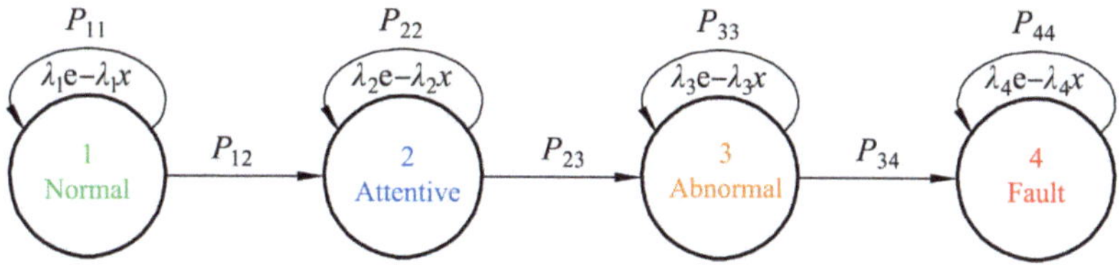

Fig. 3.4 Markov Degradation Model of Traction Power Supply Equipment

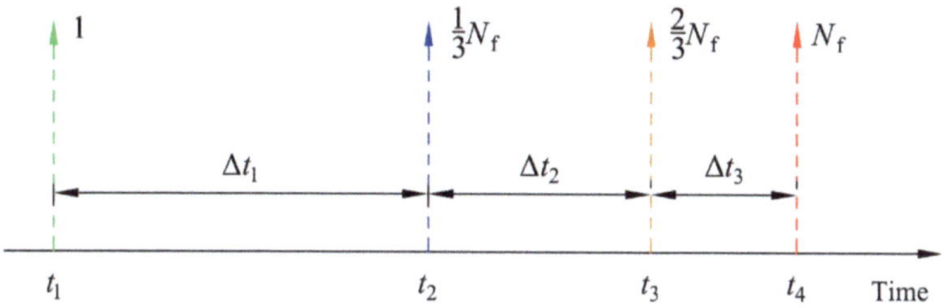

Fig. 3.5 Equipment Fault Sample Equidivision Time Series

2. All equipment starts in a normal condition (status 1), and the recorded moments are when the equipment is in a fault state (status 4). The equipment performance degradation until the fault is a step-by-step degrading process.
3. The currently observed status (fault record data) is already the equipment's fault state, which is the cumulative result of operating time and the impact of various factors up to the observation moment. This outcome can reflect the impact of operating conditions, geographic and climatic environments, and the strategies employed for daily maintenance and repair of the equipment.
4. The equipment degradation process possesses continuous-time Markov properties: under the current status (present), the future evolution of the equipment's status (future) does not depend on its past evolution (past), embodying memorylessness.

3.3.1.2 Markov State Transition Rate Calculation

For the same type of equipment, the residence time in each status can be described from a statistical probability perspective. For equipment with a substantial quantity and fault statistical sample size, the residence time for each state can be derived by equally dividing the number of fault occurrences. For instance, if the total number of a certain type of traction power supply equipment is N, and the total number of occurrences of equipment failure from the start of the statistics at time t_1 to the end at time t_4 is N_f, then the fault sample number of occurrences N_f can be equally divided to determine the residence time for each state. As shown in Fig. 3.5, t_1 marks the time of the first fault occurrence after the start of the statistics, t_2 marks the time of the N_f/

3rd fault occurrence, and t_3 marks the time of the $2N_f/3$rd fault occurrence, continuing until the N_fth fault occurrence at time t_4, which is considered the end of the statistical period. Thus, the time interval Δt_1 between t_1 and t_2 can be considered as the residence time of the equipment under normal conditions, representing the transition time from status 1 to status 2. Similarly, Δt_2 is the residence time under attentive condition, representing the transition from status 2 to status 3; Δt_3 is the residence time under abnormal conditions, representing the transition from status 3 to status 4.

The degradation law of traction power supply equipment adheres to Markov properties, and its state residence time follows an exponential distribution. If we assume that the residence time Δt_i of the equipment in the i-th state follows an exponential distribution with parameter λ_i, then the density function and the expected value of this exponential distribution are as follows:

$$f(t) = \lambda_i e^{-\lambda_i t} \tag{3.50}$$

$$E(t) = \frac{1}{\lambda_i} = \Delta t_i \tag{3.51}$$

Considering a four-status continuous-time Markov chain, it is clearly a homogeneous Markov process, thus its state transition probability is represented by:

$$\begin{cases} P_{12}(\Delta t) = \lambda_1 \Delta t + O(\Delta t) \\ P_{23}(\Delta t) = \lambda_2 \Delta t + O(\Delta t) \\ P_{34}(\Delta t) = \lambda_3 \Delta t + O(\Delta t) \end{cases} \tag{3.52}$$

where $P_{12}(\Delta t)$, $P_{23}(\Delta t)$, $P_{34}(\Delta t)$ respectively represent the state transition probabilities between *normal* and *attentive, attentive* and *abnormal,* and *abnormal* and *fault* states. $o(\Delta t)$ denotes a higher-order infinitesimal over the time interval Δt.

If q represents the equipment state transition rate, then according to the Markov process limit theorem:

$$\begin{cases} q_{11} = \lim\limits_{\Delta t \to 0} \dfrac{1 - P_{11}(\Delta t)}{\Delta t} = \lim\limits_{\Delta t \to 0} \dfrac{P_{12}(\Delta t)}{\Delta t} = \lambda_1 = q_{12} \\ q_{22} = \lim\limits_{\Delta t \to 0} \dfrac{1 - P_{22}(\Delta t)}{\Delta t} = \lim\limits_{\Delta t \to 0} \dfrac{P_{23}(\Delta t)}{\Delta t} = \lambda_2 = q_{23} \\ q_{33} = \lim\limits_{\Delta t \to 0} \dfrac{1 - P_{33}(\Delta t)}{\Delta t} = \lim\limits_{\Delta t \to 0} \dfrac{P_{34}(\Delta t)}{\Delta t} = \lambda_3 = q_{34} \end{cases} \tag{3.53}$$

where q_{11}, q_{22}, q_{33} are the self-holding rates for the *normal, attentive,* and *abnormal* states, respectively. q_{12}, q_{23}, q_{34} are the state transition rates between *normal* and *attentive, attentive* and *abnormal,* and *abnormal* and *fault,* respectively. $P_{11}(\Delta t)$, $P_{22}(\Delta t)$, $P_{33}(\Delta t)$ are the self-holding probabilities for the *normal, attentive,* and *abnormal* states, respectively.

3.3.2 Prognostics Based on Markov Degradation Process

3.3.2.1 Establishment of Markov State Set

For the same type of traction power supply equipment, each piece initially operates in status 1, the normal state, upon commencement. After a random continuous period, the equipment transitions to the next status, progressing until it reaches status 4, the fault state. In this Markov degradation process, if at time t the number of units in status i is $X_i(t)$, then the entire system's status set $\boldsymbol{S}(t)$ can be represented as $\boldsymbol{S}(t) = [X_1(t), X_2(t), X_3(t), X_4(t)]'$, where $X_i(t)$ can be obtained through the following system of integral equations:

$$\begin{cases} X_1(t) = N - Y_1\left[\int_0^t \lambda_1 X_1(\tau)\mathrm{d}\tau\right] \\ X_2(t) = Y_1\left[\int_0^t \lambda_1 X_1(\tau)\mathrm{d}\tau\right] - Y_2\left[\int_0^t \lambda_2 X_2(\tau)\mathrm{d}\tau\right] \\ X_3(t) = Y_2\left[\int_0^t \lambda_2 X_2(\tau)\mathrm{d}\tau\right] - Y_3\left[\int_0^t \lambda_3 X_3(\tau)\mathrm{d}\tau\right] \\ X_4(t) = Y_3\left[\int_0^t \lambda_3 X_3(\tau)\mathrm{d}\tau\right] \end{cases} \tag{3.54}$$

where N represents the total number of traction power supply equipment in this category; $Y_i(\cdot)(i = 1, 2, 3)$, indicates the number of equipment that has left status i and reached status $i + 1$ by time t; the integrand in $Y_i(\cdot)$ represents the rate at which equipment leaves status i to reach status $i + 1$ at time τ. For example, as expressed in the first equation of (3.54), the number of equipment in status 1 equals the initial total number of equipment N minus the number of equipment that has left status 1, calculated as $Y_1\left[\int_0^t \lambda_1 X_1(\tau)\mathrm{d}\tau\right]$; in the second and third equations, the first term on the right side of the equation, which inputs the number of equipment in the relevant status, precisely matches the number of equipment that has left the previous status; the last equation states that $X_4(t)$, the number of equipment in the fault state, equals the number of equipment that has left status 3.

For this Markov process, the matrix geometric method is typically employed to solve Eq. (3.54). However, as the number of equipment increases, the state order within the solution process also expands, making the traditional matrix geometric method impractical for large-scale systems containing numerous equipment. Therefore, for traction power supply systems with a large number of similar equipment, after establishing the Markov state set, a uniform acceleration technique is adopted to transform it into another approximable solution set of states.

3.3.2.2 Transformation of the Markov State Set

To utilize the uniform acceleration technique, an acceleration factor η is introduced to replace N (representing the total number of equipment in the system) in Eq. (3.54), thus transforming the Markov state set as follows:

$$\begin{cases} X_{1,\eta}(t) = \eta - Y_1\left[\int_0^t \eta\lambda_1 \frac{X_{1,\eta}(\tau)}{\eta}\mathrm{d}\tau\right] \\ X_{2,\eta}(t) = Y_1\left[\int_0^t \eta\lambda_1 \frac{X_{1,\eta}(\tau)}{\eta}\mathrm{d}\tau\right] - Y_2\left[\int_0^t \eta\lambda_2 \frac{X_{2,\eta}(\tau)}{\eta}\mathrm{d}\tau\right] \\ X_{3,\eta}(t) = Y_2\left[\int_0^t \eta\lambda_2 \frac{X_{2,\eta}(\tau)}{\eta}\mathrm{d}\tau\right] - Y_3\left[\int_0^t \eta\lambda_3 \frac{X_{3,\eta}(\tau)}{\eta}\mathrm{d}\tau\right] \\ X_{4,\eta}(t) = Y_3\left[\int_0^t \eta\lambda_3 \frac{X_{3,\eta}(\tau)}{\eta}\mathrm{d}\tau\right] \end{cases} \tag{3.55}$$

where $X_{i,\ \eta}(\tau)(i = 1, 2, 3, 4)$ represents the number of equipment under the transformation conditions.

Let $T > 0$ be the service time of the equipment. For $t \in [0, T]$, as η approaches infinity, the fluid limit $\overline{x}_i(t)$ of the equipment quantity $X_{i,\ \eta}(t)$ can be defined as:

$$\overline{x}_i(t) = \lim_{\eta\rightarrow} \frac{X_{i,\eta}(t)}{\eta} \tag{3.56}$$

This represents the expected value of the proportion of equipment in status i at time t relative to the total number of equipment in the system. Consequently, it is established that $\overline{x}_1(0) = 1; \overline{x}_i(0) = 0$ for $i = 2, 3, 4$, and within the service time $t \in [0, T]$,

$$\sum_{i=1}^{4} \overline{x}_i(t) = 1 \tag{3.57}$$

By differentiating both sides of eq. (3.55) with respect to t and taking the limit as η approaches infinity, we obtain:

$$\begin{cases} \frac{\mathrm{d}}{\mathrm{d}t}\overline{x}_1(t) = -\lambda_1\overline{x}_1(t) \\ \frac{\mathrm{d}}{\mathrm{d}t}\overline{x}_2(t) = \lambda_1\overline{x}_1(t) - \lambda_2\overline{x}_2(t) \\ \frac{\mathrm{d}}{\mathrm{d}t}\overline{x}_3(t) = \lambda_2\overline{x}_2(t) - \lambda_3\overline{x}_3(t) \\ \frac{\mathrm{d}}{\mathrm{d}t}\overline{x}_4(t) = \lambda_3\overline{x}_3(t) \end{cases} \tag{3.58}$$

By solving the system of differential equations in Eq. (3.58), the fluid limits $\overline{x}_i(t)$ can be obtained, and by combining this with the total number of equipment, the average number of equipment in status i can be determined.

Based on this framework, within the service time $t \in [0, T]$, the diffusion limit $D_i(t)$ and the diffusion limit cluster $\boldsymbol{D}(t)$ for the number of equipment $X_{i,\ \eta}(t)$ are defined as follows:

$$D_i(t) = \lim_{\eta \to} \sqrt{\eta}\left[\frac{X_{i,\eta}(t)}{\eta} - \overline{x}_i(t)\right], i = 1, 2, 3, 4 \tag{3.59}$$

$$D(t) = [D_1(t), D_2(t), D_3(t), D_4(t)]' \tag{3.60}$$

During $t \in [0, T]$, $T > 0$, the obtained diffusion limit $\boldsymbol{D}(t)$ follows a multivariate normal distribution, with an expectation of a 4×1 zero vector.

Construct two 4×4 matrices as follows:

$$\boldsymbol{A}(t) = \begin{bmatrix} -\lambda_1 & 0 & 0 & 0 \\ \lambda_1 & -\lambda_2 & 0 & 0 \\ 0 & \lambda_2 & -\lambda_3 & 0 \\ 0 & 0 & \lambda_3 & -\lambda_4 \end{bmatrix} \tag{3.61}$$

$$\boldsymbol{B}(t) = \begin{bmatrix} -\sqrt{\lambda_1 \overline{x}_1(t)} & 0 & 0 & 0 \\ \sqrt{\lambda_1 \overline{x}_1(t)} & -\sqrt{\lambda_2 \overline{x}_2(t)} & 0 & 0 \\ 0 & \sqrt{\lambda_2 \overline{x}_2(t)} & -\sqrt{\lambda_3 \overline{x}_3(t)} & 0 \\ 0 & 0 & \sqrt{\lambda_3 \overline{x}_3(t)} & -\sqrt{\lambda_4 \overline{x}_4(t)} \end{bmatrix} \tag{3.62}$$

The system can be established using matrix $\boldsymbol{A}(t)$, $\boldsymbol{B}(t)$, and the diffusion limit cluster $\boldsymbol{D}(t)$, forming the following system of random differential equations:

$$\begin{cases} \mathrm{d}\boldsymbol{D}(t) = \boldsymbol{A}(t)\boldsymbol{D}(t)\mathrm{d}t + \boldsymbol{B}(t)\mathrm{d}\boldsymbol{W}(t) \\ \boldsymbol{D}(0) = 0 \end{cases} \tag{3.63}$$

where $\boldsymbol{D}(t) = [\boldsymbol{D}_1(t), \boldsymbol{D}_2(t), \boldsymbol{D}_3(t), \boldsymbol{D}_4(t)]'$ is the solution to this system of random differential equations, and $\boldsymbol{W}(t)$ represents a four-dimensional standard Brownian motion. The covariance matrix $\boldsymbol{\Sigma}(t)$ of $\boldsymbol{D}(t)$ is the solution to the following ordinary differential equation system:

$$\frac{\mathrm{d}}{\mathrm{d}t}\boldsymbol{\Sigma}(t) = \boldsymbol{A}(t)\boldsymbol{\Sigma}(t) + \boldsymbol{\Sigma}(t)\boldsymbol{A}(t)' + \boldsymbol{B}(t)\boldsymbol{B}(t)' \tag{3.64}$$

Using the covariance matrix $\boldsymbol{\Sigma}(t)$ of $\boldsymbol{D}(t)$, it is possible to determine the variance $V_{ar}[\boldsymbol{D}_i(t)]$ of the diffusion limit $\boldsymbol{D}_i(t)$.

3.3.2.3 Availability Parameters Calculation

According to Eq. (3.59), the approximate value of $X_{i,\eta}(t)$ can be obtained:

$$X_{i,\eta}(t) \approx \eta\overline{x}_i(t) + \sqrt{\eta}D_i(t), i = 1, 2, 3, 4 \tag{3.65}$$

Since $\boldsymbol{D}(t)$ follows a multivariate normal distribution, the expectation and variance for random variables can be calculated:

$$E\left[X_{i,\eta}(t)\right] \approx \eta\overline{x}_i(t) \tag{3.66}$$

$$V_{ar}[X_i(t)] \approx NV_{ar}[D_i(t)] \tag{3.67}$$

Known values are represented by $\overline{x}_i(t)$, which indicates the proportion of equipment in status i relative to the total number of equipment, thus:

$$\sum_{i=1}^{4} N\overline{x}_i(t) = N \tag{3.68}$$

Using $\eta\boldsymbol{\Sigma}(t)$, the covariance matrix of the system state $\boldsymbol{S}(t)$ can be approximated. Therefore, at time t, the system state $\boldsymbol{S}(t)$ can be approximated by a multivariate normal distribution with a mean vector of $N[\overline{x}_1(t), \overline{x}_2(t), \overline{x}_3(t), \overline{x}_4(t)]'$ and a covariance matrix of $N\boldsymbol{\Sigma}(t)$, namely:

$$\boldsymbol{S}(t) \sim MVN\left\{N[\overline{x}_1(\mathrm{t}), \overline{x}_2(t), \overline{x}_3(t), \overline{x}_4(t)]', N\boldsymbol{\Sigma}(t)\right\} \tag{3.69}$$

In summary, based on the continuous-time four-state Markov process established for traction power supply equipment, and the transformation of the Markov state set using uniform acceleration techniques, the normal distribution of the number of equipment in each status can be obtained. This yields the expected values $E[X_i(t)]$ and variances $V_{\mathrm{ar}}[X_i(t)]$ for each status, enabling prognostics for this class of traction power supply equipment.

3.3.3 *Case Study of Traction Power Equipment Prognostics*

3.3.3.1 Actual Data Description

An analysis was conducted using fault statistics data from 430 units of 27.5 kV vacuum circuit breakers within the jurisdiction of a particular railway power supply section. All vacuum circuit breakers are of the same voltage level and model, and the evolution of the future state of the equipment is not influenced by its historical state, adhering to the conditions of a Markov process. The statistical duration is

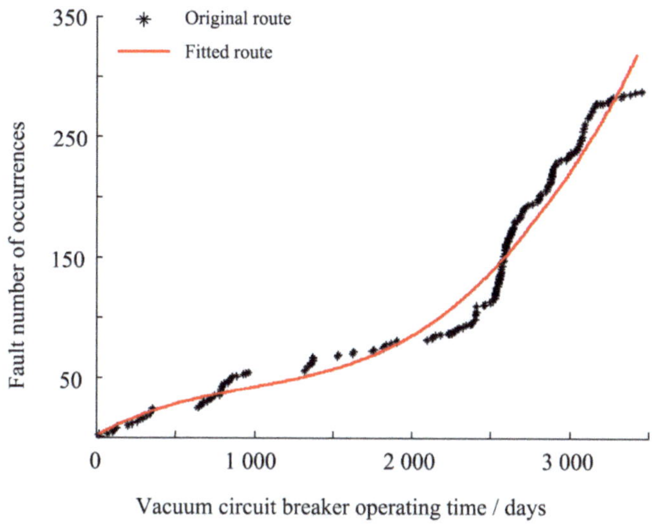

Fig. 3.6 Fault Statistics Data and Trend Curve for 430 Vacuum Circuit Breakers

116 months, during which the equipment experienced a total of 270 faults. The statistical start time is April 4, 2006, marking the first fault occurrence within this period; the statistical end time is December 4, 2015, when the 270th fault occurred.

As depicted in Fig. 3.6, the fault conditions of the 430 units of 27.5 kV vacuum circuit breakers were statistically analyzed with the horizontal axis representing time, starting from April 4, 2006, and using days as intervals. Each occurrence of equipment failure incremented the cumulative fault count, continuing until the record end date of December 4, 2015. Using the least squares method to fit the fault data reveals a growing trend in the accumulation rate of fault occurrences from the initial operation.

3.3.3.2 27.5 kV Vacuum Circuit Breaker Markov Degradation Model Actual Data Description

To validate the availability prognostics method proposed in this chapter, the statistical fault data were divided into two segments:

1. The fault data from April 4, 2006, to August 4, 2014, is selected as the first part. Based on the Markov degradation model and state division method for traction power supply equipment established in this chapter, the fault statistics data of 225 occurrences for 430 units of 27.5 kV vacuum circuit breakers are processed.
2. The second part consists of the fault data from August 4, 2014, to December 4, 2015, which will be used as verification data for the fault prediction method in the next section.

Table 3.15 Markov state division results and state residence times

Markov status	Fault number of occurrences	Fault timing interval	State residence time/d
Status 1 (normal)	1–75	2006-04-04 to 2011-06-29	1912
Status 2 (attentive)	76–150	2011-06-29 to 2013-05-21	692
Status 3 (abnormal)	151–225	2013-05-21 to 2014-08-15	451
Status 4 (fault)	Absorbing state	Statistical end time	

Using the first part of the fault statistics data, the residence time intervals for the equipment states are divided according to the time intervals experienced by equally divided fault numbers. Since there were 225 faults during this period, the time intervals for every 75 faults among similar equipment are used to define the intervals between two statuses, thereby determining each state residence time. As shown in Table 3.15, Δt_1, Δt_2, and Δt_3 represent 1939 d, 665 d, and 451 d, respectively.

Given that the state residence times follow an exponential distribution, the state transition rates of the Markov model can be calculated using Eqs. (3.50) and (3.51):

$$\begin{cases} q_{12} = \lambda_1 = \dfrac{1}{\Delta t_1} = \dfrac{1}{1912} \approx 0.0005 \\ q_{23} = \lambda_2 = \dfrac{1}{\Delta t_2} = \dfrac{1}{692} \approx 0.0014 \\ q_{34} = \lambda_3 = \dfrac{1}{\Delta t_3} = \dfrac{1}{451} \approx 0.0022 \end{cases} \tag{3.70}$$

3.3.3.3 27.5 kV Vacuum Circuit Breaker Availability Prognostics and Validation

By employing the continuous-time Markov degradation model established, along with the state transition rates determined using Eq. (3.70), and applying the uniform acceleration technique to approximate solutions for the Markov state set, prognostics can be made for any forecasted moment. This includes the expected number of equipment in each state, along with their variance and standard deviation.

Based on the records from April 4, 2006, it was found that all 430 units of the 27.5 kV vacuum circuit breakers in the traction power supply section were in operating status prior to their first fault. Due to the lack of earlier records, April 4, 2006, is considered the start date, and it is assumed that these breakers were initially in normal condition. According to relevant regulations, the service life of high-voltage vacuum interrupters generally ranges from 15 to 20 years. Therefore, calculations were performed to obtain the Markov state set for a 20-year period from the start date, showing the expected number of equipment in each state and their standard deviations as illustrated in Figs. 3.7 and 3.8.

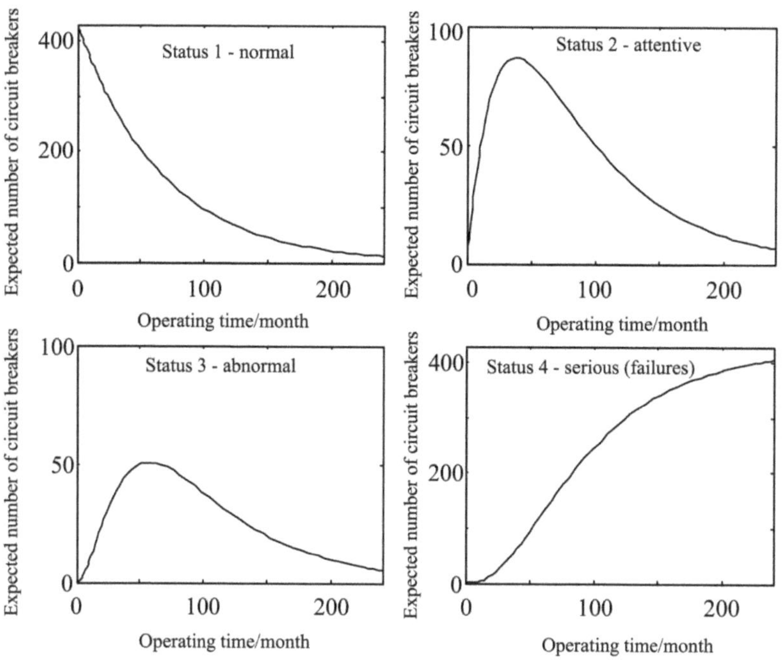

Fig. 3.7 Expected Number of Equipment in Each State Over 20 Years (240 Months)

From Figs. 3.7 and 3.8, it is evident that as the service time progresses, the number of circuit breakers in Status 1, i.e., the normal operating condition, continuously decreases, while those developing to Status 4 fault continuously increase. Without considering the maintenance process, from the start date on April 4, 2006, the number of equipment failures is expected to reach approximately 406 units after 20 years, with about 12 pieces of equipment still in normal operating condition. Furthermore, the number of breakers in Status 2 *attentive* grows to about 86 units after approximately 38 months, then shows a decreasing trend; similarly, the number of breakers in Status 3 *abnormal* peaks at about 50 units around 44 months, then gradually decreases. This observation confirms that Status 2 and Status 3 serve as transition states for these 430 vacuum circuit breakers. Once defects occur, the development from normal condition to fault state happens swiftly, aligning with the field personnel's observations of frequent operations and frequent mechanical and signal faults among these equipment.

The first part of the statistical fault data serves as the input for this method, covering a total of 100 months from April 2006 to August 2014. The prognostics based on the Markov degradation process method reflect the performance degradation law of these equipment under their operating conditions.

The subsequent section validates the prognostics results using the second part of the fault statistics data. This second part is divided into four segments, each spanning 4 months. Prognostics for the number of circuit breakers in each status were made for

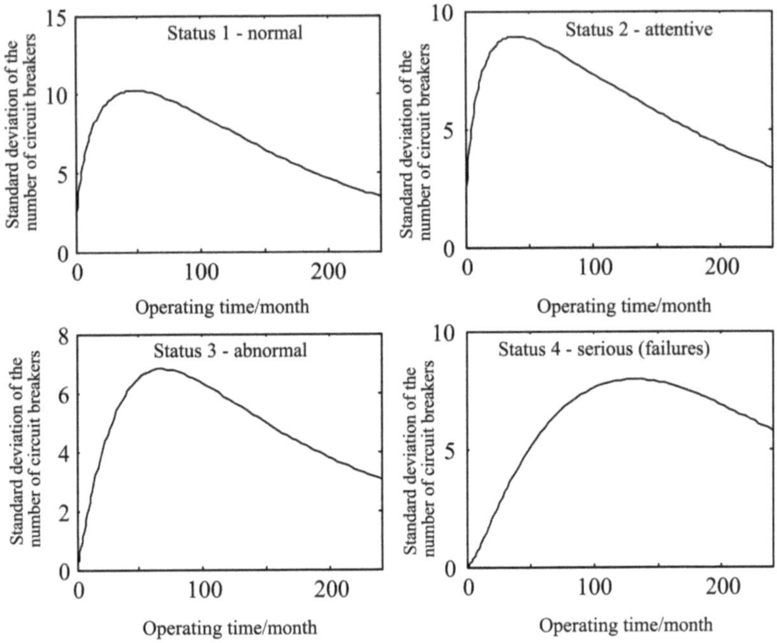

Fig. 3.8 Standard Deviations in Each State Over 20 Years (240 Months)

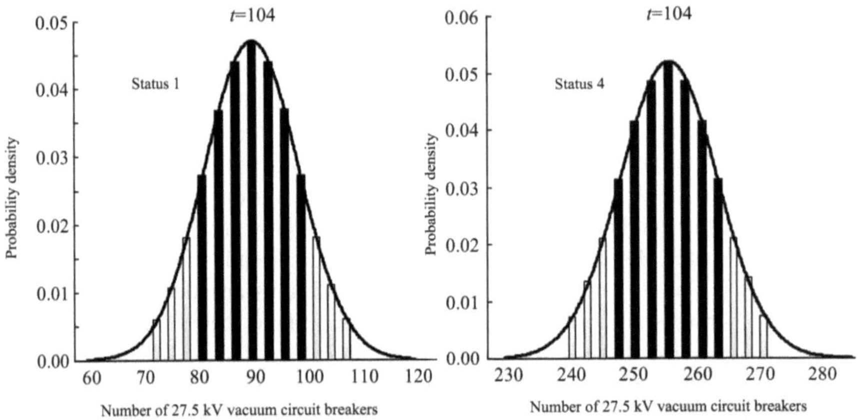

Fig. 3.9 Prognostics Results for the Number of Circuit Breakers in Status 1 and Status 4 at 104 Months

specific operational moments on December 4, 2014, April 4, 2015, August 4, 2015, and December 4, 2015. The predicted expectations of the fault state are compared with the actual number of faults. Figures 3.9, 3.10, 3.11, and 3.12 illustrate the prognostics results for the number of equipment in status 1 and status 4 from April

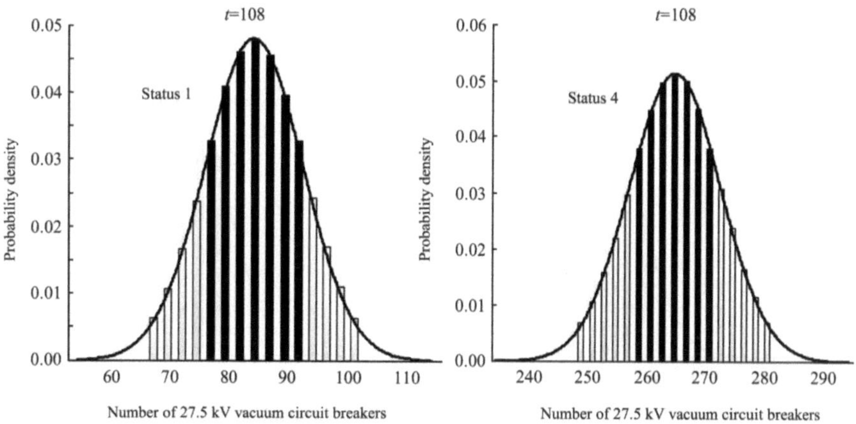

Fig. 3.10 Prognostics Results for the Number of Circuit Breakers in Status 1 and Status 4 at 108 Months

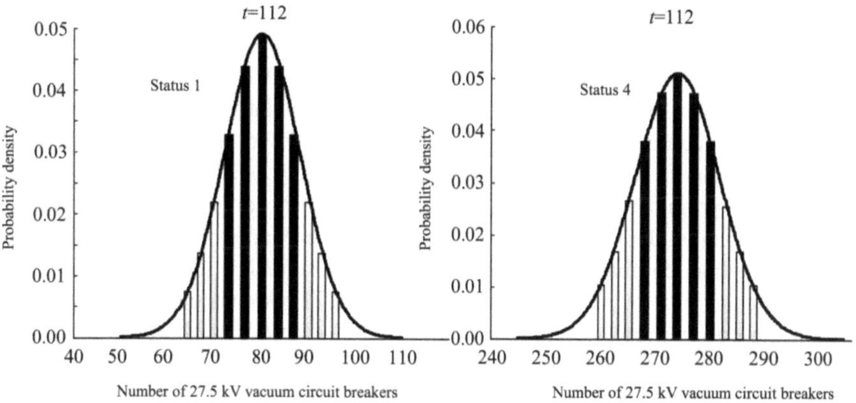

Fig. 3.11 Prognostics Results for the Number of Circuit Breakers in Status 1 and Status 4 at 112 Months

2006 through the 104th, 108th, 112th, and 116th months, respectively. The black bar intervals represent the number of equipment within one standard deviation range, while the white bar intervals represent the number within two standard deviation ranges. For example, in Fig. 3.9, the predicted expected value for equipment in status 1 is approximately 90 with a standard deviation of about 8.49; for equipment in status 4, the predicted expected value is about 256 with a standard deviation of about 7.68. In Fig. 3.10, the predicted expected value for equipment in status 1 is approximately 85 with a standard deviation of about 8.31; for equipment in status 4, it is about 265 with a standard deviation of about 7.76. The comparison of prognostics

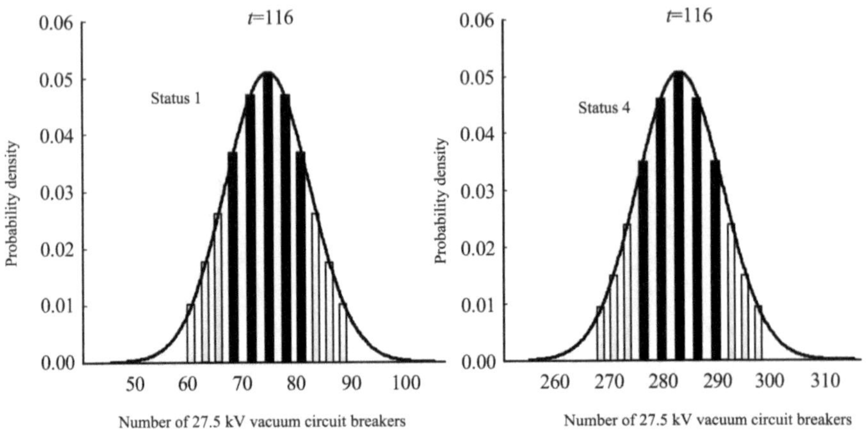

Fig. 3.12 Prognostics Results for the Number of Circuit Breakers in Status 1 and Status 4 at 116 Months

results for the 27.5 kV circuit breakers at these four operational moments with the actual number of faults is shown in Table 3.16.

From Figs. 3.9, 3.10, 3.11, and 3.12 and Table 3.16, it is evident that at the selected four operational moments, the actual cumulative number of faults for the 27.5 kV vacuum circuit breakers falls within the 95% confidence interval of the predicted results by the method proposed in this paper. Additionally, the relative error of the prognostics are within a 5% range when compared with the actual number of faults. This demonstrates that using the Markov model and its state division method, combined with the first part of the statistical fault data, effectively reflects the degradation law of states for this type of 27.5 kV vacuum circuit breakers within the power supply section. The proposed prognostics method can accurately achieve traction power equipment prognostics. The prognostics for the number of equipment in each status at future moments and their 95% confidence intervals, when compared with the second part of the fault statistics data, affirm the accuracy and effectiveness of the method.

3.4 Summary

Chapter 3 systematically describes the prognostics method for high-speed railway traction power supply equipment and systems. By analyzing the evolution of equipment failure and the relationship between equipment fault characteristic parameters and faults, and considering the influence of numerous related factors, this chapter integrates multi-parameter prognostics models and fault diagnosis models to achieve real-time prognostics of fault occurrence probability, fault type, and fault

Table 3.16 Comparison of 27.5 kV circuit breaker prognostics results with actual fault numbers

Operational moment	Duration of operation	Actual number of faults	Predicted expectation	Prognostics standard deviation	95% confidence interval	Prognostics relative error/%
2014-12-04	104 months	246	256	7.68	[241, 271]	4.06
2015-04-04	108 months	258	265	7.76	[249, 280]	2.71
2015-08-04	112 months	266	274	7.82	[258, 290]	3.01
2015-12-04	116 months	270	283	7.87	[267, 299]	4.44

location for traction power supply equipment. This section introduced two prognostics methods based on improved grey prognostics and continuous-time Markov degradation processes, and discussed prognostics processes, establishing prognostics models for high-speed railway traction power supply system equipment and exploring case studies of prognostics for such systems.

Chapter 4
Health Assessment of High-Speed Railway Traction Power Supply System

Accurately and comprehensively understanding the health condition of high-speed railway traction power supply equipment and systems is the core element and fundamental basis for formulating proactive maintenance strategies. Comprehensive analysis of health conditions can clearly reflect the actual situation of traction power supply equipment and systems, embodying the intrinsic laws of performance degradation, fault occurrence, and evolution. This facilitates the guidance of fault detection, elimination, and maintenance and repair tasks.

Health assessment, as an essential part of health diagnosis, organically integrates the service performance and degradation laws of equipment, the fault evolution process, and the intelligent early warning mechanism for safe operation. The various monitoring indicators within the intelligent early warning mechanism lay the foundation for establishing a health diagnostic indicator system, which is a prerequisite for health assessment. The service performance and the degradation law detail the operational status of the equipment (such as safe, declining) which correspond to the results of the health assessment (such as healthy, sub-healthy) and also relate to the equipment's fault evolution process (such as no fault, potential fault, functional fault). Thus, the intelligent early warning mechanism provides foundational indicators for health assessment, while health assessment offers a theoretical basis for service performance, degradation law, and fault evolution process.

Health condition assessment aims to integrate various monitoring data collected, converting them into a unified indicator—a health value—to comprehensively, accurately, intuitively, and quantitatively characterize the health condition of traction power supply equipment and systems. However, in the actual process of assessing the health condition of traction power supply equipment and systems, the variety of equipment involved, the complexity of data indicators, and the need for assessment results to guide onsite operational maintenance require a comprehensive consideration of both subjective and objective factors affecting the assessment outcomes. Under the proactive maintenance theoretical framework, it is necessary to customize multi-source heterogeneous monitoring data gradually into a health condition evaluation indicator set and health value. Furthermore, dividing the

Z. He et al., *Prognostics and Health Management of High-Speed Railway Traction Power Substations*, https://doi.org/10.1007/978-981-95-4538-4_4

corresponding health conditions and employing certain assessment methods to combine subjective and objective advantages are crucial to accurately and comprehensively understand the health condition of traction power supply equipment and systems.

4.1 Overview of Health Assessment for Traction Power Supply System

In traditional traction power supply systems, the maintenance and repair model is relatively outdated, still relying on periodic maintenance, manual inspection, and emergency fault repairs. This approach often results in a maintenance strategy that lacks specificity and involves blind operations, commonly leading to either insufficient or excessive maintenance. Moreover, fault diagnosis usually occurs after a fault has happened, which, although enabling quick identification of the time and location of the fault to expedite the resumption of system operations, the accident has already occurred by then, resulting in significant human and material losses.

The occurrence of failures in traction power supply systems or equipment is the result of numerous factors accumulating over an extended period, even if the fault appears to be sudden. There is inevitably a time process in the evolution and development of such faults. Health assessment involves summarizing the system's evolutionary laws from its real-time operational status to accurately assess the health condition of the traction power supply system. Based on these laws, the system or equipment's abnormalities are identified in advance, and preemptive maintenance is conducted to prevent accidents.

Traction power system health assessment technology leverages advanced sensors to gather relevant feature information, utilizing intelligent algorithms and models to detect, analyze, evaluate, and predict these pieces of information. By analyzing historical operating data of the equipment, current operating statuses and transformation trends are determined. Coupled with current environmental conditions and structural characteristics, precise judgments about the current health condition of the equipment are made, and reasonable prognostics about future operational trends are formulated. Utilizing the assessment results for targeted maintenance allows for early fault detection, effectively reducing the operational risk of the traction power supply system.

In the actual process of conducting health assessments, the diversity of equipment types, the complexity of the topological structure, the wide range of sources for evaluation indicators, and varied judgment standards pose significant challenges. These factors necessitate a degree of information fusion to perform a comprehensive assessment of the health condition. To clearly reflect the actual operating conditions of the traction power supply system and to illustrate the inherent evolution and development of faults, which facilitates on-site operations, health values derived from comprehensive assessment results are used to describe the health of the traction power supply system and equipment, and to classify health grades. Currently, popular health assessment methods include the AHP and fuzzy statistical methods.

The AHP organizes the traction power supply system into four hierarchical levels: the system level, subsystem layer, component layer, and indicator layer. The assessment progresses upward from the indicator layer, assessing the health conditions of the component layer, subsystem layer, and finally the system level, culminating in a comprehensive health assessment of the traction power supply system. The traction power supply system has a clear hierarchical structure. Through the analysis of connections and functional relationships between subsystems and equipment, AHP standardizes the assessment process. It determines the impact weight of each subsystem and piece of equipment, facilitating a health assessment process that combines quantitative and qualitative analysis.

Given the uncertainty of information acquired during health assessments and the subjective nature of determining weights, the health condition of the traction power supply system is not a simple binary state; that is, the results of the assessment do not have clear boundaries. Therefore, fuzzy statistical methods are commonly used. These methods employ membership functions to find the optimal assessment results, eliminating the randomness of multi-source heterogeneous data and subjective differences, and ensuring the objectivity, comprehensiveness, and accuracy of the assessment results.

Health assessment has not only garnered significant attention in traction power supply systems but has also become a hot topic of research in power systems. In power systems, health assessments typically focus on primary equipment such as transformers and generators. There is less research on health assessments for the entire power grid, and most of the techniques are consistent with those used in traction power system health assessment technology. These involve using historical and real-time information about assessment indicators to evaluate the health status of the power grid, preventing potential faults, reducing the risk of accidents, and providing a basis for maintenance and decision-making by relevant departments.

An objective and comprehensive health assessment is critically important for the reliable operation of systems and their equipment. On the one hand, by assessing the health condition of equipment and adopting state-based maintenance instead of scheduled maintenance, it reduces maintenance randomness, enhances inspection efficiency, lowers maintenance costs, and effectively minimizes additional losses caused by improper maintenance. On the other hand, by predicting future trends of systems and equipment, targeted preventive measures can be taken to avoid accidents, reduce the probability of faults, provide a basis for decision-making by relevant departments, and enhance the availability of system operations. Only when the health assessment results are accurate and reasonable can the maintenance strategy based on them be precise and complete, transforming the system and equipment towards a healthier state post-maintenance. If the assessment results are unreasonable, the system and equipment may continue to deteriorate until a fault occurs.

4.2 Health Indicators of Traction Power Supply System

The primary prerequisite for conducting a health assessment of a traction power supply system is to establish a health diagnostic indicator system. This system comprises several health indicators. Scientific and reasonable health indicators serve as the basis for health assessment and are also the foundation of the intelligent early warning mechanism for safe operation. Health indicators can be derived from traction power supply equipment status indicators (see Sect. 2.3) after appropriate screening. Therefore, the health indicators of the traction power supply system must scientifically, reasonably, and comprehensively reflect the factors affecting the health condition of the traction power supply system. The following principles should be adhered to when determining health indicators based on traction power supply system monitoring and diagnostic indicators:

1. Hierarchy: The health indicators of the traction power supply system must exhibit a hierarchical structure, where each level is interconnected and consistent with one another, with concrete indicators corresponding between the upper and lower tiers.
2. Relevance: The traction power supply system, being a composite of numerous pieces of equipment, necessitates an analysis of the interrelationships between these equipment to organize evaluation indicators rationally.
3. Holistic and Comprehensive Nature: It is essential not only to monitor the health condition of individual power supply equipment but also to consider the overall health assessment standards. Indicators should reflect not only the normal operational conditions but also potential risks and issues.
4. Feasibility and Practicality: The standards for health assessment should have clear concepts and definitions, be easily collectible and gatherable, and consider current technological conditions and capabilities.
5. Qualitative and Quantitative Combination: The indicator system for the health assessment of the traction power supply system should integrate qualitative and quantitative approaches, meaning that quantification should be based on a foundation of qualitative analysis.

The health diagnosis system of the traction power supply system gradually customizes system monitoring and diagnostic indicators into health indicators and health values, serving as the intelligent early warning mechanism for safe operation and health assessment.

The results of the health assessment of the traction power supply system are a fundamental basis for conducting health management. To more accurately reflect the actual situation of the traction power supply system and to understand the intrinsic laws of system fault occurrence and evolution, facilitating practical operations on-site, the health condition of the traction power supply system is categorized into five levels: healthy, sub-healthy, attentive, abnormal, and severely abnormal. Among them, *healthy* and *sub-healthy* belong to the system's optimal statuses. *Healthy*: This status indicates that the indicators and parameters of the traction

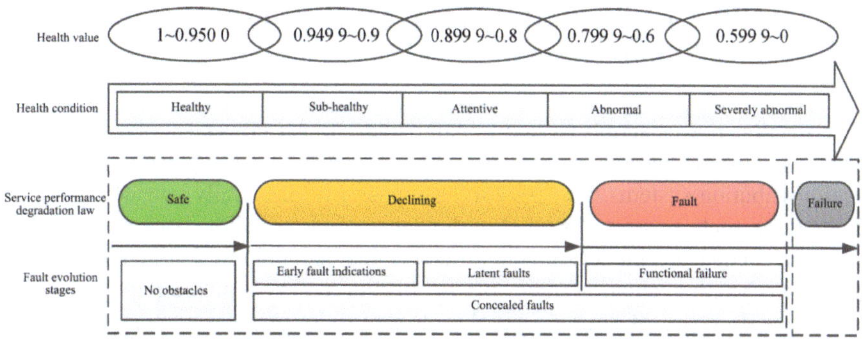

Fig. 4.1 Correspondence of Health Condition Assessment Results with Service Performance Degradation Law and Fault Evolution Process

power supply equipment are near optimal values. The operating status of the traction power supply system is normal, the performance is stable, and the likelihood of faults is extremely low, permitting long-term safe operation. *Sub-Healthy*: This status signifies that some performance parameters of the traction power supply equipment have slightly declined, yet it still ensures a stable power supply to locomotives. The power supply system's status is generally stable, with a lower possibility of faults, allowing for continued operation. *Attentive*: This level indicates that some performance metrics of the traction power supply system have significantly decreased but still maintain normal operations. The risk of faults has increased, necessitating close monitoring of the system's status and scheduling inspections by personnel. *Abnormal*: This status reveals that some indicator parameters reflect a poor operating condition of the traction power supply system, making it prone to power supply faults and requiring maintenance by personnel. *Severely Abnormal*: This level demonstrates that the overall status of the traction power supply system is poor, with some equipment parameters severely deviating from design values, making it susceptible to power supply accidents and requiring immediate attention and intervention.

To make the results of the health condition assessment more intuitive, a health value can be used to describe the various health statuses of equipment and systems. The correspondence between health condition, health value, service performance degradation law, and fault evolution process is illustrated in Fig. 4.1.

It is evident that assessing the health condition of the traction power supply system involves numerous equipment, complex evaluation indicators, and the results that need to guide on-site operational maintenance. Thus, a comprehensive evaluation must consider a multitude of factors.

4.3 Health Condition Assessment Methods for Traction Power Supply System

The traction power supply system, comprising various power supply equipment, features a distinct hierarchical structure characteristic of electric power transmission and distribution systems. The system's equipment is electrically connected in configurations such as series, parallel, redundancy, and (cold/hot) backup, making a hierarchical analysis model suitable for describing its topological structure. Furthermore, the health evaluation indicators for traction power supply equipment derive from a wide range of sources, involve diverse units of measurement, and differ greatly in standards. Data scale ranges from micro to macro, and data nature varies from qualitative to quantitative, showing significant multi-temporal and spatial scalability and multi-source heterogeneous features, with inherent stochasticity in cognition. Given that health condition assessment requires clear and definitive results, fuzzy statistical methods can be used to mitigate the randomness of multi-source heterogeneous data and subjective differences, making the assessment results more accurate, comprehensive, and objective.

The immune system, as an effective defense mechanism, can automatically defend against and repair damage when an organism is invaded by external threats or experiences internal system issues, thus ensuring the organism's health. The basic mechanism of immunity involves detecting antigens that invade the body. After identification by the antigen pattern recognizer, the results are sent to the immune management controller, where a series of complex biochemical processes produce antibodies to complete the immune process. The immunological mechanisms of organisms hold significant research value and practical significance for complex system health monitoring and state assessment. Based on these mechanisms, combined with the structure and characteristics of the traction power supply system, an immune-based health assessment mechanism as shown in Fig. 4.2 is employed.

In the immune-based health assessment mechanism illustrated in Fig. 4.2, the health diagnostic indicator system is analogous to an antigen, with health indicators acting as stimuli that trigger the entire assessment process. The multi-factorial health condition assessment model functions like an antigen pattern recognizer, where health indicators (antigens) are input into this model (recognizer) for matching and modeling. Techniques such as the AHP, entropy weighting method, and fuzzy membership function are analogous to an immune management controller. These techniques use the parameters (outputs) from the assessment model (recognizer) and combine them through a series of complex biochemical-like processes to produce antibodies, i.e., a comprehensive analysis and assessment method, facilitating a comprehensive assessment of the traction power supply system's health condition.

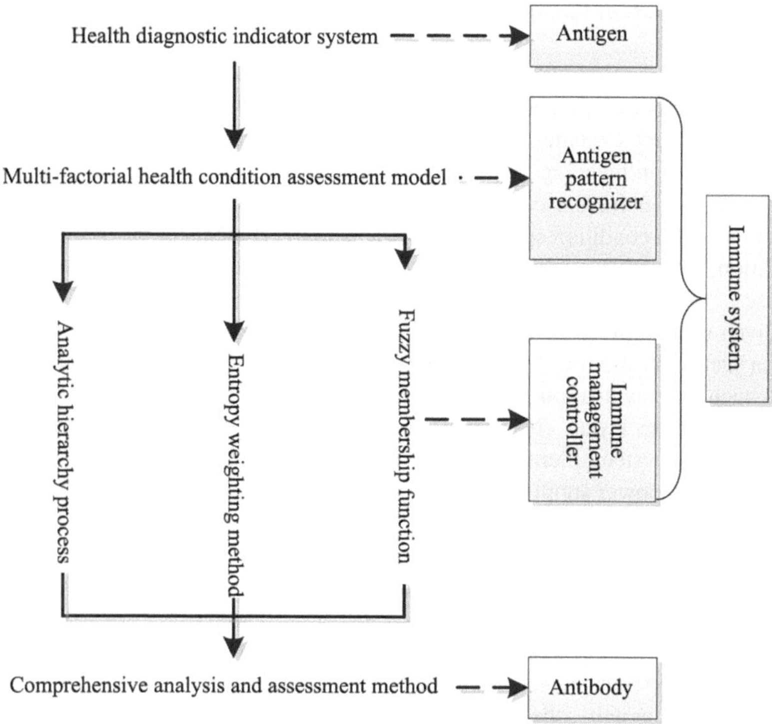

Fig. 4.2 Immune-based Health Assessment Mechanism

4.3.1 Multifactorial Assessment Model for the Health Condition of Traction Power Supply System

The traction power supply system includes numerous pieces of equipment with complex associated evaluation factors. The health condition assessment model for this system is essentially a multi-factor decision model.

For a multi-factor decision model, consider a finite goal set $I = \{I_i \mid i = 1, 2, \cdots, p\}$ and a finite alternative set $R = \{R_j \mid j = 1, 2, \cdots, n\}$. The multi-criteria evaluation issue can be resolved by determining the optimal alternative set R_x that most closely aligns with the relevant goals G_i. If represented by the fuzzy subset of R, it would be $G_i = [\mu_{i,\,1}(R_1)|R_1, \mu_{i,\,2}(R_2)|R_2, \cdots, \mu_{i,\,n}(R_n) \mid R_n]$. The essence of evaluating the health condition of the traction power supply system is to find the optimal evaluation result for the evaluation factor set from the evaluation set.

4.3.1.1 Evaluation Factor Set

Factors refer to the various attributes and performances of the evaluation object, also known as parameter indicators or quality indicators, which can comprehensively reflect the object's quality. The so-called factor set refers to a set composed of various factors influencing the evaluation object, $I = \{i_1, i_2, \cdots, i_n\}$.

The factor set for the comprehensive assessment of the traction power supply system's health condition specifically pertains to the contents of the system's health condition evaluation, such as various indicators, parameters, performances, and statuses.

Given the multitude of equipment involved and the complexity of related evaluation factors in the traction power supply system's health condition, to facilitate assessment, the evaluation factors can be appropriately stratified according to their attributes. These layers correspond to the structural composition of the traction power supply system, thereby enhancing the understanding of the health condition of the traction power supply system equipment, subsystem, and the entire system.

4.3.1.2 Evaluation Set

Let r_i be the possible evaluation result for the ith evaluation object. The set $R = \{r_1, r_2, \cdots, r_m\}$ composed of r_i is called the evaluation set.

To more accurately reflect the trend of the traction power supply system's health condition evolution, as previously described, a health status evaluation set $\boldsymbol{R} = \{Healthy,\ Sub\text{-}Healthy,\ Attentive,\ Abnormal,\ Severely\ Abnormal\}$. The corresponding health value evaluation set $\boldsymbol{R} = \{[0.9500, 1], [0.9, 0.9499], [0.8, 0.8999], [0.6, 0.7999], [0, 0.5999]\}$.

4.3.1.3 Weight Set

The importance of different evaluation factors in the assessment process varies. The weight set $\boldsymbol{W} = \{w_1, w_2, \cdots, w_n\}$ reflects the significance of each factor.

The weights should satisfy the requirement of normalization, i.e.,

$$\sum_{i=1}^{n} w_i = 1, (w_i \geq 0) \tag{4.1}$$

For different evaluation purposes, the weights of factors may differ and should be specifically analyzed according to the problem at hand. Common analytical methods include the statistical experiment method, analytical reasoning method, expert scoring method, AHP, and entropy weighting method.

Given the complexity of the traction power supply system and the multitude of evaluation factors involved, the magnitude of each factor's weight has a corresponding impact on the assessment results. A reasonable and feasible method for determining weights requires specific analysis and research.

4.3.2 AHP

The assessment process of health condition is a multi-attribute decision process. Various multi-attribute decision-making methods have different degrees of application in practice, and each method has its unique mechanism. Most methods emphasize the complexity of the decision-making method while neglecting its simplicity and scientific nature, often pursuing mathematical complexity, which can obscure the complexity of the evaluation method and the scientific nature of the evaluation conclusions. This overlooks the ease of understanding and the capability to express the evaluator's subjective judgment. Professor T. L. Saaty from the University of Pittsburgh introduced the renowned AHP in the early 1970s. AHP, as an efficient decision-making tool, is characterized by its simplicity of application, ease of understanding, and high practicality. It emphasizes the role of human judgment in the decision-making process and standardizes the decision-making process through a specific pattern, making it suitable for assessment problems that combine qualitative and quantitative approaches.

The high-speed railway traction power supply system is a vast and complex system with a clear hierarchical structure, making it suitable for layered processing to address multi-attribute decision-making problems. Using the AHP to establish a hierarchical analysis model for the traction power supply system allows for accurate and comprehensive health assessment and decision-making.

The hierarchical analysis model established for the high-speed railway traction power supply system using AHP is illustrated in Fig. 4.3. The first layer is the system level (S), used for assessing the overall health condition of the entire traction power supply system. The second layer is the subsystem layer (U), which divides the traction power supply system into major key subsystems such as the catenary, traction substation, sectioning post, and autotransformer station, and performs health assessments for each. The third layer is the component layer (E), which assesses components that make up each subsystem, such as traction transformers, circuit breakers, isolators, bus bars, cables, current transformers, GIS high-voltage cabinets in the traction substations, and contact wires, catenary wires, droppers, insulators, etc., on the catenary. The fourth layer is the indicator layer (I), where health assessment indicators are established based on the physical characteristics, operating status, and service conditions of each component. From the indicator layer, the assessment progresses upwards through the component layer, subsystem layer, and system level to finally achieve a comprehensive health assessment of the traction power supply system.

Based on the hierarchical division of the traction power system AHP model, determine the membership relationships among elements within each layer. Conduct pairwise comparisons of elements at the same layer to derive an n-order judgment

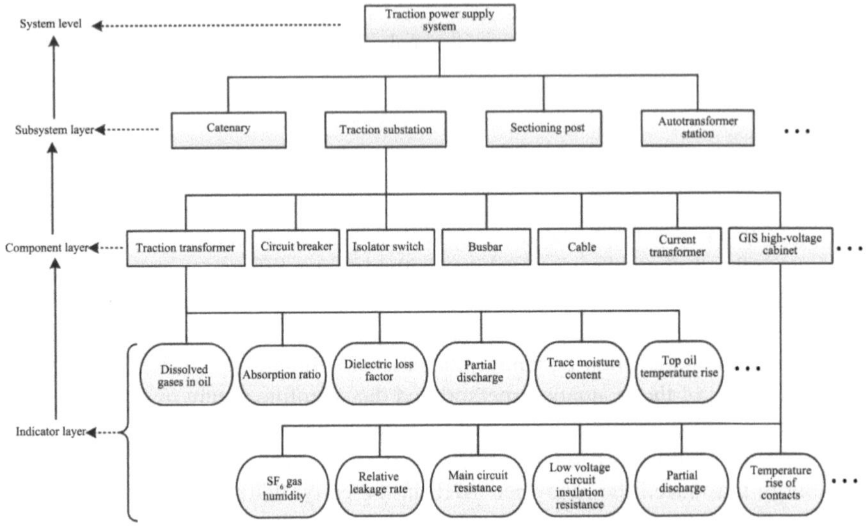

Fig. 4.3 Hierarchical Analysis Model for the High-Speed Railway Traction Power System

Table 4.1 T. L. Saaty 1–9 scaling method values and their meanings

Scale value	Meaning
1	Indicates that the two compared indicators are of equal importance
3	Indicates that one of the two compared indicators is slightly more important than the other
5	Indicates that one of the two compared indicators is significantly more important than the other
7	Indicates that one of the two compared indicators is strongly more important than the other
9	Indicates that one of the two compared indicators is extremely more important than the other
2, 4, 6, 8	The intermediate values of adjacent scales are used for compromise

matrix that indicates the relative importance of n different elements at one level to an element at the previous level, essentially constructing a pairwise comparison matrix:

$$\boldsymbol{A} = \left(a_{ij}\right)_{n \times n} \tag{4.2}$$

where $a_{ij} > 0$, $a_{ij} = 1/a_{ji}$, $a_{ii} = 1$. Use the T. L. Saaty 1–9 scale method to represent the results of pairwise comparisons between different elements, where the value a_{ij} is as shown in Table 4.1.

Calculate the maximum eigenvalue of the pairwise comparison matrix $\boldsymbol{A}$ and its corresponding eigenvector. Normalize this to obtain the initial relative weight vector.

Table 4.2 RI values for pairwise comparison matrices of order 1–11

Matrix order n	1	2	3	4	5	6	7	8	9	10	11
RI values	0.00	0.00	0.58	0.90	1.12	1.24	1.32	1.41	1.45	1.49	1.51

$$\boldsymbol{w} = (w_1, w_2, \cdots, w_n) \tag{4.3}$$

During the construction of the pairwise comparison matrix, the matrix's order and variations in subjective thinking can impact the matrix created. Therefore, it is necessary to perform consistency verification on the initial relative weights. The formula for consistency verification is:

$$CR = CI/RI \tag{4.4}$$

where CI is the general consistency deviation index of the pairwise comparison matrix $\boldsymbol{A}$, calculated by $CI = (\lambda_{max} - n)/(n - 1)$; λ_{max} is the maximum eigenvalue of $\boldsymbol{A}$; RI is the average random consistency index of the pairwise comparison matrix $\boldsymbol{A}$, used to eliminate the effect of matrix order and to correct the CI value, with its values shown in Table 4.2; CR is the random consistency ratio of the pairwise comparison matrix $\boldsymbol{A}$.

Generally, for $n \geq 3$, if $0 < CR < 0.10$, it indicates that the consistency deviation of the pairwise comparison matrix is small, and the weights assigned to each indicator are reasonable; otherwise, the values in the pairwise comparison matrix need to be adjusted according to Table 4.1 until the consistency meets the requirements.

4.3.3 Entropy Weighting Method

In information theory, entropy is used to represent the disorder degree of a system. If an indicator's information entropy is smaller, the uncertainty is less, which means the indicator provides more information and has a greater impact on comprehensive evaluation, thus warranting a higher weight. Conversely, greater uncertainty, less information, and higher entropy imply a lower weight in the comprehensive evaluation. This is the fundamental idea behind the entropy weighting method.

The definition of entropy is given by:

$$H_i = -\sum_{i=1}^{m} P_{ij} \ln P_{ij} \tag{4.5}$$

Where $P_{ij} = h_{ij} / \sum_{i=1}^{m} h_{ij}$ is the probability of the variable of the indicator to be determined; h_{ij} is the normalized value of the i-th measurement of the j-th indicator.

The entropy weight of the j-th indicator is

$$w_H^j = (1 - H_j) / \sum_{j=1}^{n} (1 - \mathrm{H}_j) \tag{4.6}$$

Thus, the entropy-based evaluation indicator weight vector can be obtained as $\boldsymbol{W} = \left(w_H^1, w_H^2, \cdots, w_H^n\right)$.

For n evaluation objects and m indicators, the steps to determine the indicator weights using the entropy method are:

1. Calculating the proportion of the i-th evaluation object indicator value under the j-th indicator as

$$P_{ij} = h_{ij} / \sum_{i=1}^{m} h_{ij} \tag{4.7}$$

2. Calculating the entropy value of the indicator as

$$H_j = -\sum_{i=1}^{m} P_{ij} \ln P_{ij} \tag{4.8}$$

3. Calculating the weight of the indicator as

$$w_H^j = (1 - H_j) / \sum_{j=1}^{n} (1 - H_j) \tag{4.9}$$

In Eq. (4.9), $1 - H_j$ is called the difference coefficient. The smaller the entropy value of an indicator, the larger its difference coefficient, indicating a greater degree of variation in the indicator value and a larger proportion in the evaluation.

4.3.4 *Fuzzy Statistical Method*

Building upon the foundation of the traction power system hierarchical analysis model, relying solely on initial relative weights for health assessment can introduce significant subjectivity. Changes in matrix order and differences in subjective thinking can significantly affect the assessment results, impacting the accuracy of the health assessment. Fuzzy statistics can unfold a multi-indicator object in a one-dimensional space to achieve an objective comprehensive assessment. Therefore, by employing the fuzzy statistical method and introducing a fuzzy membership function to adjust the initial relative weights, the influence of matrix order and subjective thinking on weight calculation results can be effectively eliminated, ensuring the objectivity and accuracy of the health condition assessment results.

For the indicators collected at the indicator layer $I_1, I_2, \cdots, I_n$, establish the factor set as follows:

$$\boldsymbol{I} = \{I_1, I_2, \cdots, I_n\} = \{\text{ Indicator 1, Indicator 2, } \cdots \text{, Indicator } n\} \tag{4.10}$$

The health condition of the traction power supply system is divided into five levels, with the corresponding assessment set defined from highest to lowest as:

$$\begin{aligned} \boldsymbol{R} &= \{R_1, R_2, R_3, R_4, R_5\} \\ &= \{\mathit{Healthy}, \mathit{Sub-Healthy}, \mathit{Mildly\ Unhealthy}, \mathit{Moderately\ Unhealthy}, \mathit{Severely\ Unhealthy}\} \end{aligned} \tag{4.11}$$

Using the fuzzy statistical method, involve m experts in the assessment who categorize each health indicator according to the given levels in the assessment set R. Subsequently, count the frequency δ_{is} of each health indicator I_i belonging to each assessment level $R_s (s = 1, 2, \cdots, 5)$. The membership degree of health assessment indicator I_i belonging to health grade R_s is then calculated as:

$$\lambda_i^{(s)} = \frac{\delta_{is}}{n} \tag{4.12}$$

Since fuzzy sets can represent a collection of elements that possess certain properties to varying degrees, the membership function can reflect the extent to which these elements belong to the fuzzy set, with values ranging from 0 to 1. A higher value of the membership function indicates a higher degree of belonging to the fuzzy set. Based on the hypothesis in reference [135], the membership degree values within the fuzzy distribution interval are determined using an equal division method. This facilitates automatic processing by computer programming and has a minimal impact on assessment accuracy.

Based on the physical meaning of the health indicator and combined with the analysis of the assessment system, two types of fuzzy distribution forms, benefit-based type (the larger the better) and cost-based type (the smaller the better), can be considered. According to the design standards, detection specifications, and operating characteristics of each component of the traction power supply system, the indicator value range is set to (a, b). Within this interval, three equally spaced points c_2, c_3, and c_4 are inserted to divide it into four equal parts. Assuming the actual value of any indicator I is z, the expressions of these two fuzzy membership functions are given.

4.3.4.1 Benefit-Type Fuzzy Distribution Membership Function

$$\lambda_i^{(1)} = \begin{cases} 1, & z \geqslant b \\ (z - c_4)/d, & c_4 \leqslant z < b \\ 0, & z < c_4 \end{cases}$$

$$\lambda_i^{(s)} = \begin{cases} 0, & z \geqslant c_{5-s+2} \\ (c_{5-s+2} - z)/d, & c_{5-s+1} \leqslant z < c_{5-s+2} \\ (z - c_{5-s})/d, & c_{5-s} \leqslant z < c_{5-s+1} \\ 0, & z \leqslant c_{5-s} \end{cases} \tag{4.13}$$

$$\lambda_i^{(5)} = \begin{cases} 0, & z \geqslant c_2 \\ (c_2 - z)/d, & a \leqslant z < c_2 \\ 1, & z < a \end{cases}$$

where a is the worst critical value, b is the best critical value.

$d = (b - a)/4$, $s = 2, 3, 4$, Let $a = c_1$, $b = c_5$. The visual representation of the benefit-based type fuzzy distribution is shown in Fig. 4.4.

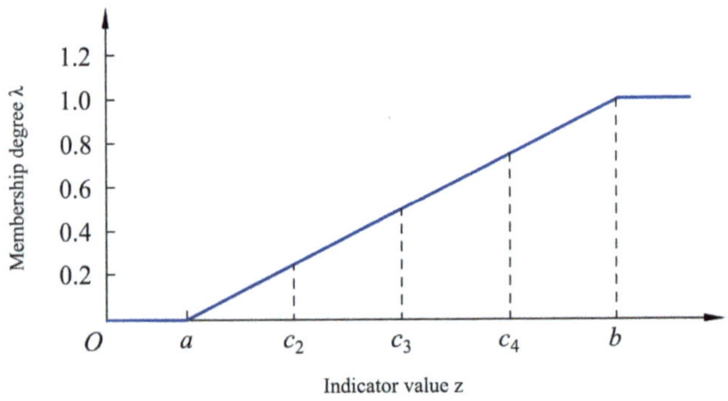

Fig. 4.4 Benefit-Based Type Fuzzy Distribution Form

4.3.4.2 Cost-Type Fuzzy Distribution Membership Function

$$\lambda_i^{(1)} = \begin{cases} 1, & z < a \\ (c_2 - z)/d, & a \leqslant z < c_2 \\ 0, & z \geqslant c_2 \end{cases}$$
$$\lambda_i^{(s)} = \begin{cases} 0, & z < c_{s-1} \\ (z - c_{s-1})/d, & c_{s-1} \leqslant z < c_s \\ (c_{s+1} - z)/d, & c_s \leqslant z < c_{s+1} \\ 0, & z \geqslant c_{s+1} \end{cases} \tag{4.14}$$
$$\lambda_i^{(5)} = \begin{cases} 0, & z < c_4 \\ (c_5 - z)/d, & c_4 \leqslant z < b \\ 1, & z \geqslant b \end{cases}$$

where a is the best critical value, b is the worst critical value.

$d = (b - a)/4$, $s = 2, 3, 4$, Let $a = c_1$, $b = c_5$. The visual representation of the cost-based type fuzzy distribution is shown in Fig. 4.5.

Based on the membership degrees of each health indicator across five health grades at the indicator layer, construct the indicator assessment matrix:

$$\boldsymbol{D}_i = \left[\lambda_i^{(1)}, \lambda_i^{(2)}, \lambda_i^{(3)}, \lambda_i^{(4)}, \lambda_i^{(5)}\right] \tag{4.15}$$

The health indicator status monitoring information can be output based on the membership degree value.

Let:

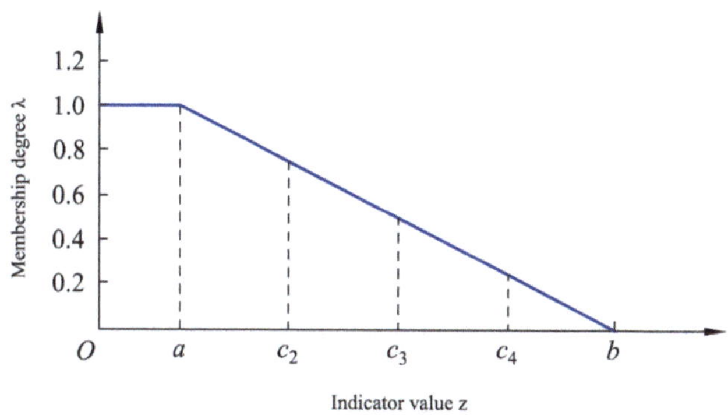

Fig. 4.5 Cost-Based Type Fuzzy Distribution Form

$$e_i = max\left\{\lambda_i^{(s)}\right\}, s = 1, 2, \cdots, 5 \tag{4.16}$$

If the indicator I_j belongs to *abnormal* or *severely abnormal* (i.e., $e_j = \lambda_j^{(4)}$ or $e_j = \lambda_j^{(5)}$), output an alarm signal 1; if I_j belongs to *healthy*, *sub-healthy*, or *attentive* (i.e., $e_j = \lambda_j^{(1)}$, $e_j = \lambda_j^{(2)}$, or $e_j = \lambda_j^{(3)}$), output a normal signal 0.

Thus, the objective correction coefficient of each indicator can be obtained:

$$\alpha_i = \frac{1/e_i}{\sum(1/e_i)} \tag{4.17}$$

where $\sum(1/e_i)$ is the cumulative sum of the reciprocals of the e_i values corresponding to all health indicators of a component.

Finally, the composite weight of each indicator in the indicator layer is obtained as:

$$\beta_i = \frac{\alpha_i w_i}{\sum(\alpha_i w_i)} \tag{4.18}$$

where $\sum(\alpha_i\omega_i)$ is the cumulative sum of the $\alpha_i\omega_i$ values corresponding to all health indicators of a component.

By introducing the objective correction coefficient to adjust the initial relative weights, the influence of subjective factors in determining the initial relative weights is effectively eliminated. This allows for the efficient, convenient, and rigorous determination of accurate composite weight values in health assessment.

State monitoring information obtained from the indicator layer is aggregated to form a state sequence vector as follows:

$$\boldsymbol{K}_I = K_1, K_2, \cdots, K_n \tag{4.19}$$

where n is the total number of indicators used for health condition assessment, and $K_i(i = 1, 2, \cdots, n)$ is the status monitoring information output for each indicator, with values being either 0 or 1.

The composite weight vector of the indicator layer $\boldsymbol{\beta}_I = (\beta_i)_{1 \times n}(i = 1, 2, \cdots, n)$ is constituted by all indicators under a component. The impact factor of the indicator layer on the component layer is given by:

$$\gamma_E = \boldsymbol{K}_I \boldsymbol{\beta}_I^{\mathrm{T}} \tag{4.20}$$

Progressing upwards hierarchically, the impact factor of the component layer on the subsystem layer γ_U and the impact factor of the subsystem layer on the system level γ_S are calculated as:

$$\gamma_U = \sum(\gamma_E w_E) \quad (4.21)$$

$$\gamma_S = \sum(\gamma_U w_U) \quad (4.22)$$

where w_E and w_U are the relative weights calculated per Eq. (4.3) for each component of the component layer and each subsystem of the subsystem layer, respectively.

Ultimately, the health values for components, subsystems, and the system are computed as follows:

$$H_E = 1 - \gamma_E \quad (4.23)$$

$$H_U = 1 - \gamma_U \quad (4.24)$$

$$H_S = 1 - \gamma_S \quad (4.25)$$

where $H \in [0, 1]$, and a higher H indicates a better health condition of the system. By calculating the health value, a quantitative health assessment of individual equipment, subsystems, and the entire traction power supply system is achieved. Referring to Fig. 4.1, the current health condition, service performance degradation, and fault evolution stage for each can be determined.

4.3.5 Fuzzy Comprehensive Evaluation Method

Firstly, starting from individual factors, establish the membership degree of each evaluation factor to the evaluation set, i.e., establish the fuzzy mapping from U to V as follows:

$$\underset{\sim}{f} : U \to F(V), \forall u_i \in U, u_i \to \underset{\sim}{f}(u_i) = \frac{r_{i1}}{v_1} + \frac{r_{i2}}{v_2} + \cdots + \frac{r_{im}}{v_m} \quad (4.26)$$

In the equation, $r_{ij} - u_i$ belongs to the membership degree of v_j.

From $\underset{\sim}{f}(u_i)$, obtain the evaluation set $\boldsymbol{R}_i = \{r_{i1}, r_{i2}, \cdots, r_{im}\}$ for a single factor. The evaluation sets of each factor form the rows of the single-factor evaluation matrix as follows:

$$\boldsymbol{R} = \begin{bmatrix} r_{11} & r_{12} & \cdots & r_{1m} \\ r_{21} & r_{22} & \cdots & r_{2m} \\ \vdots & \vdots & \cdots & \vdots \\ r_{n1} & r_{n2} & \cdots & r_{nm} \end{bmatrix} \quad (4.27)$$

The i-th row of the evaluation matrix reflects the support level of the i-th evaluation factor for various evaluation conclusions, while the j-th column reflects the support level of all evaluation factors for the j-th evaluation conclusion. When considering the impact of all factors simultaneously, represent the effect of each factor with a corresponding weight coefficient, enabling the derivation of a comprehensive assessment result considering multiple factors as follows:

$$\boldsymbol{B} = W \circ \boldsymbol{R} = (w_1, w_2, \cdots w_n) \begin{bmatrix} r_{11}, & r_{12}, \cdots, & r_{1m} \\ r_{21}, & r_{22}, \cdots, & r_{2m} \\ \vdots & \vdots & \vdots \\ r_{n1}, & r_{n2}, \cdots, & r_{nm} \end{bmatrix} = (b_1, b_2, \cdots b_m) \tag{4.28}$$

where b_{j} indicates the support level for the *j*-th evaluation conclusion when considering the impact of all factors.

The fuzzy operator $\circ$, is often divided into the following types:

4.3.5.1 M(∧, ∨), Principal Factor Determinant Type

$$b_{\mathrm{j}} = \vee_{i=1}^{P} (w_i \wedge r_{ij}) = \max_{1 \leqslant i \leqslant P} \left\{ \min_{1 \leqslant j \leqslant m} (w_i, r_{ij}) \right\}, (j = 1, 2 \cdots m) \tag{4.29}$$

In this operator, the result b_j is determined only by the smaller of w_i and r_{ij} (first taking the minimum, then the maximum). This algorithm considers only the principal factor, with minor influences from other factors.

4.3.5.2 M(·, ∨), Principal Factor Prominent Type

$$b_{\mathrm{j}} = \vee_{i=1}^{P} (w_i \cdot r_{ij}) = \max_{1 \leqslant i \leqslant P} (w_i, r_{ij}), (j = 1, 2 \cdots m) \tag{4.30}$$

Similar to M(∧, ∨), this operator involves multiplying r_{ij} by a weight w_i less than 1, indicating a correction value when considering multiple factors. It retains all information but focuses on the principal factors, neglecting minor ones.

4.3.5.3 M(∧, ⊕), Type of Unbalanced Averaging

$$b_{\mathrm{j}} = \bigoplus_{i=1}^{P} (w_i \cdot r_{ij}) = \min \left\{ 1, \sum_{i=1}^{P} (w_i \wedge r_{ij}) \right\}, (j = 1, 2 \cdots m) \tag{4.31}$$

$\oplus$ is the bounded sum, i.e., $a \oplus b = \min(1, a + b)$.

Algorithms (1), (2), and (3) all emphasize the dominant role of the main factor, often resulting in the loss of some useful information in the evaluation results.

4.3.5.4 M(·, ⊕), Weighted Average Type

$$b_{\mathrm{j}} = \left(1 \wedge \sum_{i=1}^{P} \left(w_i \cdot r_{ij}\right), (j = 1, 2 \cdots m)\right. \tag{4.32}$$

The weighted average algorithm balances all factors according to the size of their weights, making it more suitable for the health condition assessment of traction power supply systems, which require consideration of numerous evaluation factors.

The comparison of the above four fuzzy operators is shown in Table 4.3.

The outcome of the fuzzy evaluation is the membership degree to various health grades, forming a fuzzy vector. When assessing multiple states, further processing of the fuzzy comprehensive evaluation result vector is essential to derive the final outcome. Common processing methods include:

1. Maximum Membership Degree Principle Method

   If $b_r = \max\limits_{1 \le j \le m} \{b_j\}$, then the evaluation set v_{r} corresponding to b_{r} is the final result of the health condition evaluation. This method is simple and intuitive, and is commonly used in practice. However, it may lose a significant amount of information in some cases, and its application requires meeting certain conditions. The threshold principle method is an improvement based on this foundation.
2. Threshold Principle Method

   To overcome the drawbacks of the maximum membership degree method, which may lose information and yield unreasonable conclusions in some cases, a threshold is set. The maximum membership principle is only applicable under certain conditions.

   Definition:

Table 4.3 Comparison of several fuzzy operators

	M(∧, ∨)	M(·, ∨)	M(∧, ⊕)	M(·, ⊕)
	Principal factor determinant type	Principal factor prominent type	Type of unbalanced averaging	Weighted average type
Degree of weight coefficient manifestation	Not evident	Evident	Not evident	Evident
Comprehensiveness	Weak	Weak	Strong	Strong
Utilization of R's information	Insufficient	Insufficient	Comparatively sufficient	Sufficient

$$\alpha = \frac{m\beta - 1}{2\gamma(m-1)} \tag{4.33}$$

Where:

$$\begin{aligned} \beta &= \max(b_j) = \max_{1 \leqslant j \leqslant m}\{b_j\} / \sum\nolimits_{j=1}^{m} b_j \\ \gamma &= \sec_{1 \leqslant j \leqslant m}\{b_j\} / \sum\nolimits_{j=1}^{m} b_j \end{aligned} \tag{4.34}$$

Using α as the threshold for evaluating the applicability of the maximum membership principle, the larger the α, the more effective the maximum membership principle.

3. Weighted Average Principle Method

   This method considers each evaluation level as a relative position and calculates the evaluated object's relative position by weighting and summing their ranks, expressed as:

$$C = \frac{\sum_{j=1}^{m} b_j^{\,k} * l}{\sum_{j=1}^{m} b_j^{\,k}} \tag{4.35}$$

where k is a coefficient to be determined, generally set to 1 or 2. As k approaches infinity, the weighted average principle converges to the maximum membership degree principle.

4. Fuzzy Vector Scalarization Method

   The calculation formula is:

$$C = \frac{\sum_{j=1}^{m} b_j^{\,k} * c_j}{\sum_{j=1}^{m} b_j^{\,k}} \tag{4.36}$$

To achieve the purpose of health condition assessment for the traction power supply system, the health grade of the traction power supply system can be determined based on the maximum membership principle. The health value corresponding to the traction power supply system can be determined using the weighted evaluation method.

4.4 Health Condition Assessment of Traction Substation

The traction power supply system is composed of subsystems such as the traction substation, catenary, sectioning post, and autotransformer station. A traction substation is selected as a typical subsystem for case analysis. The traction substation consists of equipment such as traction transformers, circuit breakers, disconnect switches, busbars, cables, voltage transformers, current transformers, and GIS

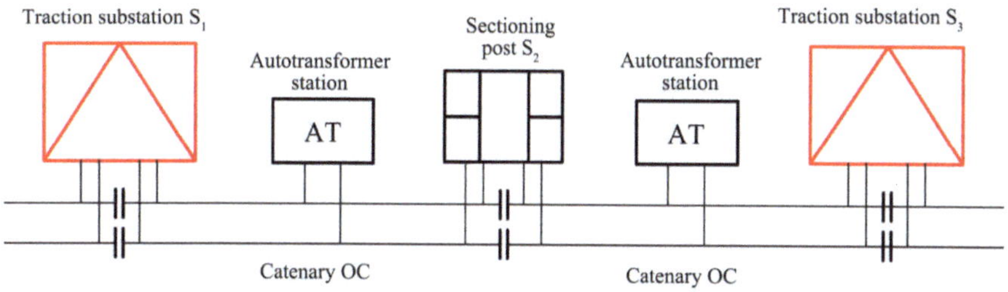

Fig. 4.6 Schematic of a Traction Substation and Its Power Supply Section

high-voltage cabinets. Traction transformers and GIS high-voltage cabinets are selected as typical equipment for case analysis.

An analysis is conducted using a traction substation and its power supply arm, as shown in Fig. 4.6. In Fig. 4.6, the catenary OC has no backup, and substations S_1 and S_3 can achieve cross-region power supply through sectioning post S_2, thus serving as each other's backup. The impact of an autotransformer station fault on the system is relatively minor.

Inside traction substation S_1, there are four single-phase transformers T_1 to T_4. T_1 and T_3 are configured as a Vx-connected traction transformer, while the other two, T_2 and T_4, constitute its 100% backup. The power flow from the main transformer's 220 kV incoming side passes sequentially through a circuit breaker DL, three sets of four current transformers CT, traction transformers T_1 and T_3, six feeder cables CA, four traction side busbars M, two 27.5 kV feeder side GIS high-voltage cabinets, and two voltage transformers PT.

4.4.1 Health Condition Assessment of Traction Transformers

For traction transformer T_1, monitor its dissolved gases in oil (total hydrocarbon content) I_1, absorption ratio I_2, dielectric loss factor I_3, partial discharge I_4, trace moisture content I_5, and top oil temperature rise I_6. According to the following standards: *DL/T 596-1996 Preventive Test Code for Electric Power Equipment, JB/T 501-2006 Test Guide for Power Transformers, GB/T 7252-2001 Guide to the Analysis and the Diagnosis of Gases Dissolved in Transformer Oil, GB 1094.2-1996 Power Transformers—Part 2: Temperature Rise*, the prescribed critical values for these six indicators can be determined. They are classified according to their fuzzy distribution as shown in Table 4.4.

At a given moment, the values collected from traction transformer T_1 are {130, 1.8, 0.65%, 190, 10, 23}.

Using the AHP, a pairwise comparison matrix is constructed for the factor set $\{I_1, I_2, I_3, I_4, I_5, I_6\}$. The matrix elements are determined based on the importance derived from pairwise comparisons of these six indicators, followed by a consistency check.

Table 4.4 Indicator layer parameters T_1 of traction transformers

Health indicator	Total hydrocarbon content	Absorption ratio	Dielectric loss factor	Partial discharge	Trace moisture content	Top oil temperature rise
Critical value	<150 μL/L	≥1.3	≤0.8%	<500 pC	≤25 mg/L	<60 K
Distribution type	Cost-based type	Benefit-based type	Cost-based type	Cost-based type	Cost-based type	Cost-based type

$$A_{T1} = \begin{bmatrix} 1 & 2 & 2 & 3 & 3 & 0.33 \\ 0.5 & 1 & 2 & 3 & 4 & 0.5 \\ 0.5 & 0.5 & 1 & 2 & 3 & 0.5 \\ 0.33 & 0.33 & 0.5 & 1 & 2 & 0.25 \\ 0.33 & 0.25 & 0.33 & 0.5 & 1 & 0.2 \\ 3 & 2 & 2 & 4 & 5 & 1 \end{bmatrix} \tag{4.37}$$

The maximum eigenvalue is calculated as $\lambda_{max} = 6.2206$, hence $CI = (6.2206 - 6)/(6 - 1) = 0.0441$. According to Eq. (4.4), the consistency verification is $CR = 0.0441/1.24 = 0.0356 < 0.1$, which meets the requirement. The feature vector corresponding to λ_{max} is obtained and normalized to get the initial relative weight vector $\boldsymbol{w}_I = (0.2137, 0.1846, 0.1290, 0.0737, 0.0509, 0.3481)$.

For indicators I_1, $I_3 \sim I_6$, a cost-type fuzzy distribution membership function (Eq. 4.14) is employed, and for indicator I_2, a benefit-type fuzzy distribution membership function (Eq. 4.13) is applied. Calculate their membership degrees belonging to the five health grades and construct the indicator assessment matrix according to Eq. (4.15). Then, calculate e_i using Eq. (4.16) as follows:

$$\begin{aligned} \boldsymbol{D}_1 &= \begin{bmatrix} 0 & 0 & 0 & 0.5333 & 0.5333 \end{bmatrix}, & e_1 &= 0.5333 \\ \boldsymbol{D}_2 &= \begin{bmatrix} 0 & 0.8571 & 0.1429 & 0 & 0 \end{bmatrix}, & e_2 &= 0.8571 \\ \boldsymbol{D}_3 &= \begin{bmatrix} 0 & 0 & 0 & 0.7500 & 0.7500 \end{bmatrix}, & e_3 &= 0.7500 \\ \boldsymbol{D}_4 &= \begin{bmatrix} 0 & 0.4800 & 0.5200 & 0 & 0 \end{bmatrix}, & e_4 &= 0.5200 \\ \boldsymbol{D}_5 &= \begin{bmatrix} 0 & 0.4000 & 0.6000 & 0 & 0 \end{bmatrix}, & e_5 &= 0.6000 \\ \boldsymbol{D}_6 &= \begin{bmatrix} 0 & 0.4667 & 0.5333 & 0 & 0 \end{bmatrix}, & e_6 &= 0.5333 \end{aligned}$$

Using Eqs. (4.17) and (4.18), the composite weight vector of the indicator layer is obtained: $\boldsymbol{\beta}_{IT1} = (0.2403, 0.1291, 0.1031, 0.0850, 0.0509, 0.3915)$. According to Eq. (4.19), the status sequence vector of traction transformer T_1 is determined as $\boldsymbol{K}_{IT1} = (1, 0, 1, 0, 0, 0)$. From Eq. (4.20), $\gamma_{ET1} = 1 \times 0.2403 + 1 \times 0.1031 = 0.3434$. Then, using Eq. (4.23), the health value of traction transformer T_1 is obtained as $H_{ET1} = 1 - 0.3434 = 0.6566$.

Referring to Fig. 4.1, the current status of this traction transformer is classified as abnormal, indicating a significant decline in service performance and potential internal degradation leading to functional failure. A maintenance strategy should be promptly devised to address these issues and prevent failure.

4.4.2 Condition Assessment of GIS High-Voltage Cabinets

For the 27.5 kV feeder side GIS high-voltage cabinet, monitor the SF_6 gas humidity (converted to 20 °C) I_1, relative leakage rate I_2, main circuit resistance I_3, low voltage circuit insulation resistance I_4, partial discharge I_5, and temperature rise of contacts I_6. According to the standards *GB/T 28428-2012 27.5 kV and 2 × 27.5 kV AC metal-enclosed switchgear and controlgear for electrification railway*, *GB/T 11022-2011 Common specifications for high-voltage switchgear and controlgear standards*, *GB/T 3906-2006 Alternating-current metal-enclosed switchgear and controlgear for rated voltages above 3.6 kV and up to and including 40.5 kV*, *DL/T 617-2010 Specification for Gas-insulated metal-enclosed switchgear*, and *DL/T 603-2006 Regulation of operation and preventive maintenance for gas-insulated metal-enclosed switchgear*, the specified critical values for the above six indicators can be determined and classified according to fuzzy distribution forms, as shown in Table 4.5.

At a certain time, the indicator values collected from the GIS high-voltage cabinet are $\{150, 0.08\%, 30, 1.8, 46, 21\}$.

Using the AHP to construct the pairwise comparison matrix for the factor set $\{I_1, I_2, I_3, I_4, I_5, I_6\}$, each element of the matrix is valued based on the results of the importance pairwise comparison of these six indicators, and the consistency verification is obtained as follows:

$$A_{\text{GIS}} = \begin{bmatrix} 1 & 0.33 & 3 & 4 & 2 & 0.5 \\ 3 & 1 & 5 & 6 & 4 & 2 \\ 0.33 & 0.2 & 1 & 2 & 0.5 & 0.25 \\ 0.25 & 0.17 & 0.5 & 1 & 0.33 & 0.2 \\ 0.5 & 0.25 & 2 & 3 & 1 & 0.33 \\ 2 & 0.5 & 4 & 5 & 3 & 1 \end{bmatrix} \tag{4.38}$$

The maximum eigenvalue is calculated as $\lambda_{max} = 6.1221$, hence $CI = (6.1221 - 6)/(6 - 1) = 0.0244$. According to Eq. (4.4), the consistency verification is $CR = 0.0244/1.24 = 0.0197 < 0.1$, which meets the requirement. The feature vector

Table 4.5 Parameters of the indicator layer for the 27.5 kV feeder side GIS high-voltage cabinet

Health indicator	SF_6 gas humidity	Relative leakage rate	Main circuit resistance	Low voltage circuit insulation resistance	Partial discharge	Temperature rise of contacts
Critical value	≤500 μL/L	≤0.5%	≤100 μΩ	≥1 MΩ	≤50 pC	≤65 K
Distribution form	Cost-based type	Cost-based type	Cost-based type	Benefit-based type	Cost-based type	Cost-based type

corresponding to λ_{max} is obtained and normalized to get the initial relative weight vector $\boldsymbol{w}_I = (0.1594, 0.3826, 0.0640, 0.0430, 0.1005, 0.2505)$.

For indicators $I_1 \sim I_3$, I_5, and I_6, the cost-type fuzzy distribution membership function (Eq. 4.14) is used. For indicator I_4, the benefit-type fuzzy distribution membership function (Eq. 4.13) is used. The membership degrees belonging to the five health grades are calculated and the indicator assessment matrix is constructed according to Eq. (4.15). Then, e_i is calculated using Eq. (4.16).

$$
\begin{aligned}
\boldsymbol{D}_1 &= \begin{bmatrix} 0 & 0.8000 & 0.2000 & 0 & 0 \end{bmatrix}, \quad e_1 = 0.8000 \\
\boldsymbol{D}_2 &= \begin{bmatrix} 0.3600 & 0.6400 & 0 & 0 & 0 \end{bmatrix}, \quad e_2 = 0.6400 \\
\boldsymbol{D}_3 &= \begin{bmatrix} 0 & 0.8000 & 0.2000 & 0 & 0 \end{bmatrix}, \quad e_3 = 0.8000 \\
\boldsymbol{D}_4 &= \begin{bmatrix} 0.2000 & 0.8000 & 0 & 0 & 0 \end{bmatrix}, \quad e_4 = 0.8000 \\
\boldsymbol{D}_5 &= \begin{bmatrix} 0 & 0 & 0 & 0.3200 & 0.3200 \end{bmatrix}, \quad e_5 = 0.3200 \\
\boldsymbol{D}_6 &= \begin{bmatrix} 0 & 0.7077 & 0.2923 & 0 & 0 \end{bmatrix}, \quad e_6 = 0.7077
\end{aligned}
$$

Using Eqs. (4.17) and (4.18), the composite weight vector of the indicator layer is obtained as $\boldsymbol{\beta}_{I\mathrm{GIS}} = (0.1246, 0.3738, 0.0501, 0.0336, 0.1965, 0.2213)$. According to Eq. (4.19), the status sequence vector of the GIS high-voltage cabinet is determined as $\boldsymbol{K}_{I\mathrm{GIS}} = (0, 0, 0, 0, 1, 0)$. From Eq. (4.20), $\gamma_{E\mathrm{GIS}} = 1 \times 0.1965 = 0.1965$. Then, using Eq. (4.23), the health value of the GIS high-voltage cabinet is obtained as $H_{E\mathrm{GIS}} = 1 - 0.1965 = 0.8035$.

Referencing Fig. 4.1, the current GIS high-voltage cabinet is in an attentive condition, indicating some deterioration in service performance with potential or concealed faults present. Further testing is necessary to ascertain the type of fault and to formulate a maintenance plan to prevent further degradation.

4.4.3 Health Condition Assessment of the Traction Substation

For the subsystem, traction substation S_1, the AHP was applied to the internal component set $\{\mathrm{DL}, \mathrm{CT}, \mathrm{T1}, \mathrm{T3}, \mathrm{CA}, \mathrm{M}, \mathrm{GIS}, \mathrm{PT}\}$. Each element of the matrix is valued based on the results of the pairwise comparison of the importance of these eight components, and the consistency verification is obtained as follows:

$$
\boldsymbol{A}_{\mathrm{S1}} = \begin{bmatrix}
1 & 8 & 2 & 2 & 6 & 5 & 0.5 & 4 \\
0.13 & 1 & 0.14 & 0.14 & 0.5 & 0.33 & 0.11 & 0.25 \\
0.5 & 7 & 1 & 1 & 5 & 3 & 0.33 & 2 \\
0.5 & 7 & 1 & 1 & 5 & 3 & 0.33 & 2 \\
0.17 & 2 & 0.2 & 0.2 & 1 & 0.5 & 0.14 & 0.33 \\
0.2 & 3 & 0.33 & 0.33 & 2 & 1 & 0.17 & 0.5 \\
2 & 9 & 3 & 3 & 7 & 6 & 1 & 5 \\
0.25 & 4 & 0.5 & 0.5 & 3 & 2 & 0.2 & 1
\end{bmatrix} \tag{4.39}
$$

The maximum eigenvalue is calculated as $\lambda_{max} = 8.1723$, hence $CI = (8.1723 - 8)/(8 - 1) = 0.0246$. According to Eq. (4.4), the consistency verification is $CR = 0.0246/1.41 = 0.0174 < 0.1$, which meets the requirement. The feature vector corresponding to λ_{max} is obtained and normalized to get the initial relative weight vector $\boldsymbol{w}_E = (0.2251, 0.0215, 0.1346, 0.1346, 0.0321, 0.0496, 0.3275, 0.0750)$.

At the system level, for the traction power supply system S of the power supply section, a pairwise comparison matrix was created for the subsystem set $\{S_1, S_2, S_3, OC, AT\}$. The elements of this matrix were determined from pairwise comparisons of the importance of these five subsystems, as shown in Table 4.1, yielding the following:

$$A_S = \begin{bmatrix} 1 & 0.5 & 1 & 0.33 & 3 \\ 2 & 1 & 2 & 0.5 & 4 \\ 1 & 0.5 & 1 & 0.33 & 3 \\ 3 & 2 & 3 & 1 & 5 \\ 0.33 & 0.25 & 0.33 & 0.2 & 1 \end{bmatrix} \tag{4.40}$$

The maximum eigenvalue is calculated as $\lambda_{max} = 5.0496$, hence $CI = (5.0496 - 5)/(5 - 1) = 0.0124$. According to Eq. (4.4), the consistency verification is $CR = 0.0124/1.12 = 0.0111 < 0.1$, which meets the requirement. The feature vector corresponding to λ_{max} is obtained and normalized to get the initial relative weight vector $\boldsymbol{w}_U = (0.1426, 0.2491, 0.1426, 0.4069, 0.0321, 0.0587)$.

4.4.3.1 Considering Only the Health Condition of the Traction Transformer

Focusing solely on the health condition of the traction transformer T_1, if T_1's health value, previously calculated as $H_{ET1} = 0.6566$, and assuming the health values of other equipment in the component set are 1, then using eq. (4.21), the impact factor of the component layer (traction transformer T_1) on the subsystem layer (Shijiazhuang traction substation S_1) is $\gamma_{US1} = \gamma_{ET1} \times w_{ET1} = 0.3434 \times 0.1346 = 0.0462$. Consequently using eq. (4.24), the health value of the traction substation S_1 is calculated as $H_{US1} = 1 - 0.0462 = 0.9538$, indicating a healthy condition.

Using eq. (4.22), the impact factor of the subsystem layer (traction substation S_1) on the system layer (the power supply section) is $\gamma_S = \gamma_{US1} \times w_{US1} = 0.0462 \times 0.1426 = 0.0066$. Therefore, using eq. (4.25), the health value of the traction power supply system S in that power supply section is $H_S = 1 - 0.0066 = 0.9934$, also indicating a healthy condition.

Despite the traction transformer T_1 having a lower health value and being in an abnormal condition, its impact on the overall health of the traction substation is minimized due to the presence of 100% backup with two other transformers, T_3 and T_4, within the facility. This redundancy ensures that the entire traction power supply system maintains a high health value and remains in good condition.

4.4.3.2 Considering Only the Health Condition of the GIS High-Voltage Cabinet

If only the health condition of the GIS high-voltage cabinet is considered, with the health value of the GIS high-voltage cabinet being the previously calculated result $H_{E\text{GIS}} = 0.8035$, and the health value of all other equipment in the component set being 1, then using eq. (4.21), the impact factor of the component (GIS high-voltage cabinet) layer on the subsystem (traction substation S_1) layer is $\gamma_{US1} = \gamma_{E\text{GIS}} \times w_{E\text{GIS}} = 0.1965 \times 0.3275 = 0.0644$. Therefore, using eq. (4.24), the health value of the traction substation (S_1) is obtained as $H_{US1} = 1 - 0.0644 = 0.9356$, which is in a sub-health condition.

Using eq. (4.22), the impact factor of the subsystem layer (traction substation S_1) on the system layer (the power supply section) is $\gamma_\text{S} = \gamma_{US1} \times w_{US1} = 0.0644 \times 0.1426 = 0.0092$. Therefore, using eq. (4.25), the health value of the traction power supply system S in that power supply section is $H_\text{S} = 1 - 0.0092 = 0.9908$, also indicating a healthy condition.

Comparative analysis with the previous scenario, where only the health condition of the traction transformer was considered, reveals that although the health value of the GIS high-voltage cabinet is higher than that of the traction transformer and its health condition is better, the traction substation is in a sub-health condition in this case. The health value is lower than in the previous scenario. This is because the 27.5 kV feeder side GIS high-voltage cabinet does not have a backup, and its critical condition has a greater impact on the health condition of the traction substation, resulting in the overall traction power supply system's health value being slightly lower than in the previous scenario.

4.4.3.3 Considering Both the Health Condition of the Traction Transformer and the GIS High-Voltage Cabinet

If the health conditions of both the traction transformer T_1 and the GIS high-voltage cabinet are considered simultaneously, i.e., the health values of T_1 and the GIS high-voltage cabinet are the previously calculated results $H_{E\text{T1}} = 0.6566$ and $H_{E\text{GIS}} = 0.8035$, respectively, while the health values of all other equipment in the component set are 1, then using eq. (4.21), the impact factor of the component (traction transformer T_1 and GIS high-voltage cabinet) layer on the subsystem (traction substation S_1) layer is calculated as: $\gamma_{US1} = \gamma_{E\text{T1}} \times w_{E\text{T1}} + \gamma_{E\text{GIS}} \times w_{E\text{GIS}} = 0.3434 \times 0.1346 + 0.1965 \times 0.3275 = 0.1106$. $\text{US1} = \gamma_{E\text{T1}} \times w_{E\text{T1}} + \gamma_{E\text{GIS}} \times w_{E\text{GIS}} = 0.3434 \times 0.1346 + 0.1965 \times 0.3275 = 0.1106$. Therefore, using eq. (4.24), the health value of traction substation S_1) is obtained as $H_{US1} = 1 - 0.1106 = 0.8894$, which is in an attentive condition.

Using eq. (4.22), the impact factor of the subsystem layer (traction substation S_1) on the system layer (the power supply section) is $\gamma_\text{S} = \gamma_{US1} \times w_{US1} = 0.1106 \times 0.1426 = 0.0158$. Therefore, using eq. (4.25), the health value of the traction power supply system S in that power supply section is $H_S = 1 - 0.0158 = 0.9842$, also indicating a healthy condition.

Comparative analysis with scenarios that consider only the health conditions of individual equipment shows that when both equipment are simultaneously in a certain degree of distress, the health value of the traction substation significantly decreases, and the health condition deteriorates further. However, the overall decrease in the health value of the entire traction power supply system is not pronounced. This resilience is attributed to the ability of traction substations S_1 and S_3 to provide a cross-sectional power supply, with sectioning post S_2 acting as a backup. This configuration mitigates the impact of one substation's distressed condition on the entire system, thus maintaining a relatively high health value for the entire system.

4.5 Health Assessment of Catenary System

Catenary systems, which are responsible for providing energy to electric locomotives, are crucial for the operation of electrified railways. Electric locomotives rely on the sliding contact between the locomotive's pantograph and the catenary for power, as they do not carry their own energy sources. Therefore, maintaining the catenary in good condition is essential for ensuring the safe and punctual operation of electrified railways. Due to their outdoor installation, lack of backup, and the strong impacts they endure from pantographs, catenaries are prone to faults and have become vulnerable points in electrified railway systems.

The catenary has become one of the weak links in electrified railways. Strengthening the monitoring of the catenary status is an important prerequisite for realizing the high-speed railway catenary maintenance and repair concept of *preventive priority, thorough inspection, and cautious repair*. Currently, railway departments commonly employ two types of catenary testing methods: static and dynamic testing. Static testing involves routine maintenance checks using portable testing equipment performed manually, while dynamic testing involves installing testing equipment on specialized inspection vehicles to measure parameters under actual operating conditions. However, further research is needed on how to appropriately evaluate the health status of catenaries from the testing parameters obtained and use these evaluations to guide the management and maintenance of catenary equipment.

Advancements in testing techniques have provided more high-precision measurement data for determining the catenary status. However, the current methods for analyzing measurement data and evaluating catenary status are relatively simple, often employing the standard comparison method. This method involves comparing the measured values with predefined standard values to determine whether the catenary is in a fault condition. This approach offers only two evaluations of the catenary's health condition: 'normal' and 'fault', which are binary evaluations ('either/or'). The results are simplistic and fail to reflect the development and changes in the catenary network's health condition. They cannot capture the trend of the catenary network's health condition development, nor can they identify deteriorating operating trends or provide early warnings for impending faults. Consequently, this method has limited effectiveness in guiding state-based maintenance and repair of the catenary. As the traction power supply 6C system is gradually implemented in the future, the amount of detection data related to the catenary's operational status will increase. Faced with a massive amount of catenary monitoring data, there is currently a lack of effective data processing and suitable status evaluation methods. With the accumulation of 6C monitoring over time, there will be an explosion of data but a lack of knowledge. Therefore, it is crucial to research data processing and state assessment methods suitable for catenary management.

Catenary systems feature mutually influencing and interrelated structural parameters, where changes in one parameter can affect others. Therefore, a holistic assessment of the catenary's health condition is necessary rather than relying on individual parameters. This section introduces a comprehensive assessment model for catenary network health conditions using the theory of membership degrees from fuzzy mathematics. This model translates qualitative evaluations of catenary status into quantitative, comprehensive assessments of catenary health value. By employing entropy weighting theory to mine the information inherent in measurement data, this approach more objectively and rigorously determines the weights of different indicator parameters during integration. The quantitative assessment of health condition will provide a basis for state-based maintenance and future health management of catenaries.

4.5.1 Assessment Indicator Set for Catenary System

For an electric locomotive to operate normally, it is crucial to maintain good current collection quality between the pantograph and the catenary. The degree of current collection quality and the interaction between the contact suspension and the pantograph relate to the catenary's structural integrity, which includes various mechanical and electrical parameters. These parameters are essential indicators for evaluating the operational quality of catenaries.

Based on the structural characteristics of catenaries and current detection methods, a fuzzy comprehensive assessment indicator system for catenary status has been established, as depicted in Fig. 4.7. This system incorporates sub-indicators commonly used in current catenary testing, ensuring the data is readily accessible. With the anticipated future implementation of the 6C detection system, this indicator system is designed to be expandable to include additional testing items.

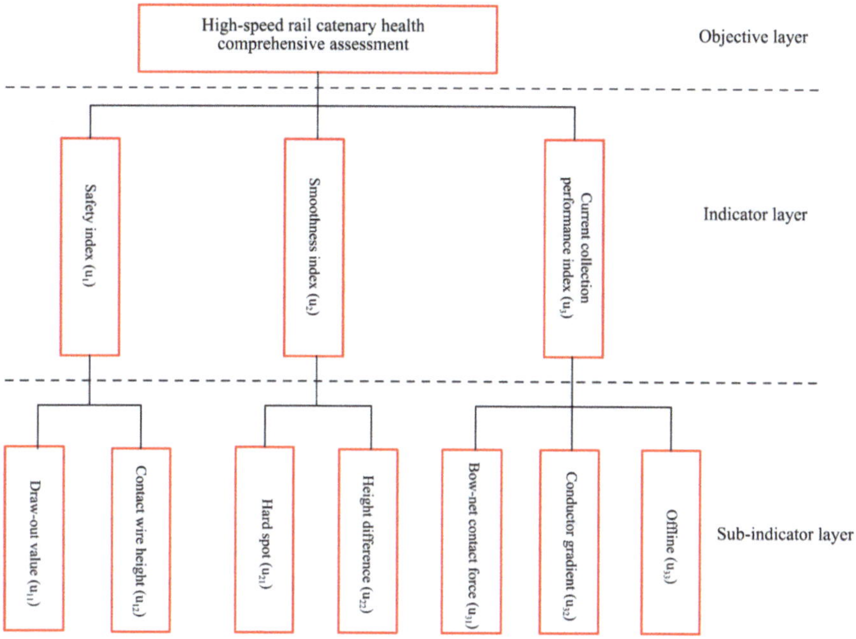

Fig. 4.7 Catenary Status Comprehensive Assessment Indicator System

4.5.2 Health Condition Judgment Set for Catenary System

After determining the influencing factors for judging the health condition of the catenary network, it is necessary to categorize the different statuses of the catenary health levels, i.e., to establish a judgment set for the catenary network health condition. Unlike the current binary success-failure judgment, to better reflect the actual situation of the catenary and to illustrate the intrinsic patterns of catenary fault occurrence and evolution, as well as to facilitate practical on-site operations, the catenary health condition is divided into five levels, namely:

$$V = \left\{\text{Healthy}(v_1), \text{Sub}-\text{healthy}(v_2), \text{Minor Unhealthy}(v_3), \right.$$
$$\left. \text{Moderate Unhealthy}(v_4), \text{Severe Unhealthy}(v_5)\right\}$$

These divisions are based on the *Catenary Operation Maintenance Regulations* and practical evaluation experience, with specific standard threshold values for each indicator outlined in Table 4.6.

If the five health grades of the catenary are represented by values between [0, 1], where values close to 1 indicate good status and values close to 0 indicate deteriorating status, the catenary network health condition can be quantified. Using

Table 4.6 Catenary network health condition evaluation standard limit values

Indicator	Grade				
	v_1	v_2	v_3	v_4	v_5
Contact wire draw-out value (mm)	400	450	500	550	600
Contact wire height (mm)	6415	6415 ± 15	6415 ± 45	6415 ± 85	6415 ± 135
Hard spot (g)	30	40	50	60	70
Height difference (mm)	0	25	50	100	150
Pantograph-catenary pressure (N)	120	120 ± 30	120 ± 60	120 ± 80	120 ± 100
Gradient (‰)	0	0.8	1	2	3
Pantograph offline time (ms)	0	20	40	60	80

$V = \{1, 0.8, 0.6, 0.4, 0.2\}$, the health value for the catenary state assessment can be ultimately determined. This evaluation result will be more intuitive and can help establish a historical data repository for catenary network health condition assessment, making it easier to reveal its development and evolution patterns.

After determining the assessment indicators and judgment set, processing the obtained monitoring data based on fuzzy theory to determine the influence weights of different indicators is a key component for conducting a comprehensive assessment of the catenary status.

4.5.3 *Fuzzification of Indicator Parameters for Catenary System*

Upon defining the indicator set for the catenary system, the parameters obtained from the tests require normalization due to their varying dimensions. As all the indicators in the health assessment of the catenary system adhere to a *smaller-the-better* type, the normalization formula is defined as follows:

$$x_i = \left(x_a - x_i'\right)/\left(x_a - x_b\right), i = 1, 2, \cdots, m \tag{4.41}$$

where x_i' is the measured value, x_a is the limit value of the indicator, and x_b is the optimum value of the indicator, x_i is the normalized value of the indicator.

Given that the catenary functions as a mechanical structure, its availability indicators and fuzzy set levels utilize a triangular membership function, which is illustrated in Fig. 4.8.

The corresponding membership function is defined as:

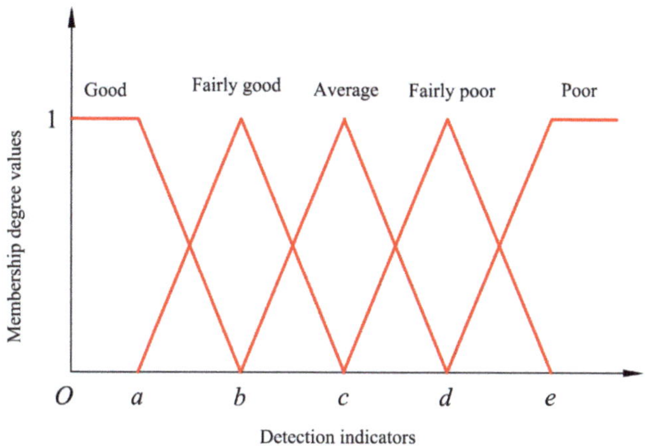

Fig. 4.8 Triangular Membership Function Diagram

$$\mu_1(x)=\begin{cases}1 & x\leqslant a\\(b-x)/(b-a), & x\in(a,b)\\0 & x\geqslant b\end{cases}$$

$$\mu_2(x)=\begin{cases}0 & x\leqslant a \text{ or } x\geqslant c\\(x-a)/(b-a), & x\in(a,b)\\(c-x)/(c-b), & x\in(b,c)\end{cases}$$

...

$$\mu_5(x)=\begin{cases}0 & x\leqslant d\\(x-d)/(\mathrm{e}-d) & x\in(d,\mathrm{e})\\1 & x\geqslant \mathrm{e}\end{cases} \tag{4.42}$$

where a, b, c, d, e are the normalized values derived from the limit values listed in Table 4.6.

4.5.4 Fuzzy Comprehensive Evaluation of Catenary Health Condition

Based on the membership functions, the degree to which various indicators belong to the judgment set can be assessed. If u_j is the j-th factor in the assessment indicator set, and v_l is the l-th (where $l=1, 2, \cdots, L$) object in the judgment set, then the degree of membership of u_j belonging to v_l is represented as r_{jl}. Thus, $\boldsymbol{R}_j=(r_{j1}, r_{j2}, \cdots, r_{jl})$

Table 4.7 Dynamic testing data of a catenary segment

Date	Draw-out value (mm)	Contact wire height (mm)	Hard spot (g)	Height difference (mm)	Contact force (N)	Gradient (‰)	Offline (ms)
10.15	405	6537	45	146	140	2.5	44
10.14	412	6450	48	120	150	1.5	32

forms the single-factor assessment set for u_j, and the matrix $\boldsymbol{R} = (\boldsymbol{R}_1\boldsymbol{R}_2\cdots\boldsymbol{R}_n)$ composed of these row vectors constitutes the fuzzy judgment matrix.

The comprehensive evaluation of the catenary's health condition can be achieved through the following formula:

$$\boldsymbol{B} = \boldsymbol{W} \circ \boldsymbol{R} \tag{4.43}$$

where $\circ$ is the fuzzy operator. Due to the numerous influencing factors on the catenary network health condition, to avoid losing information and to comprehensively consider the impact of various factors, the weighted average operator $\mathrm{M}(\cdot, \bigoplus)$ is selected for the calculation. Thus,

$$b_l = \sum_{j=1}^{n} w_j r_{jl} \tag{4.44}$$

If $b_r = \max\limits_{1 \leqslant l \leqslant L}(b_l)$ is the maximum element of the fuzzy comprehensive result $\boldsymbol{B}$, then according to the maximum membership principle, the evaluation v_r corresponding to b_r can be taken as the final assessment result.

Table 4.7 provides data from the inspection of a segment of catenary on October 14 and 15, 2010, with inspection vehicle speeds of 247 km/h and 253 km/h respectively.

After normalizing the values for each indicator using Formula (4.41) into a standardized matrix:

$$\begin{bmatrix} 0.975 & 0.096 & 0.625 & 0.027 & 0.800 & 0.167 & 0.450 \\ 0.940 & 0.881 & 0.550 & 0.200 & 0.700 & 0.500 & 0.600 \end{bmatrix} \tag{4.45}$$

Combining the standard threshold values for each level listed in Table 4.6 and the membership function of Eq. (4.42), the single factor judgment set for the data measured on October 15 is obtained. The fuzzy judgment matrix composed of the single factor judgment sets of the 7 detection indicators is as follows:

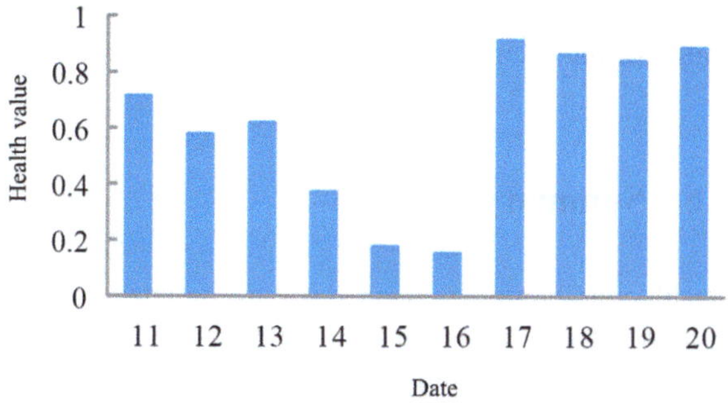

Fig. 4.9 Health Assessment Status Chart for a Segment of Catenary

$$
\boldsymbol{R} = \begin{bmatrix} 0.900 & 0.100 & 0 & 0 & 0 \\ 0 & 0 & 0 & 0.259 & 0.741 \\ 0 & 0.500 & 0.500 & 0 & 0 \\ 0 & 0 & 0 & 0.081 & 0.919 \\ 0.333 & 0.667 & 0 & 0 & 0 \\ 0 & 0 & 0 & 0.500 & 0.500 \\ 0 & 0 & 0.800 & 0.200 & 0 \end{bmatrix} \tag{4.46}
$$

Calculating the entropy for the time series formed by the two sets of monitoring data:

$$
\boldsymbol{H} = [\ 0.356\ 0.165\ 0.355\ 0.188\ 0.356\ 0.289\ 0.351\ ] \tag{4.47}
$$

Substituting into Eq. (4.6), the final weight vector is obtained as:

$$
\boldsymbol{W}_H = [\ 0.130\ 0.168\ 0.131\ 0.164\ 0.130\ 0.144\ 0.131\ ] \tag{4.48}
$$

Using Eqs. (4.43) and (4.44) for comprehensive assessment, the results are:

$$
\boldsymbol{B}_{C1} = [\ 0.160\ 0.166\ 0.170\ 0.156\ 0.348\ ] \tag{4.49}
$$

At this point, according to the maximum membership principle, it can be seen that the final assessment result is *severely abnormal*. Using $V = \{1, 0.8, 0.6, 0.4, 0.2\}$ to replace the scores corresponding to the five grades, and applying the weighted average according to Formula (4.44), the catenary's health assessment value is $v = 0.1756 < 0.2$, still classifying it as *severely abnormal*. Figure 4.9 presents the health value assessment results of this catenary segment from October 11 to October 20, showing a deterioration in the catenary's health condition starting on the 14th, with worse conditions on the 15th and 16th. Following maintenance on the 16th, the catenary's status significantly improved. This quantitative assessment result is more intuitive and better reflects the development and variation trends of the catenary's health condition. After accumulating data over time, future development trends can be predicted, facilitating early prevention, which is crucial for ensuring the availability of the catenary and reducing potential losses due to faults.

4.6 Summary

Health condition assessment involves the integration and transformation of various monitoring data into a unified health value indicator, providing a comprehensive, accurate, intuitive, and quantitative characterization of the health condition of traction power supply equipment and systems. This chapter introduces health assessment methods for high-speed railway traction power supply systems, including the AHP, entropy weight method, fuzzy statistics, and fuzzy comprehensive evaluation method. By combining these methods with the service performance and degradation laws of equipment and the fault evolution process, a multifactorial model of traction power supply system health condition is established. This model customizes multi-source heterogeneous monitoring data into a set of health condition evaluation indicators and health values. Using health condition assessment methods, the advantages of both subjective and objective assessments are integrated. Comprehensive health assessments are conducted on traction transformers, GIS high-voltage cabinets, traction substations, and catenary systems to fully and accurately grasp the health condition of the traction power supply equipment and systems.

Chapter 5
Reliability and Risk Assessment of High-Speed Railway Traction Power Supply System

The operational environment of China's high-speed railways is highly variable, and the traction power supply system installed along the tracks exhibits diverse comploutageies over spatial and temporal scales. Since its commissioning, the traction power supply system, which has been continuously in service status, needs to withstand the changes of the four natural seasons and occasionally suffers from the devastating impacts of lightning, rain, snow, and strong winds. Covering more than 30 provinces and cities, the high-speed railway network shows significant spatial variability, especially for individual high-speed railways lines that exceed 1000 kilometers in travel distance, along which the traction power supply system spans multiple regions and natural climate zones, potentially experiencing different environmental impacts at the same time, such as the Beijing-Harbin railway reaching cold regions and the Lanzhou-Xinjiang railway crossing two major wind zones. The complex and variable external service environment poses considerable challenges in establishing a unified risk assessment model for the entire line's traction power supply system. Additionally, the internal service environment of the traction power supply system is equally complex. According to the different transport capacity requirements and train operation schedules from origin to destination across regions, the density of traffic varies from section to section, causing diverse impacts on the traction substation equipment due to varying traction load characteristics. Moreover, due to the different characteristics of regional power grids, including adequacy, power supply capacity, and power quality, the reliability of the external power source for the high-speed railway traction power supply system also varies. Therefore, most existing studies focus on analyzing the reliability of power supply equipment, traction substation connections to the power system, and capacity configurations individually. This approach leads to assessment results that only reflect a single aspect of security risk and fail to evaluate the overall risks faced by the traction power supply equipment or the system as a whole. To ensure the safe, reliable, and continuous operation of high-speed railways, the traction power supply system requires extremely high operational reliability. Traction power supply equipment, such as traction transformers and circuit breakers, is the core component of the

Z. He et al., *Prognostics and Health Management of High-Speed Railway Traction Power Substations*, https://doi.org/10.1007/978-981-95-4538-4_5

high-speed railway traction power supply system, and its safety and reliability are critically important for the normal operation of the entire high-speed railway. Additionally, the reliability and risk assessment of traction power supply equipment provide a basis for operational maintenance decisions and are key topics in current research on traction power supply system maintenance technology.

In the traction power supply system, many factors will disrupt normal operation, induce faults, or even lead to safety accidents. By employing security risk assessment theories to analyze and assess the risks of the power supply system, it is possible to comprehensively understand the system's risk status from both macroscopic and microscopic perspectives. This enables the evaluation of the security risks associated with different equipment components or systems under various operating conditions and stages, facilitating timely measures to enhance safety management and reduce or even prevent safety incidents, thereby ensuring the safe and stable supply of power. According to security risk assessment theories, the risk assessment of the power supply system generally includes four steps: risk identification, risk analysis, risk quantification, and comprehensive risk assessment, detailed as follows:

(1) Risk Identification. By reviewing the historical fault records of the power supply system and considering its own characteristics and those of the operating environment, identify sources of risk within various components, operations, and maintenance activities and their processes; recognize risk factors such as component aging, disaster weather, and human activities; confirm risk accidents, such as equipment damage and system power outages; and screen for types of risks that are of concern to onsite maintenance personnel.
(2) Risk Analysis. Clearly define and describe the identified risks and their characteristics, segment the power supply system into regions and units, and establish models for reliability and risk analysis of the power supply system, such as failure mode and effect analysis and fault tree analysis models. Qualitatively describe and analyze the conditions under which risks occur and the potential scenarios of these risks.
(3) Risk Quantification. Determine the outage model for components and systems to calculate the probability of risk; ascertain the severity of risk for components and systems to calculate risk loss; and quantify the risks faced by the power supply system from both the probability and severity of consequences, establishing a comprehensive risk indicator system for components and systems.
(4) Comprehensive Risk Assessment. Based on the quantitative results of the risk indicators, identify the level of risk within the system, comprehensively assess and evaluate the risk conditions faced by the power supply system, to facilitate risk prioritization, establish risk acceptance criteria, and optimize risk management activities.

5.1 Overview of Reliability and Risk Assessment in High-Speed Railway Traction Power Supply Systems

In the field of power systems, AC power systems were among the first subjects of reliability studies, and the results of these studies are now widely applied at various stages of power system planning, design, and operation. Today, the methods for reliability assessment in power systems can primarily be divided into two types. One type is the analytical method, which includes techniques such as failure mode and effect analysis (FMEA), fault tree analysis (FTA), the network equivalent method, and the minimal path method. The main approach is to explore and analyze the failure modes and causes of power supply equipment like transformers, circuit breakers, and disconnecting switches, establishing analytical models of equipment and system reliability to calculate corresponding reliability indicators. The other type is the simulation method, such as the Monte Carlo simulation method, which uses reliability data and distribution characteristics of power supply equipment to generate a series of random numbers, simulating the state of equipment and related time sequences, thus approximating the calculation of reliability indicators. The development of these two methods is relatively mature, and existing research often aims to improve traditional methods by considering more factors. The reliability of power systems is closely related to factors such as the aging failure of power equipment, external environmental interference, and human activities.

At the beginning of the twenty-first century, international standards concerning the reliability and security risk of traction power supply systems were proposed, notably the *IEC 62278:2002 – Railway Applications: Specification and Demonstration of Reliability, Availability, Maintainability, and Safety (RAMS)*. In 2004, American engineer Sergo Sagareli first introduced a conceptual framework for traction power supply system reliability. Although railways transport played a crucial role during the economic development period in the United States and still accounts for about 35% of the world's total railway length today, its railways transport is primarily freight-based, consisting of a mix of diesel and electric locomotives, with high-speed railways development lagging behind. Consequently, there has been little direct research related to the reliability of traction power supply systems in the US. In contrast, countries like Japan and France, which developed high-speed railways technology earlier, designed their traction power supply systems with reference to the RAMS standards and enhanced system reliability using methods such as the FTA, redundancy analysis, and multi-path strategies. Additionally, researchers in France and the Netherlands have studied the impact of human factors on railway system reliability.

Research on the reliability of traction power supply systems in China began in the 1990s, and in 2008, the standard *GB/T 21562-2008: Railway Applications: Specification and Demonstration of Reliability, Availability, Maintainability and Safety* was issued. However, qualitative and explanatory standards are increasingly difficult to meet the demands of quantitative reliability assessments. Therefore, exploration of the reliability of traction power supply systems in China needs to focus on the following three aspects:

(1) Reliability evaluation considering substation wiring and structures. Assess the reliability of substations by considering different main wiring structures and transformer backup methods.
(2) Reliability evaluation considering OCS components. Assess the reliability of the OCS by considering factors such as different hanging methods of OCS, the lifespan of components, and maintenance methods that may cause OCS failures.
(3) Reliability evaluation considering external factors. Focus on traction power supply equipment or power supply lines, taking into account external factors such as meteorological and geographical environments, and human maintenance activities, and conduct reliability assessments primarily using data and model-based methods.

5.1.1 Reliability Assessment Indicators for High-Speed Railway Traction Power Supply System

As a fundamental element within the reliability assessment framework, the reliability indicators for the traction power supply system are a key focus of research. Existing studies commonly base their approach on the operational characteristics of the traction power supply system, centering on system or equipment failures, and define reliability assessment indicators in terms of equipment reliability, functional reliability, and system integrity. In the initial stage of researching high-speed railway traction power supply system reliability in China, researchers commonly used the failure rate (i.e., the ratio of the number of incorrect actions to the total number of actions of the traction power supply system equipment protection within a certain period.) to describe the reliability of the equipment. This approach has effectively guided the improvement of China's traction power supply systems. However, it does not take into account factors such as equipment usage time, external environmental conditions related to the equipment, and the operational lifespan of the equipment. Therefore, it does not comprehensively reflect the reliability of the traction power supply system and its equipment. For instance, if the equipment's fault frequency is very low, or if no faults have occurred during a statistical period, the result obtained by the traditional method of calculating the failure rate would inevitably be very low. In this case, the reliability cannot be adequately reflected using traditional probabilistic statistical methods. Additionally, the impact of the number of equipment included in the statistical data cannot be reflected in the indicator, making it difficult to effectively represent the reliability level of the traction power supply system. To address these issues, subsequent research suggests using indicators such as mean time to failure (MTTF), mean time to repair (MTTR), and mean time between failures (MTBF) to evaluate the reliability of traction power supply equipment and systems. It is proposed to base the traction power supply system reliability indicators on the operational and working characteristics of the traction power supply equipment and system, using them as the foundation and starting point for both historical and future reliability evaluations and analyses. These reliability indicators should be subdivided into various aspects such as power supply quality indicators, fault outage indicators, scheduled outage indicators, external influence indicators, and equipment performance indicators. This approach allows the reliability indicators to reflect the characteristics, structure, operational conditions, and impact on the normal operation of trains of the traction power supply equipment and its system. These indicators also serve as a basic measure of the equipment and system's reliability level.

Furthermore, from a quantitative statistical and calculation perspective, reliability indicators are proposed for equipment, components, and system-level entities, reflecting their operation and service status. Common assessment metrics include the failure rate, repair rate, and mean time to repair. By defining and analyzing real-world data according to the following indicator parameters, the main equipment reliability indicators of the traction power supply system can be calculated.

5.1.1.1 Performance Indicators for Traction Substation Equipment

(1) Failure Rate (λ): The failure rate, denoted as λ, is the average frequency at which equipment malfunctions occur over a specified period, generally quantified in terms of incidents per year.

$$\text{Failure rate } (\lambda) = \frac{\text{Total number of equipment failures}}{\text{Number of units}} \ (\text{incidents } per \text{ year}) \tag{5.1}$$

Additionally, this rate can be ascertained through statistical reliability analysis, incorporating reliability figures obtained from railway operations and equipment manufacturers. For a 365-day year, an alternate equation for computing the failure probability of traction substation equipment is given by:

$$\text{Failure rate } (\lambda) = \frac{\text{Annual failure duration of equipment}}{365\text{x}24} \tag{5.2}$$

(2) MTTF: MTTF is defined as the average duration before power restoration after an equipment failure, typically measured in hours per incident.
(3) MTTR: MTTR is defined as the average time required for personnel to conclusively rectify (including diagnosis and repair) an equipment failure, typically measured in hours per incident.
(4) MTBF: MTBF is measured as the average interval between the completion of a repair and the occurrence of the next failure. Assuming an exponential distribution for reliability, MTBF is inversely proportional to the failure rate, mathematically expressed as:

$$MTBF = \frac{1}{\lambda} \tag{5.3}$$

(5) Repair Rate (μ): The repair rate, denoted as μ, measures the frequency at which equipment can be repaired within a specified timeframe, often cited as occurrences per hour. Assuming an exponential distribution for the repair rate, it is inversely related to MTTR as:

$$MTTR = \frac{1}{\mu} \tag{5.4}$$

(6) Annual Repair Time (U): The annual repair time is a measure of the average time spent on repairs annually, calculated by the product of the failure rate and MTTR:

$$U = \lambda \times MTTR \tag{5.5}$$

(7) Available Rate (R)/Unavailable Rate (F): The available rate quantifies the probability that equipment is in a functional state, providing a comprehensive evaluation of equipment availability by integrating factors such as failure rate, impact of failures, and maintenance efficiency. The equation for the available rate is:

$$R = \frac{365 \times 24 - U}{365 \times 24} \times 100\% \tag{5.6}$$

Accordingly, the unavailable rate is derived as follows:

$$F = 1 - R \tag{5.7}$$

5.1.1.2 Reliability Indicators for High-Speed Railway Traction Power Supply Systems

Power Supply Quality Indicators

(1) Power Supply Availability: Power supply availability refers to the ratio of the effective power supply time to the total time during a statistical period for a particular high-speed railway line.

$$\text{Power Supply Availability} = 1 - \left(\frac{\text{Fault Outage Time}}{\text{Statistical Period Time}}\right) \times 100\% \tag{5.8}$$

(2) Train Delay Time: Train delay time refers to the total time of train delays over the course of a year on a high-speed railway line.

$$\text{Train Delay Time} = \sum \text{Each Train Delay Time (min)} \tag{5.9}$$

(3) Number of Delayed Trains: Number of delayed trains refers to the total number of trains that experienced delays over a year on a high-speed railway line.

$$\text{Number of Delayed Trains} = \sum \text{Each Occurrence of Delayed Trains} \tag{5.10}$$

(4) Average Train Delay Time: This indicates the average delay time per train, calculated over a year for a specific high-speed railway line.

$$\text{Average Train Delay Time} = \frac{\text{Total Train Delay Time}}{\text{Number of Delayed Trains}} (\text{min/trains}) \tag{5.11}$$

(5) Average Number of Delayed Trains per Power Supply Interruption: This indicator measures the average number of train delays caused by each power outage over a year.

$$\text{Average Number of Delayed Trains } \textit{per} \text{ Power Supply Interruption} = \frac{\text{Number of Delayed Trains}}{\text{Number of Fault Power Outages}} (\text{trains/incident}) \tag{5.12}$$

Fault Outage Indicators

(1) Fault Outage Time: This refers to the total time of power outages due to faults over a year.

$$\text{Fault Outage Time} = \sum \text{Each Fault Outage Time (min)} \tag{5.13}$$

(2) MTTF

$$\text{Mean Time to Failure} = \frac{\text{Fault Outage Time}}{\text{Number of Fault Power Outages}} (\text{min}) \tag{5.14}$$

5.1.2 High-Speed Railway Traction Power Supply System Reliability Assessment Model

In modeling and solving indicators for the reliability assessment of high-speed railway traction power supply systems, the approaches used at the system level and equipment level are similar and mainly include the simulation method, the analytical method, and a hybrid modeling method combining both simulation and analytical methods. The simulation method is intuitive but has a distinct statistical nature, and the required computation time and precision are closely related. This method involves selecting and estimating states through probability distribution sampling and obtaining reliability indicators using statistical methods, including the Monte Carlo simulation method and non-exponential distribution method. The analytical method constructs a reliability probability model based on the structure of the system, the functions of the system and equipment, and their logical relationships, and precisely solves the model through recursive or iterative processes to calculate the system's reliability indicators. Its advantages include clear physical concepts and high model accuracy, while its downside is that the computational load soars with the size of the system. Leveraging the precision and clarity of physical concepts inherent in the analytical method, the hybrid method utilizes the analytical method where applicable and employs the simulation method when the problem scale exceeds the solving capacity of the analytical method.

1. Simulation Method

   The simulation method involves establishing an accurate mathematical model of a system and simulating various system states that randomly occur on a computer. This approach uses numerical methods to simulate an actual process, statistically deriving the system's reliability indicators from a large number of simulation trials. Additionally, it can generate probability distribution-based reliability indicators. The number of occurrences in random simulations is independent of system size, making the method highly adaptable with a simple algorithm and program structure. It is particularly suitable for calculating the reliability of complex systems. However, the precision of this calculation often faces limitations due to various factors. Constructing an accurate mathematical model for China's high-speed railway traction power supply system is labor-intensive and challenging. The simulation method, which inherently has statistical characteristics, lacks sufficient accuracy for systems like the traction power supply system that demand high reliability. This method is more appropriate for systems with many low-reliability components, whereas the components of the traction power supply system are generally reliable. Therefore, relying solely on the simulation method for reliability analysis of the high-speed railway traction power supply system is inappropriate. In recent years, to address the lack of systemic empirical information, an improved method, the Bayesian network method, has emerged.
2. Analytical Method

Currently, the analytical method is widely used in the reliability analysis of China's traction power supply system. Commonly employed analytical methods include the reliability block diagram method, Markov process model, network model method based on feeder segmentation, Go method, and FTA method. Each of these reliability models has distinct characteristics suited for specific applications.

The reliability block diagram method is one of the most fundamental reliability assessment models, primarily aimed at the reliability study of simple two-state systems. It represents the logical relationships between various components of a system in the form of a diagram. By using the reliability diagram and the reliability parameters of each component, the system's overall reliability parameters are quantified.

The Markov process model is mainly used in the reliability analysis of repairable systems within reliability assessments. The study typically employs a state space method to describe the states at different stages of a repairable system and the transitions between these states. An important premise in the application of Markov random processes is that the transition probabilities between states are memoryless, independent of the system's historical states.

The network model method based on feeder segmentation proposes segmenting the system into automatic isolation zones and manual isolation zones, organized around feeders. This is determined by the positions of automatic and manual switching equipment. On this foundational segmentation, feeders are defined as regional nodes and switching arcs that form a network model, allowing for reliability analysis by region. This significantly reduces the analysis compared to conducting searches on each component or equipment individually. This model is suitable for reliability studies of complex power distribution systems.

The Go method is a "goal-oriented" modeling approach (goal oriented, GO). Compared to traditional analytical methods, the Go method directly uses system flowcharts and schematic diagrams, transforming them into GO diagrams according to specific rules. GO diagrams represent the structural, functional, and logical relationships between systems and equipment using GO symbols. Focusing on system simulation and emulation, this method is particularly applicable to reliability studies of systems with clear structures and well-defined relationships between equipment.

The FTA model starts from the least desirable events, focusing on the root causes and intermediate processes that lead to system faults. It uses a fault tree to express the logical relationships of the system, providing a clear and intuitive visualization. On this foundation, the structure function of the fault tree's top event can be quickly written, facilitating the reliability analysis of the system. This model, based on graphical representation, employs a top-down approach to system and equipment failure analysis. It begins by identifying the top event (typically a system fault event), then determines the root causes or basic events that lead to the failure of the top event, and constructs a fault tree diagram model. The consistency of the fault tree's form with the actual system is challenging to ascertain; it usually requires identifying the minimal cut sets first, and then

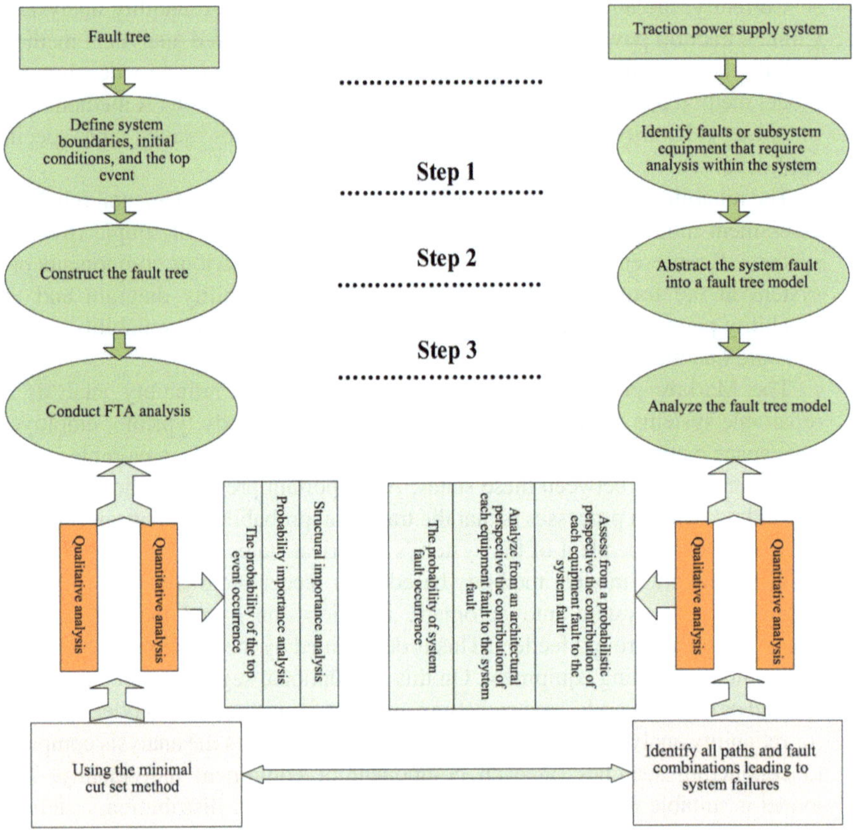

Fig. 5.1 The FTA process

evaluating and quantifying the overall system's reliability based on the failure rates of each basic event, their logical relationships, and the probability of the minimal cut sets. FTA enables analysts to gain a deeper understanding of the system, particularly in terms of structure, functionality, failures, and maintenance, thereby making reliability improvement efforts more effective. It is suitable for systems with numerous electrical components, high component reliability, complex structures, and predominantly redundant configurations. The specific process for analyzing the reliability of the traction power supply system using the fault tree model is shown in Fig. 5.1. Compared to traditional reliability analysis methods, FTA is one of the most commonly used methods in reliability analysis, but once a fault tree is constructed, it is not easily modified.

3. Hybrid Method

   Currently, there are two approaches to using the hybrid modeling method for system reliability assessment: The first approach alternates or sequentially uses the simulation method and analytical method to reduce the variance of the results obtained by integrating both methods. The second and more common approach involves a simple analytical judgment of the system state to shorten the duration of each state assessment, thereby increasing simulation speed.

Given the second approach, for complex power distribution network systems, the network is first simplified into a basic main feeder system using the analytical method. Then, the simplified main feeder system is analyzed using the simulation method to obtain the reliability probability distribution indicators at each load point, thus deriving a probability distribution calculation method for various reliability indicators of the complex distribution network. The hybrid method enriches the reliability indicator system.

4. Bayesian Network Method

   Although typical reliability assessment methods such as the FTA method, Markov process model, GO method, and hybrid method have been applied in practical engineering, these models or methods lack bidirectional inference capability. The Bayesian network, however, possesses unique bidirectional inference capabilities, overcoming the limitation that once established, the models mentioned above are difficult to modify. When using Bayesian networks for reliability analysis of traction substations, one can conduct forward reasoning from top to bottom. Starting from the causes of faults, it quickly deduces the outcomes, allowing the inference of the entire substation system failure probability from unit fault indicators within the traction substation, which is commonly used in reliability analysis. Moreover, it also supports backward reasoning from bottom to top, where conclusions can lead to the identification of causes. Based on the system fault in the traction substation, the probability of contribution by each subsystem or component fault can be determined, commonly used to identify weak links in system faults. Bayesian networks can handle incomplete information and data, allowing for the construction of relatively accurate Bayesian network models even in situations with incomplete data, thus facilitating system diagnostics.

5.1.3 Risk Sources in High-Speed Railway Traction Power Supply Systems

Once high-speed railway traction power supply equipment is operational, whether in continuous service or as standby equipment, it remains in service status. Service equipment such as OCS, due to having no backups, being exposed to the elements, spanning long supply intervals, and enduring long-term friction with pantographs, is highly prone to failures. Additionally, traction substations have numerous high-voltage insulation equipment, which are susceptible to damage due to the random shocks and imbalance characteristics of traction loads, thus also posing a risk of failure. Should a failure occur in the traction power supply system or its equipment, it could likely cause train stoppages or reduced speeds, resulting in delays or cancellations, and might even lead to accidents involving personal and property safety, with severe consequences. Therefore, analyzing and measuring the likelihood and severity of failures in high-speed railway traction power supply systems and equipment, predicting their potential development trends based on current operating conditions, and assessing their security risks are crucial to make sure that dispatch and maintenance personnel can take timely measures to reduce or avoid significant incidents.

Security risk assessment involves the comprehensive application of safety systems engineering methods to measure and predict risks. This process includes identifying, qualitatively and quantitatively analyzing the hazardous or unsafe factors within a system, and determining the likelihood and severity of potential dangers. Since the 1980s, the electric power sector has gradually adopted risk assessment techniques to address its safety issues, with well-developed analysis methods and assessment models for transmission and transformation equipment, as well as entire systems. These models are applied in scenarios such as operational dispatch, maintenance planning, and prevention of cascading failures. Objectively speaking, the traction power supply system can be considered a special user within the large electrical power system. However, the high-speed railway traction power supply system differs from conventional power systems in several aspects: it has distinct load characteristics, such as high train power, high traffic density, and strong impact loads; in terms of equipment, besides differences in voltage levels and wiring configurations, the traction power supply system also includes dynamically worn components like the OCS; compared to electric power distribution networks, it can achieve cross-border power supply, among other capabilities. These unique characteristics make the high-speed railway traction power supply system distinct and directly impact the safe and reliable operation of high-speed trains. Therefore, with increasing demands for safety and reliability, risk assessment of the high-speed railway traction power supply system has become a focus of attention.

In high-speed railways, the traction power supply system carries the critical task of supplying energy to high-speed trains, and its safety and reliability are paramount for the normal operation of the railway. The traction power supply system primarily consists of two parts: the traction substation and the traction network, which act like the 'heart' and 'blood vessels' of the system, responsible for high-voltage power transformation, train power supply, and traction return.

Based on field surveys and analyses of historical accident records, the main risks in the high-speed railway traction power supply system stem from equipment performance degradation, service environment, and human maintenance activities. Equipment performance degradation primarily involves issues such as insulation fatigue, aging, and mechanical wear of the equipment. Service environment factors include external influences like climate and geographical conditions, as well as internal influences such as traction load impact and nonlinearity. Human maintenance activities are discussed in terms of the maintenance mode and the security risks arising from human operations.

5.1.3.1 Equipment Performance Degradation Factors

The traction power supply system comprises numerous components such as OCS, cables, transformers, circuit breakers, disconnecting switches, current transformers, lightning arresters, and secondary equipment. The root cause of system failures is equipment failure due to shutdowns, which result from gradual performance degradation. The occurrence and progression of equipment failure inevitably go through a temporal process. Seemingly random failures have inherent mechanisms and regularities; even sudden failures have a gestation and development phase, albeit very brief. While reliability issues such as the overload capacity of traction transformers and the wear resistance of OCS are considered during design, manufacturing, and factory testing, equipment begins to age or wear from the moment it is commissioned, thereby increasing the likelihood of failures. Excluding sudden incidents, this likelihood tends to have a cumulative effect, generally increasing over time. When equipment performance degradation reaches a certain level, the equipment transitions from a safe and reliable period to a 'diseased' operational phase. During this phase, equipment performance continues to decline, and parameters such as insulation, wear resistance, and durability approach critical values, significantly increasing the likelihood of failure due to external factors or spontaneous events. Accidents caused by equipment's own performance degradation, such as explosions due to insulation aging in traction transformers and GIS equipment, or pantograph snagging incidents due to wear between the pantograph and OCS, not only require costly maintenance for major repairs or complete replacements but also lead to prolonged downtime, preventing normal railway operations and causing immeasurable adverse social impacts with severe security risk consequences.

Equipment performance degradation, as a factor influencing the security risk of the traction power supply system, requires special attention. Therefore, in practical applications, models that reflect the status of equipment performance degradation are more reasonable, and the results of security risk assessments obtained are also more realistic. Sudden equipment failures and outage are often induced by their service environment and are closely related to their operating conditions and external environment.

5.1.3.2 Service Environment Risk Factors

The conditions of the equipment service environment directly affect the performance of the equipment. Harsher service environment conditions will accelerate aging and increase wear, and they are also common triggers for sudden failures, impacting the safety of the traction power supply system. The service environment is divided into external service environment (geographical conditions, climate, etc.) and internal service environment (traction load characteristics) for separate analysis of their impact on security risk.

Traction power supply equipment is installed along the tracks, with numerous components and complex structures, and is mostly placed outdoors, making them highly susceptible to external environmental influences. Existing research indicates that the failure rate of most components varies with changes in climate conditions, and climate factors significantly affect the reliability indicators of the system and equipment. For example, the OCS of high-speed railways not only experience frequent high-speed dynamic contact and friction with pantographs but are also mostly mounted on elevated bridges. Being exposed and without backups, the traction power supply system is more vulnerable to natural disasters such as lightning, which can cause tripping faults and lead to economic losses. The comprehensive grounding system for traction power supply is buried underground, serving both as a railways return and a discharge path for lightning and short-circuit currents. It is severely affected by lightning strikes and chemical corrosion. Once damaged, it incurs high maintenance costs and lengthy repair times. In high-altitude regions, frequent strong winds pose significant security risks to OCS and outdoor power supply and transformation equipment.

Moreover, due to the drastic changes and frequent impacts of the high-speed railway traction load, the internal service environment also affects the safety and reliability of the traction power supply system and its components. Taking the transformer as a case study, compared to a power transformer, a traction transformer during operation, in addition to being influenced by multiple physical fields and environmental factors such as electric fields, magnetic fields, temperature fields, and stress fields, frequently suffers impacts from traction load and OCS short-circuit currents, which introduces specific challenges. These challenges are characterized by: (1) Load imbalance: Imbalance in the loads of the two power supply arms of the traction network leads to differences in temperature rise and aging of different windings. (2) Load shock: Sudden load changes, frequent periodic variations, and high overload requirements place enormous electromagnetic stress on windings, cores, and insulation, making them susceptible to damage. (3) Load non-linearity: AC-DC-AC high-speed electric multiple unit have a broader harmonic output spectrum than conventional railways, producing high-order characteristic harmonics. These harmonics cause additional losses in the core and windings of traction transformers, exacerbating their aging.

5.1.3.3 Human Maintenance Activities Factors

In addition to the impact of equipment performance degradation and service environment, the security risk of the traction power supply system also stems from human maintenance activities, primarily reflected in the following two aspects:

(1) Problems such as insufficient or excessive maintenance caused by issues with the maintenance mode. Currently, various railway bureaus' power supply sections and maintenance workshops primarily adopt traditional maintenance methods for the traction power supply system and key equipment, which involve

scheduled maintenance combined with routine inspections and organizing emergency repairs immediately after accidents. Maintenance mode face challenges such as broad maintenance scopes, heavy maintenance workloads, and unclear objectives, preventing timely detection and resolution of potential system or equipment risks. This can also lead to wastage of human and material resources to some extent. In terms of fault handling, especially during emergency repairs, the fault diagnosis methods are somewhat outdated, relying heavily on manual experience to investigate the causes of failures. This approach is somewhat random and struggles to provide timely and accurate fault warnings and targeted diagnostics, potentially leaving original security risks unaddressed. Overall, problems with the maintenance mode fundamentally introduce a degree of blindness and speculation to maintenance work, often leading to situations of insufficient or excessive maintenance. This may temporarily eliminate risk factors or resolve issues on the surface but can plant seeds for future safety hazards in the equipment's operation and may even increase the probability of future failures.

(2) Risk expansion caused by human operation. The execution of tasks such as scheduled inspection planning, issuing dispatch control commands, line fault localization, and maintenance during "curfew maintenance" all require the involvement of maintenance crew members. Due to limitations in knowledge, external pressure, and psychological factors, these personnel might not be able to perform the operations quickly and accurately, leading to equipment damage and the exacerbation of faults. Thus, human activities inevitably introduce risk factors associated with manual operations. These human risk factors can be mainly divided into two parts: (1) Human misjudgment: Inability to correctly identify system or equipment states may lead to delayed or incorrect handling of faults, expanding the scope of failures in the traction power supply system and causing additional losses. (2) Staff have insufficient understanding of the system and equipment operation mechanisms, misoperations or incorrect dispatch commands can cause equipment damage or system faults, worsening the risk consequences.

5.1.4 *Challenges in Research on High-Speed Railway Traction Power Supply System Reliability and Risk Assessment*

Current research on the reliability of high-speed railway traction power supply systems has achieved a series of results, yet some issues remain. The traction power supply system is crucial for ensuring the safe and reliable operation of high-speed railways, and its reliability analysis is vital for the secure and stable operation of the entire railway line. FTA, based on Boolean algebra and probability theory, is one of the most effective tools in the field of reliability analysis, with system components' failure rates and the relationships between events being essential data inputs. However, as the scale and comploutagey of systems and equipment continue to increase, the limitations of traditional FTA have become more apparent, restricting its application in practical engineering: on one hand, for many systems and equipment, obtaining fault data is challenging, and the fault probability of components is difficult to determine precisely, making it hard to establish a comprehensive reliability indicator system; on the other hand, in practical engineering, the failure mechanisms of systems are often unclear, leading to uncertainties in the relationships between events. Additionally, there are several other issues in reliability and risk assessment:

5.1.4.1 Incompleteness and Uncertainty of Data and Information

The high-speed railway traction substation online monitoring system has progressively enhanced its functionality based on traction characteristics, and corresponding monitoring and recording equipment have been developed for traction substation equipment. Some primary high-voltage equipment online monitoring systems have matured. In accordance with China's high-speed railway traction substation maintenance regulations, which combine real-time online monitoring with periodic condition-based maintenance, most equipment in traction substations undergoes minor maintenance and preventive insulation testing annually, yielding related offline testing data. Consequently, the high-speed railway traction power supply system contains a vast amount of online monitoring data, offline testing records, and inspection work ledgers. These are collected and stored through various equipment or means to reflect the service status of different equipment or subsystems. However, some fault events still lack systematic recording, categorization, and analysis. The collection periods, unit dimensions, precision requirements, and data formats of recorded data vary greatly, and the information has not been properly managed and preserved. This results in missing accident samples and historical records, or issues with record formats that prevent timely uploads for assessment and analysis, causing incompleteness of data and information. Moreover, information on accident analysis is primarily obtained from safety logs or subjective descriptions by maintenance personnel, leading to increased uncertainty in data and information for the high-speed railway traction power supply system.

The incompleteness of data and information can lead to a lack of critical data sources for some advanced risk assessment methods, preventing further deductions and thus hindering the realization of risk quantification work. Additionally, issues of uncertainty also pose obstacles in grasping the status of equipment and systems and describing the likelihood and severity of failures, affecting the accurate estimation of fault probabilities and risk consequences.

5.1.4.2 Comploutagey of the Traction Power Supply System's Service Environment

China's high-speed railway operates in a highly variable environment, and the traction power supply system installed along the routes exhibits complex diversity across the spatial and temporal scales of the external service environment. Since its inauguration, the continuously operational traction power supply system has weathered the changes of the four seasons and irregular onslaughts of disastrous climatic events such as lightning, snow, and strong winds. The high-speed railways network covers more than 30 provinces and cities, with significant spatial scale variations, especially for single high-speed railways lines that travel over 1000 kilometers. These systems traverse multiple regions and natural climate zones and may experience different environmental impacts simultaneously. For example, the Beijing-Harbin railway reaches cold regions from Beijing, and the Lanzhou-Xinjiang railway crosses two major wind zones along its route. The complex and variable external service environment makes it difficult to establish a unified risk assessment model for the entire line's traction power supply system.

Additionally, the internal service environment of the traction power supply system also exhibits comploutagey. The train operation schedules, determined by varying transport capacity requirements and departure-to-arrival timings across different regions, result in varying traffic densities across sections, thereby affecting the traction load impact on the traction substation equipment of each section differently. Moreover, the regional power grids in different areas vary in their adequacy, power supply capacity, and power quality, leading to differences in the reliability of the external power sources for the high-speed railway traction power supply system. Consequently, most existing studies analyze the reliability of power supply equipment, traction substations, and their connection to the power system configuration separately, resulting in assessment outcomes that only reflect unilateral security risks and fail to evaluate the overall risk of the traction power supply system.

5.1.4.3 Insufficiencies in Fault Handling and Maintenance Modes

Regarding fault handling for the high-speed railway traction power supply system, apart from routine monitoring and testing, post-fault handling is primarily used, which involves diagnosing and resolving faults after their occurrence. However, this approach can lead to serious consequences such as equipment damage, line shutdowns, or even personnel injuries. Not only does it result in high costs for equipment repair or replacement, but it may also cause considerable negative impacts on service speed, personal safety, and social reputation, leading to an expansion of the risk scope and worsening of risk consequences.

Maintenance and repair modes also require improvements. Currently, the maintenance mode still adopts a passive approach, combining scheduled periodic maintenance, manual routine inspections, and emergency repairs after an accident. This approach is prone to either insufficient or excessive maintenance. The maintenance scheduling system of the traction power supply system is based on referencing maintenance procedures, guidelines, and technical standards from the electric power system sector, with some content directly adopted from existing conventional railway protocols without fully considering the structural characteristics and service conditions of the high-speed railway traction power supply system. This results in maintenance activities lacking specificity, directly manifesting as inefficiencies or even misguided repair work implementation. Moreover, in most cases, maintenance personnel rely on experience to bring equipment out of a fault state. While such knowledge and experience can be considered expert in the field, it is undeniable that human subjectivity introduces a degree of blindness and speculation to the maintenance process, which might temporarily resolve issues superficially but could lay down risk hazards for future operations. The problems with the aforementioned maintenance mode make the system and equipment more prone to new security risks after undergoing maintenance activities, and these factors are difficult to define, leading to challenging quantitative risk assessment issues.

Given the existing problems in the security risk assessment of the traction power supply system, the future of high-speed railway traction power supply system security risk assessment will face the following challenges:

(1) Addressing the issues of incompleteness and uncertainty of data and information. Establish targeted security risk assessment methods to reduce incompleteness and uncertainty in data collection and processing, analysis and calculation models, and interpretation of risk results. This approach should enable a more accurate estimation of the likelihood of failures when foundational data is missing or insufficient.
(2) The comploutagey and variability of the service environment pose challenges to risk modeling. Explore wide-area spatio-temporal identification, real-time online monitoring, and offline historical data to uniformly establish a traction power supply system risk assessment model based on railway lines, forming a risk indicator system to describe the security risks of the system and equipment.
(3) How to establish a maintenance effectiveness evaluation model and a human reliability model. Improve the shortcomings of fault handling and maintenance modes, reducing the potential risk consequences brought about or triggered by improper maintenance and human factors, and from a risk control perspective, study maintenance decisions and their optimization.
(4) Risk assessment is a multifaceted, comprehensive issue. How to integrate risks induced by equipment performance degradation, service environment, and human maintenance activities factors to achieve a more comprehensive security risk assessment, and effectively apply the results of risk assessments.

5.2 Reliability Assessment of High-Speed Railway Traction Power Supply Equipment Considering the Operating Characteristics

Currently, there is extensive research on the reliability of power transformers. Initially, a constant failure rate was used to measure equipment reliability, reflecting long-term operating conditions based on statistical averages, but this approach does not account for the impact of operating conditions on reliability. With the acquisition and processing of load information, equipment monitoring data, and testing and maintenance records, more factors have been incorporated into the reliability assessment process for transformers. For example, an aging failure model has been developed based on the cumulative thermal effect causing a decrease in transformer insulation strength, and a health index model based on health condition assessment. For traction transformers, some scholars have conducted studies on temperature rise under traction loads, used thermal circuit models to calculate transformer temperature rises for several common wiring methods, and analyzed the impact of temperature rise on the aging of traction transformers, such as calculating the temperature rise and life loss of V/X connection transformers based on actual loads. However, these studies have not considered the impact of the nonlinearity of traction loads on transformer temperature rise and reliability.

This section drives research considering operating characteristics of high-speed railway traction power supply equipment reliability assessment methods using field monitoring data and historical fault records of traction power supply equipment. Initially considering the operating characteristics of high-speed railway traction power supply equipment, taking the world's first 220 kV level midpoint-drawn Scott traction transformer at the Shangyu North traction substation on China's Hangzhou-Ningbo high-speed railway as a case study, this section proposes a reliability assessment method that accounts for the load characteristics of high-speed railways.

5.2.1 High-Speed Railway Traction Load Characteristics Analysis

5.2.1.1 Impulsiveness of High-Speed Railway Traction Load

Unlike typical electric power loads, high-speed railway traction loads have their unique characteristics. Certain special operating conditions of high-speed trains, such as phase mismatch, and ascending or descending gradients, can cause sudden increases or decreases in load. The sectioned traction power supply mode results in a highly impulsive nature of the traction load.

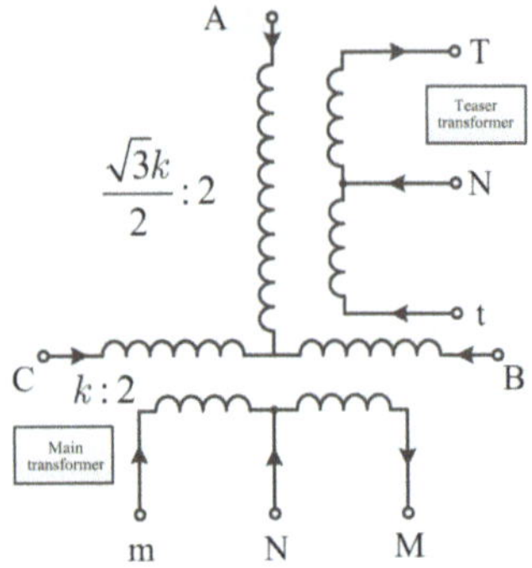

Fig. 5.2 Wiring principle diagram of a center-drawn scott traction transformer

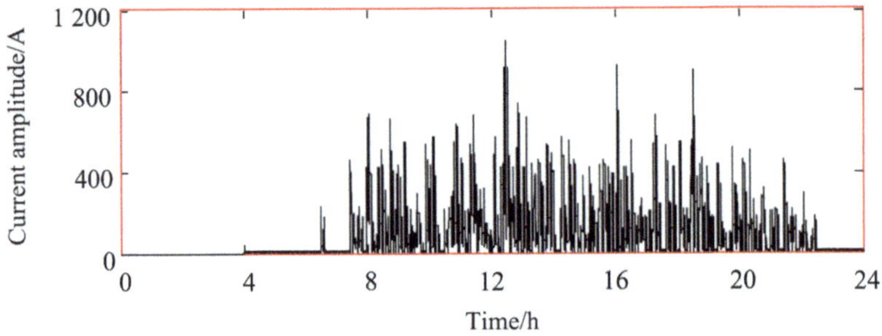

(a) teaser transformer load current curve

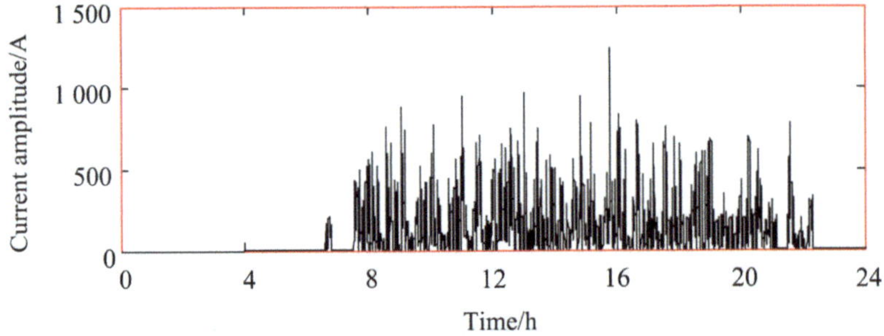

(b) main transformer load current curve

Fig. 5.3 Daily Traction Load Current Curve. (**a**) Teaser transformer load current curve. (**b**) Main transformer load current curve

As shown in Fig. 5.2, the wiring principle of a center-tapped Scott transformer is illustrated, and the daily load current curves for the teaser and main transformer of the center-drawn Scott I main transformer are shown in Fig. 5.3.

To ensure safety, high-speed railway train operation schedules generally set a 4-h comprehensive maintenance window at night, managed by dispatch to control power outages and restoration times. During this period, there is no load until morning trains start operating; however, when high-speed trains are running, the value of current can be quite large. From Fig. 5.3, it is evident that the current amplitude for the teaser transformer can reach up to 1050 A, and for the main transformer, up to 1240 A, causing certain overload conditions. However, the average load rate is quite low, only about 0.13 and 0.15, indicating that the traction load varies significantly and has a strong impulsiveness.

5.2.1.2 Imbalance of High-Speed Railway Traction Load

Traction load is a single-phase load, with the traction transformer converting three-phase electricity into single-phase for its use. Due to varying load sizes and densities on the two power supply arms, there are differences in the loads on the windings on both sides of the traction transformer. Figure 5.4 illustrates the daily load rate difference curve between the teaser and main transformer.

As shown in Fig. 5.4, there is a significant difference in feeder current between the teaser and main transformer of the transformer, and due to the transformation ratio, there is also a substantial difference in the currents between the primary side windings. The imbalance of high-speed railway traction load reflects this disparity in load currents on both sides' windings. Existing studies have shown that the imbalance of traction load can cause significant differences in the temperature rise across different windings, with insulation aging varying among them. The consequence of this imbalance is that one side of the traction transformer's windings always experiences a higher temperature rise and faster insulation aging. If this imbalance persists over the long term, it leads to uneven utilization of the traction transformer windings. The overall reliability condition of the traction transformer is primarily affected by the side of the windings that experience higher temperature rises and more severe insulation aging.

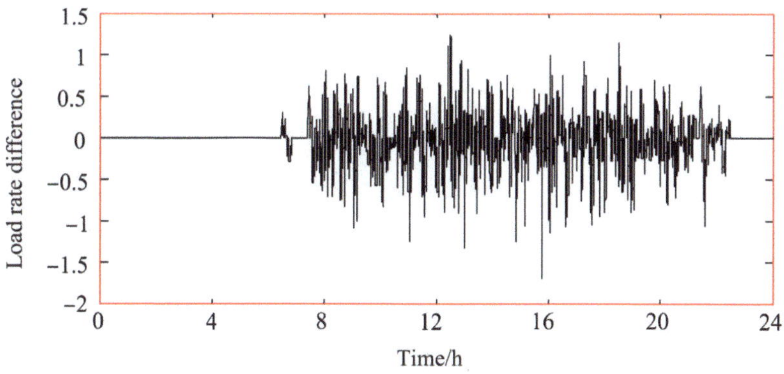

Fig. 5.4 Daily Load Rate Difference Curve Between teaser and main transformer

5.2.1.3 Non-linearity of High-Speed Railway Traction Load

The operation of high-speed trains not only introduces impulsiveness and imbalance but also exhibits non-linear characteristics. The total demand distortion (TDD) is the ratio of the root mean square value of the harmonic content to the rated or maximum demand fundamental component. Using actual daily load currents and harmonic current statistics from Fig. 5.3, the TDD results of harmonics on both sides of the traction transformer windings are shown in Fig. 5.5, indicating the impact of certain non-linear harmonics brought by the high-speed railway traction load.

Compared to conventional AC-DC electric locomotives, AC-DC-AC high-speed electric multiple unit improve the grid-side power factor and reduce the content of low-order harmonics, but they produce a broader harmonic output spectrum, generating high-order characteristic harmonics. These higher-order harmonic currents are transmitted through the traction network and flow into the traction transformer, where they cause skin and proximity effects in the transformer. These effects lead to additional losses in the windings, causing overheating or localized overheating.

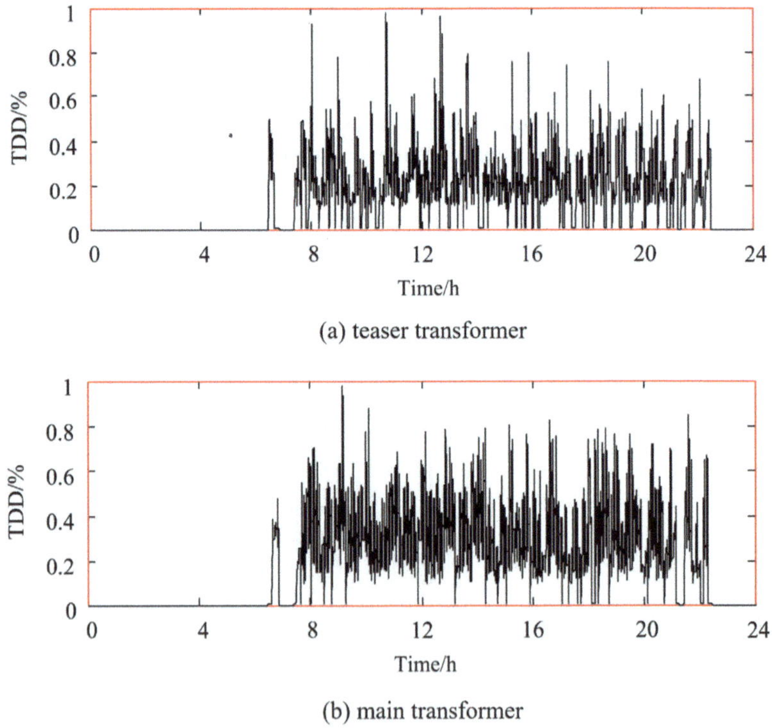

Fig. 5.5 Statistics of the TDD on both sides of the traction transformer windings. (**a**) Teaser transformer. (**b**) Main transformer

5.2.2 *Traction Transformer Reliability Assessment Considering Load Characteristics*

5.2.2.1 Non-linear Impact Quantification and Calculation

Without considering the impact of extreme weather and active tripping, internal losses and reduced insulation strength are the main causes of failure in oil-immersed transformers. This process accumulates over the service time and is irreversible.

To quantify the non-linear effects of the traction load, this section introduces the harmonic loss factor based on the IEEE C57.110 standard to calculate the transformer derating factor caused by non-linear harmonics. By adjusting the rated current values of the low-voltage side windings of the transformer in real-time using the derating factor, the harmonic impact is quantified in the calculations of transformer temperature rise, ultimately reflecting the impact of traction load non-linearity on transformer reliability.

The total loss P_{TL} of a traction transformer can be divided into no-load loss P_{NL} and load loss P_{LL}, where the additional loss caused by harmonic currents primarily affects the load loss P_{LL}. The load loss P_{LL} consists of DC resistance loss P_R, winding eddy current loss P_{EC}, and stray loss P_{OSL}, related as follows:

$$P_{LL} = P_R + P_{EC} + P_{OSL} \tag{5.15}$$

Under rated load conditions (with the rated current I_N as the current base and the DC resistance loss at the rated current $I_N^2 R$ R as the baseline power), the per unit value calculation of load loss is as follows:

$$P_{LL-N}^* = P_{R-N}^* + P_{EC-N}^* + P_{OSL-N}^* \tag{5.16}$$

In the actual operation of the transformer, the per-unit values $P_R^*, P_{EC}^*, P_{OSL}^*$ for DC resistance loss, eddy current loss, and stray loss are defined respectively:

$$P_R^* = \frac{I^2 R}{I_N^2 R} = I^{*2} \tag{5.17}$$

$$P_{EC}^* = \sum_{h=1}^{h_{max}} \left(\frac{I_h}{I_N}\right)^2 h^2 P_{EC-N}^* \tag{5.18}$$

$$P_{OSL}^* = \sum_{h=1}^{h_{max}} \left(\frac{I_h}{I_N}\right)^2 h^{0.8} P_{OSL-N}^* \tag{5.19}$$

where I_h represents the effective value of harmonic current; $I = \sqrt{\sum_{h=1}^{h_{max}} I_h^2}$, where I is the total effective current.

Through the above analysis, the per unit value of the total load loss under harmonic impact for the traction transformer can be obtained as follows:

$$P_{LL}^{*}=P_{R}^{*}+P_{EC}^{*}+P_{OSL}^{*}=I^{*2}+\sum\nolimits_{h=1}^{h_{max}}\left(\frac{I_h}{I_N}\right)^2h^2P_{EC-N}^{*}$$
$$+\sum\nolimits_{h=1}^{h_{max}}\left(\frac{I_h}{I_N}\right)^2h^{0.8}P_{OSL-N}^{*}=I^{*2}+\frac{I^{*2}\cdot\sum\nolimits_{h=1}^{h_{max}}I_h^2h^2}{I^2}P_{EC-N}^{*}$$
$$+\frac{I^{*2}\cdot\sum\nolimits_{h=1}^{h_{max}}I_h^2h^{0.8}}{I^2}P_{OSL-N}^{*}=I^{*2}\left(1+\frac{\sum\nolimits_{h=1}^{h_{max}}I_h^2h^2}{\sum\nolimits_{h=1}^{h_{max}}I_h^2}P_{EC-R}^{*}+\frac{\sum\nolimits_{h=1}^{h_{max}}I_h^2h^{0.8}}{\sum\nolimits_{h=1}^{h_{max}}I_h^2}P_{OSL-R}^{*}\right) \tag{5.20}$$

According to the IEEE C57.110 standard, the harmonic loss factors for winding eddy current loss and stray loss due to harmonics are defined as follows:

$$F_{HL}=\frac{\sum\limits_{h=1}^{h_{max}}I_h^2h^2}{\sum\limits_{h=1}^{h_{max}}I_h^2},\ F_{HL-OSL}=\frac{\sum\limits_{h=1}^{h_{max}}I_h^2h^{0.8}}{\sum\limits_{h=1}^{h_{max}}I_h^2} \tag{5.21}$$

where F_{HL} and $F_{HL\text{-}OSL}$ represent the impacts of harmonic current on winding eddy current loss and stray loss, respectively. By substituting equation (5.21) into Eq. (5.20), we can obtain:

$$P_{LL}^{*}=P_{R}^{*}+P_{BC}^{*}+P_{OSL}^{*}=I^{22}\left(1+F_{HL}P_{EC-R}^{*}+F_{HL-OSL}P_{OSL-R}^{*}\right) \tag{5.22}$$

Using the factors F_{HL} and F_{HL-OSL}, combined with actual operating voltages, currents, and the harmonic current spectrum recorded at each time point, the derating factor induced by harmonic currents on the traction transformer can be calculated. To keep the total load loss below the rated value P_{LL-N}^{*}, the per unit value of the maximum allowable current is determined as follows:

$$I_{max}^{*}=\sqrt{\frac{P_{LL-N}^{*}}{1+F_{HL}P_{EC-N}^{*}+F_{HL-OSL}P_{OSL-N}^{*}}} \tag{5.23}$$

Derating of the traction transformer:

$$R_{APR}=\frac{S_N-S}{S_N}\times 100\%=1-\frac{V_{rms}}{V_{rms-N}}\cdot I_{max}^{*} \tag{5.24}$$

where S_N and V_{rms-N} are the rated capacity and rated voltage of the traction transformer; S and V_{rms} are the operating capacity and effective voltage under harmonic conditions.

With the obtained derating factor value, the rated current of the low-voltage side windings of the transformer, $I_{h\cdot R}$, can be adjusted. As the harmonic current changes each moment, $I_{h\cdot R}$ will also change with the capacity reduction. The specific calculation equation is as follows:

$$I_{h\cdot R} = \frac{S_N \times (1 - R_{APR})}{V_{ms-N}} \tag{5.25}$$

5.2.2.2 Internal Temperature Rise Calculation of the Traction Transformer

Studies on center-tapped Scott transformers influenced by traction load characteristics have not yet been reported. This section models and analyzes it using the thermal circuit model and its differential equations to calculate the hotspot temperature rise of the transformer, characterizing both the traction load characteristics and the internal heat generation and transfer processes within the transformer.

The internal structure of a center-tapped Scott connection transformer only differs from a standard Scott connection in having a secondary side center tap, which has a negligible impact on the oil flow cooling circuit. Therefore, it can be assumed that its thermal circuit model is consistent with the Scott connection. The thermal capacities of the transformer's T and M winding positions, and the core and oil are denoted as $C_{th,\ M}$, $C_{th,\ T}$, $C_{th,\ fe}$, and $C_{th,\ oil}$. Since the heat generated by iron loss is relatively small and stable, $C_{th,\ fe}$ and the thermal resistance from the core to the oil can be neglected in calculations. Based on this, the thermal circuit model is shown in Fig. 5.6.

Based on the rules of the thermal circuit model, the internal transformer temperature rise equation is: derived as follows:

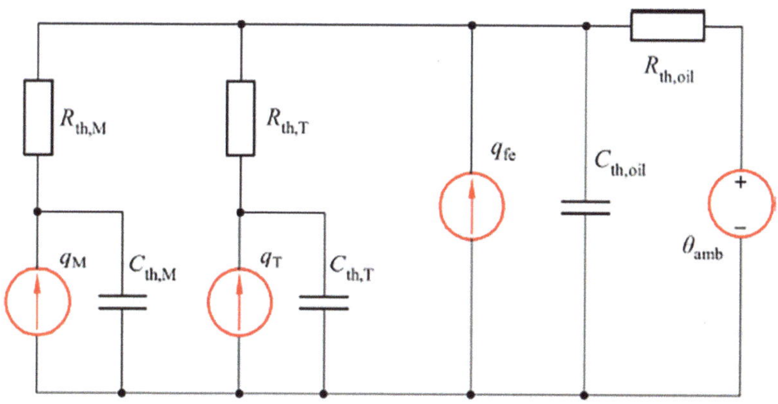

Fig. 5.6 The thermal circuit model of a center-tapped scott connection traction transformer

$$\begin{cases} q_M = \dfrac{(\theta_M - \theta_{oil})^{1/n}}{R_{th,M}} + C_{th,M}\dfrac{d\theta_M}{dt} \\ q_T = \dfrac{(\theta_T - \theta_{oil})^{1/n}}{R_{th,T}} + C_{th,T}\dfrac{d\theta_T}{dt} \\ q_{fe} = \dfrac{(\theta_{oil} - \theta_{cmb})^{1/n}}{R_{th,oil}} + C_{th,oil}\dfrac{d\theta_{oil}}{dt} - \left[\dfrac{(\theta_M - \theta_{oil})^{1/n}}{R_{th,M}} + \dfrac{(\theta_T - \theta_{oil})^{1/n}}{R_{th,T}}\right] \end{cases} \tag{5.26}$$

where q_M, q_T, and q_{fe} represent the heat generated by the copper losses of the M windings, T windings, and the core iron loss, respectively; θ_M, θ_T, θ_{oil}, and θ_{cmb} represent the hot-spot temperatures of the M and T windings, the top oil temperature, and the environmental temperature, respectively; $R_{th,\ M}$ and $R_{th,\ T}$ are the thermal resistances of the M and T windings relative to the oil, and $R_{th,\ oil}$ is the oil-to-air thermal resistance (including the transformer casing to air thermal resistance); n is a parameter characterizing the cooling method, which is 0.8 under natural cooling conditions.

By analyzing the relationship between the copper losses of the M windings or T windings under any load condition and those under rated conditions, and denoting the temperature difference between the transformer hotspot and top oil temperature under rated conditions as $\Delta\theta_{\text{kndR}}$, and the difference between the top oil temperature and environmental temperature as $\Delta\theta_{\text{oilR}}$, let $\gamma = \frac{q_{fe}}{q_M + q_T}$, $k = \frac{\gamma}{2(1+\gamma)}\left(\frac{\Delta\theta_{\text{oilR}}}{\Delta\theta_{\text{vndR}}}\right)^{1/n}$, the temperature rise calculation model is established as follows:

$$\begin{aligned} (\theta_M - \theta_{\text{oil}})^{1/n} + \Delta\theta_{vmRR}^{1/n-1} \cdot \tau_{vnd} \cdot \frac{d\theta_M}{dt} &= \frac{\left[pI_\alpha^2 + 3(1+K) \times I_\beta^2\right]\Delta\theta_{vrnR}^{1/n}}{(3K+4p)I_{hR}^2} \\ (\theta_T - \theta_{\text{oil}})^{1/n} + \Delta\theta_{wndR}^{1/n-1} \cdot \tau_{knd} \cdot \frac{d\theta_T}{dt} &= \frac{(2\sqrt{3}p + 3K)I_\alpha^2\Delta\theta_{winR}^{1/n}}{(3K+4p)I_{k,R}^2} \\ \frac{\Delta\theta_{\text{oilR}}^{1/n}}{1+\gamma} + \Delta\theta_{\text{oilR}}^{1/n-1} \cdot \tau_{\text{oil}} \cdot \frac{d\theta_{\text{oil}}}{dt} &= (\theta_{\text{oil}} - \theta_{cub})^{1/n} - k(\theta_M - \theta_{\text{oil}})^{1/n} - k(\theta_T - \theta_{\text{oil}})^{1/n} \end{aligned} \tag{5.27}$$

where τ_{wndR} and τ_{oilR} denotes the time constants for the windings and oil, respectively; I_α, I_β are the load current values of the α and β arms of the T line; taking parameters $p = 1.1$, $\gamma = 4.7$, $K = 2$.

It is important to note that in Eq. (5.27), the rated current $I_{l\downarrow R}$ of the low-voltage side windings of the transformer is adjusted in real-time using Eq. (5.25), quantifying the imbalance of the high-speed railway traction load, impulsiveness, and non-linearity into the transformer's temperature rise calculation. The resulting Eq. (5.27) calculates the hot-spot temperatures of the windings and the oil temperature of the traction transformer.

5.2.2.3 Failure Rate Calculation and Reliability Assessment

The aging process of transformers is commonly described using the Weibull distribution, which has been successfully applied to transformer fault diagnosis and life information analysis. The failure rate and cumulative probability distribution function are defined as follows:

$$\lambda_{Te\ ibull}(t) = \frac{\beta}{\eta}\left(\frac{t}{\eta}\right)^{\beta-1} \tag{5.28}$$

$$F_{\text{Weibull}}(t) = 1 - \exp\left(-\frac{t}{\eta}\right)^{p} \tag{5.29}$$

where β is the shape parameter, and η is the characteristic life parameter, which is the scale parameter in the Weibull distribution.

Experimental evidence shows that the aging of insulation paper follows the Arrhenius law, relating the transformer insulation lifetime T_j to the hot-spot temperature θ_H as follows:

$$T_j = C\exp\left(\frac{B}{\theta_H + 273}\right) \tag{5.30}$$

where B and C are empirical constants.

To reflect the impact of temperature on transformer reliability, the Arrhenius-Weibull model is established by setting $\eta = T_j$, substituting the equation (5.30) into Eqs. (5.28) and (5.29), to derive:

$$\lambda_{AW}(t\,|\,\theta_H) = \frac{\beta}{Ce^{\frac{B}{\theta_H+273}}}\left(\frac{t}{Ce^{\frac{B}{\theta_H+273}}}\right)^{\beta-1} \tag{5.31}$$

$$F_{AW}(t\,|\,\theta_H) = 1 - \exp\left(\frac{t}{Ce^{\frac{B}{\theta_H+273}}}\right)^{\beta} \tag{5.32}$$

Assuming that parameters β and C remain constant at different temperatures, these parameters can be estimated using the least squares method or maximum likelihood method.

Due to transformers potentially operating under various load conditions, their internal hotspot temperature rises vary with the load, and the temperature rises influenced by the traction load impact exhibit more stochastic fluctuations. Therefore, describing the service conditions of a transformer over a period or at equal time intervals poses certain challenges. Utilizing existing research results on the aging mechanism of thermally upgraded insulation paper, the life loss of the transformer under any temperature rise can be uniformly converted to a reference temperature, and the relative aging rate V is related to the hottest spot temperature of the windings θ_H as follows:

$$V = \mathrm{e}^{\frac{15000}{\theta_0+273} - \frac{15000}{\theta_H+273}} \tag{5.33}$$

For thermally upgraded insulation paper, the reference temperature θ_0 is taken as 110 °C.

For clarity of the model, let's divide the service time T into n intervals t_1, t_2, L, t_n, choosing appropriate interval spacing such that the hot-spot temperature of the windings, $\theta_H(t_i)$, remains constant for each interval t_1. Based on Eq. (5.30), the cumulative calculation of insulation lifetime loss within the interval t_i yields the total insulation life loss over the service time T:

$$T_e = \sum_{i=1}^{n} t_i V(t_i) = \sum_{i=1}^{n} t_i \mathrm{e}^{\frac{15000}{\theta_0+273} - \frac{15000}{\theta_H(t_i)+273}} \tag{5.34}$$

where $V(t_i)$ is the relative aging rate within the interval t_i; T_e represents the equivalent operating time of the transformer at the reference temperature θ_0 over service time T, with subsequent time intervals Δt calculated similarly and denoted as Δt_e.

By the definition of conditional probability, the probability of aging failure for the traction transformer after having effectively operated for time T and within the subsequent time interval Δt is:

$$\begin{aligned} P_{AW} &= P(T \leqslant t \leqslant T + \Delta t \mid t > T) = P(T_e \leqslant t \leqslant T_e + \Delta t_e \mid t > T_e, \theta_0) \\ &= \frac{F_{AT}(T_e + \Delta t_e \mid \theta_0) - F_{AW}(T_e \mid \theta_0)}{1 - F_{AW}(T_e \mid \theta_0)} \end{aligned} \tag{5.35}$$

By substituting equation (5.32) into Eq. (5.35), we obtain the failure probability due to transformer aging:

$$P_{AT} = 1 - \exp\left[\left(\frac{T_\mathrm{e}}{\mathrm{Ce}^{\left(\frac{B}{\theta_0+273}\right)}}\right)^{\beta} - \left(\frac{T_\mathrm{e} + \Delta t_\mathrm{e}}{\mathrm{Ce}^{\left(\frac{B}{\theta_0+273}\right)}}\right)^{\beta}\right] \tag{5.36}$$

The principle process flow for failure probability calculation and reliability assessment is illustrated in Fig. 5.7.

By quantifying the losses caused by load non-linearity harmonics through the harmonic loss factor into the derating factor, and incorporating the high-speed railway load characteristics into the internal temperature rise calculation of the traction transformer, the impact of insulation aging caused by internal temperature rise on the reliability of the traction transformer during service is considered. This leads to the calculation and assessment analysis of reliability indicators.

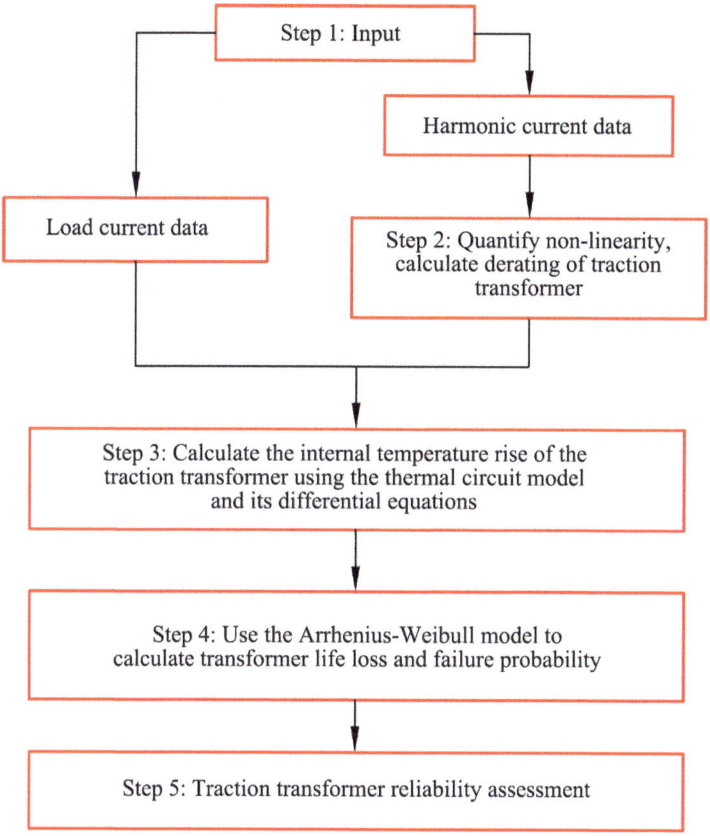

Fig. 5.7 Principle flow diagram of traction transformer reliability assessment method

5.2.3 *Case Study of Traction Transformer Reliability Assessment*

5.2.3.1 Case Study Introduction

From the nameplate of the center-tapped Scott connection transformer at the Shangyu North traction substation on the Hangzhou-Ningbo high-speed line, we gather the following data: the rated capacity of the traction transformer is (31.5 + 31.5)/(20 + 20)/(20 + 20) MVA, rated voltage is 220/2 × 27.5.2 × 27.5 kV, no-load losses for the teaser and main transformer are respectively 20.6 kW and 20.1 kW, and load losses are 83.8 kW and 138.2 kW, with natural cooling method employed. Meteorological data indicates that the average annual environmental temperature in the area is 16.7 °C, which, combined with actual measured traction load data, is used for simulation analysis.

Table 5.1 Traction transformer reliability model parameters

$\Delta\theta_{oilR}$	$\Delta\theta_{wndR}$	τ_{oilR}	τ_{wndR}	n	β	B	C
55 °C	23 °C	3 h	3 min	0.8	5.9	15,000	1.903×10^{-12}

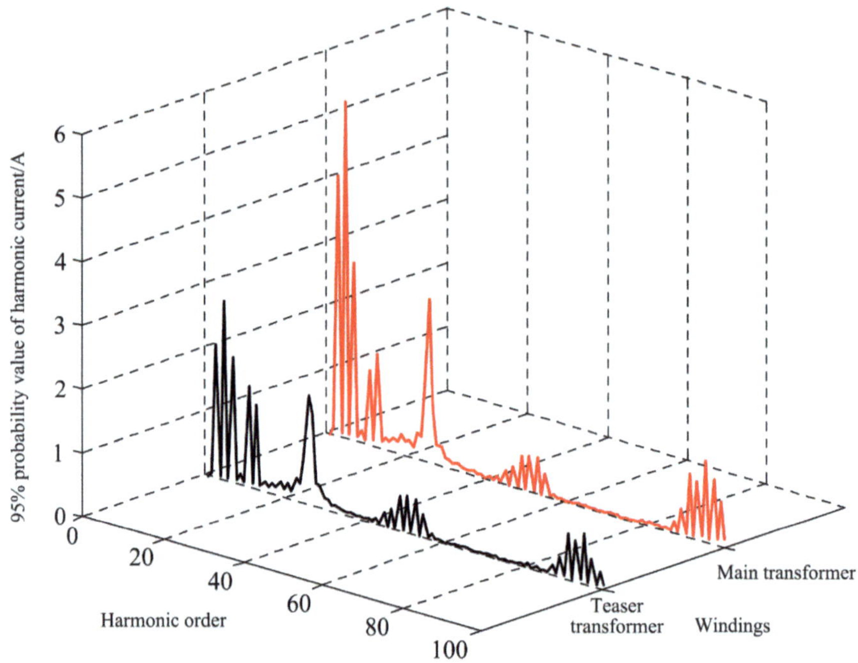

Fig. 5.8 Daily 95% probability value spectrum chart for harmonics on T and M sides

Traction transformer reliability model parameters considering load characteristics are shown in Table 5.1.

Measured load data, analyzed through Fourier analysis, provides harmonic current data. Figure 5.8 shows the 95% probability values of the 2-100th order harmonics throughout the day for both teaser and main transformer.

5.2.3.2 Calculation of Failure Rate of Traction Transformer Considering Load Characteristics

Using harmonic data, the derating of the traction transformer is calculated for different times of the day, and the quantified results of the impact on the power quality index are shown in Figs. 5.9 and 5.10. Taking the teaser transformer as a case study, the 95% probability maximum derating factor is 26.67%, with an average of 5.33% throughout the day.

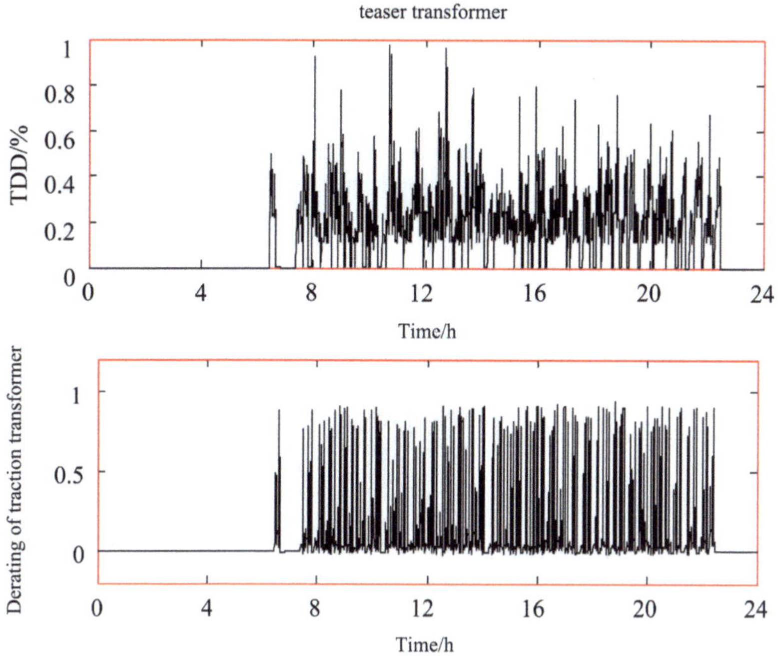

Fig. 5.9 Quantified results of the impact of the power quality index on the traction transformer teaser transformer

The results of transformer derating due to harmonics are integrated into the calculations for traction transformer temperature rise under harmonic impact. The resulting curves of windings hot-spot and oil temperature variations are derived, and life loss is calculated based on the reliability model, as shown in Fig. 5.11.

Considering the impacts of traction load imbalance and impulsiveness on the transformer, based on the hotspot temperature rise calculation results, a case study day after 20 years of equivalent operation is taken to calculate the short-term failure probability of the traction transformer both with and without the impact of non-linear harmonics. Predictions assess the transformer's reliability every $\Delta t = 1\text{h}$ interval within that day. Figures 5.12 and 5.13 respectively show the 24-h failure probability of the teaser and main transformer of the traction transformer, without and with considering harmonic impact.

5.2.3.3 Traction Transformer Reliability Assessment Analysis Considering Load Characteristics

On the foundation of considering the imbalance and impulsiveness of traction load, the load rate, internal temperature rise, and failure probability of windings before and after considering non-linear effects on the transformer are analyzed. Results are shown in Table 5.2.

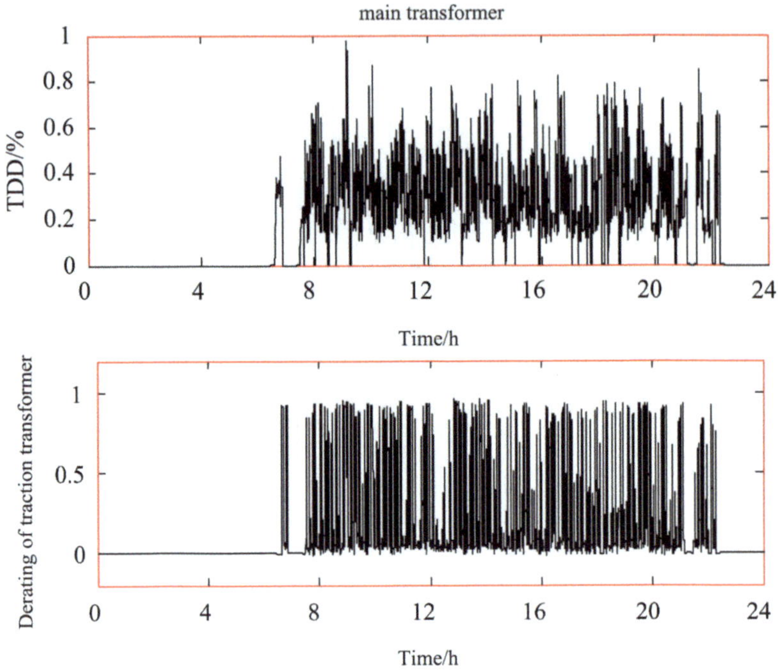

Fig. 5.10 Quantified results of the impact of the power quality index on the traction transformer main transformer

From the analysis and statistical results, the following conclusions can be drawn.

(1) The imbalance of traction load leads to uneven current distribution among the windings, resulting in significant differences in hotspot temperature rise between the teaser and main transformer of the transformer. As indicated in Table 5.2, the main transformer's hotspot temperature can exceed the teaser transformer by about 20 °C at maximum and about 3 °C on average per day. The side with the higher hotspot temperature also has a higher failure probability, with the main transformer reaching a maximum failure probability of 0.0045. The overall failure probability of the transformer is equivalent to that of the side with the higher rate. If the windings on the side with higher failure rates continuously carry heavier loads, this will accelerate aging and lead to a decrease in the utilization efficiency of the traction transformer.

(2) Unlike conventional electric power loads, the frequent impulsive nature of traction loads causes severe fluctuations in transformer windings' hotspot and oil temperatures, with temperature overshoots occurring during peak loads. As shown in Fig. 5.11, around the 16th hour, the temperature of the main transformer windings overshoots due to a significant overload, exceeding 120 °C, causing a steep increase in insulation life loss. Figures 5.12 and 5.13 show that the failure probability during this hour spikes dramatically, with the maximum

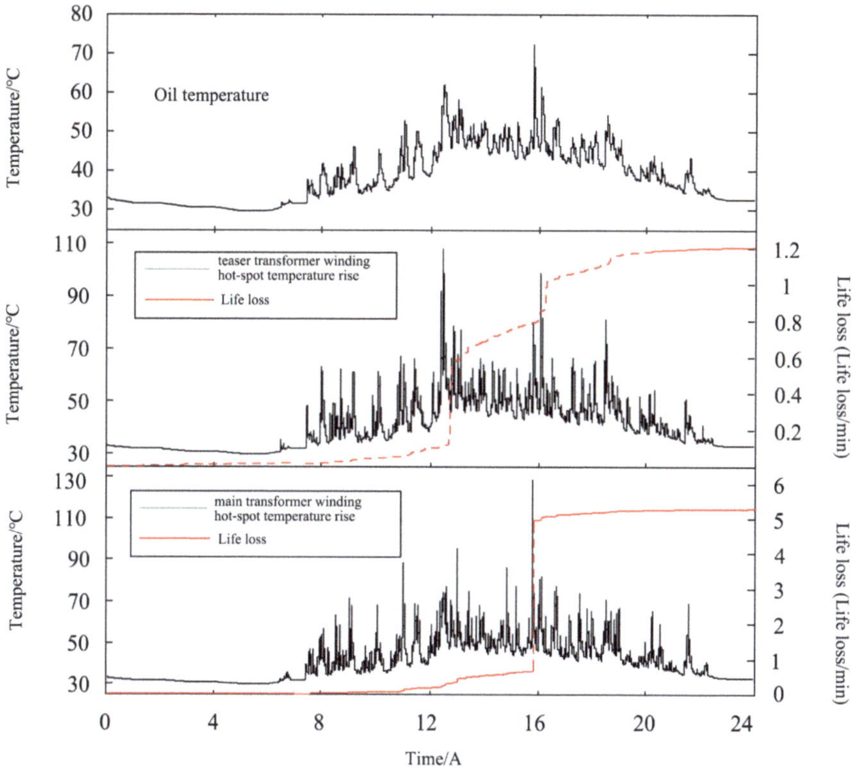

Fig. 5.11 Temperature rise and life loss curve considering harmonic impact

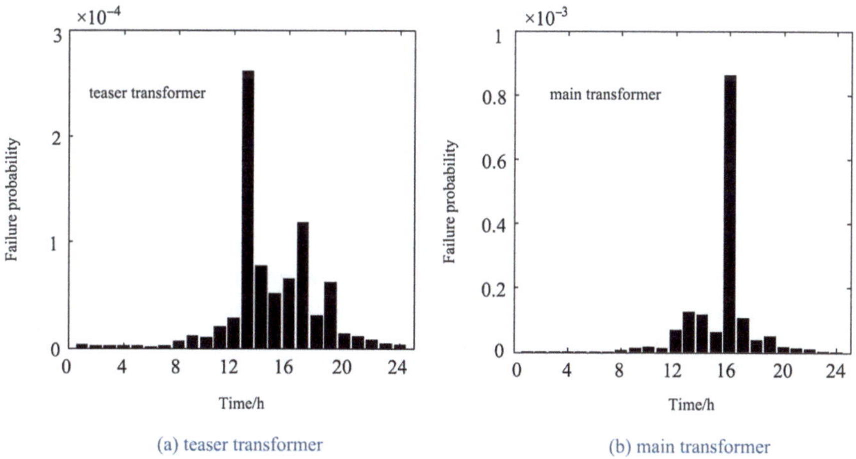

Fig. 5.12 Failure probability of traction transformer T and M windings without considering harmonic impact. (**a**) Teaser transformer. (**b**) Main transformer

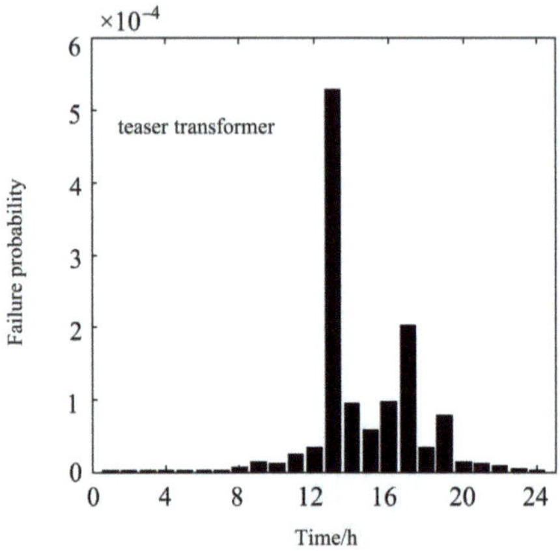

(a) teaser transformer

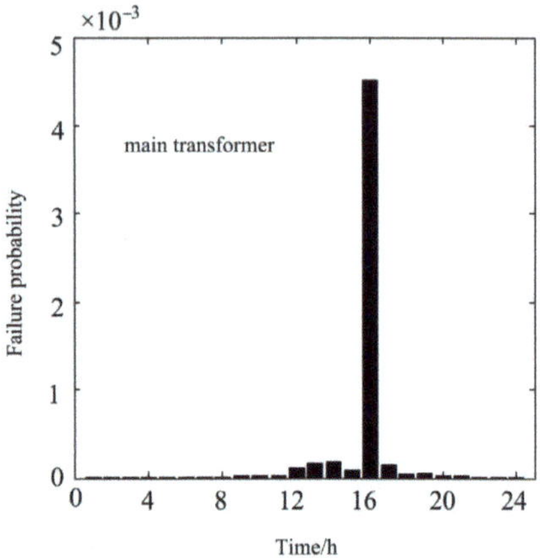

(b) main transformer

Fig. 5.13 Failure probability of traction transformer T and M windings considering harmonic impact

failure probability of the main transformer being about 20 times higher than the daily average, indicating a substantial change.

(3) The non-linear harmonics generated by high-speed railways operations introduce additional losses to the transformer. By calculating the derating factor to adjust the rated current values needed for calculating temperature rise, changes in the internal hotspot temperature of the traction transformer are reflected, influencing its failure probability. Comparisons between Figs. 5.12 and 5.13 show that considering the impact of non-linear harmonics, the maximum failure probability of the traction transformer can increase by an order of magnitude compared to when it is not considered. Both the 95% probability values and the average values are higher than those without considering harmonic impact, indicating that the effects of non-linearity on transformer reliability cannot be ignored.

(4) Although the hot-spot temperatures of the traction transformer windings and oil temperature fluctuations are substantial, their daily average temperature and cumulative life loss are generally less than those of conventional power transformers. The situations with sharp temperature rises and higher failure probabilities primarily occur during overload and when the effects of non-linear harmonics are pronounced. It is evident that the factors influencing the reliability of the traction transformer that can be controlled by humans mainly involve the traction load current and harmonic current.

In conclusion, it is recommended to balance the load on the various windings of the traction transformer by reasonably adjusting the train operation schedule and installing harmonic suppression equipment to reduce harmonic current. This will lower the operational risks of the traction transformer and enhance its reliability.

Table 5.2 Load Rate, internal temperature rise, and failure rates of both sides of windings before and after considering non-linear effects on transformer

		Teaser transformer load rate/pu	Main transformer load rate/pu	Teaser transformer hot-spot temperature of windings/°C	Main transformer hot-spot temperature of windings/°C	Oil temperature/°C	Teaser transformer windings failure probability	Main transformer windings failure probability
Without harmonic impact	Maximum value	1.4445	1.7151	97.0170	109.8478	65.1837	2.6197×10^{-4}	8.6063×10^{-4}
	95% probability maximum value	0.5875	0.6110	60.6703	63.2004	52.3833	1.1794×10^{-4}	1.2527×10^{-4}
	Average value	0.1311	0.1472	45.3818	46.5085	42.4732	3.316×10^{-5}	6.4663×10^{-5}
With harmonic impact	Maximum value	1.6259	2.0094	107.8644	128.5111	72.1637	5.2768×10^{-4}	4.5000×10^{-3}
	95% probability maximum value	0.6056	0.6362	61.6546	64.9426	53.1235	2.0165×10^{-4}	1.8566×10^{-4}
	Average value	0.1385	0.1649	45.6226	46.9866	42.7212	5.1626×10^{-5}	2.2692×10^{-4}

5.3 Reliability Assessment of High-Speed Railway Traction Power Supply System with Relay Protection

Once a fault occurs in the traction power supply system, it may lead to delays or even cancellations of high-speed trains, directly affecting the entire train operation schedule and causing immeasurable losses. Therefore, whether the traction power supply system can operate safely and reliably is directly related to the safety and stability of high-speed railway operations. Based on the characteristics of the high-speed railway traction power supply system, and through on-site practical observations and literature research, the following difficulties and issues still exist in the system-level reliability assessment:

(1) Within the traction power supply system, the structural and functional connections between various equipment, especially the coupling relationships and coordination logic between primary equipment and relay protection equipment, significantly increase system comploutagey, making the assessment of system reliability more challenging.
(2) The outcomes representing the reliability of the traction power supply system are not merely binary (success or failure); they relate to the performance states of various equipment and require the quantification of reliability indicators.
(3) The operation and maintenance processes of the traction power supply system generate some equipment reliability statistical data. However, the sample size is small, and the issue of assessing system reliability based on small sample data urgently needs to be addressed.

Given these reasons, there is limited overall research on the reliability assessment of the traction power supply system, and comprehensive studies considering the overall reliability assessment of the traction power supply system with relay protection are rare. To fill this gap, this section, driven by existing statistical data on the reliability of equipment and relay protection within the traction power supply system, proposes a method for considering the reliability assessment of the high-speed railway traction power supply system with relay protection. Initially, a qualitative analysis of the impact of relay protection failure on the reliability of the traction power supply system is conducted. Then, based on the actual system composition (including primary equipment of the traction substation, OCS equipment, and appropriately configured relay protection), a typical traction power supply system failure mode and effect analysis is performed to organize and establish an FMEA table. According to the FMEA, integrated primary and secondary equipment traction substation and OCS fault tree models are established, and Monte Carlo simulations are used to solve the system reliability indicator, achieving a quantification and assessment of the overall reliability of the high-speed railway traction power supply system.

5.3.1 *Analysis of the Impact of Relay Protection Failure on Traction Power Supply System Reliability*

In the high-speed railway traction power supply system, relay protection is primarily divided into transformer protection and feeder protection. Transformer protection is mainly responsible for protecting the transformers within the traction substation and consists of electrical protection (differential current protection, low voltage startup overcurrent protection for high/low side, overload protection) and non-electrical protection (temperature protection, oil level abnormality protection, gas protection, pressure relief protection). Feeder protection functions to protect the feeder cables or overhead lines from the traction substation to the OCS, and it also acts on OCS line protection, mainly comprising distance protection (impedance protection), low voltage start-up instantaneous current break protection, overcurrent protection, and incremental current protection. During the long-term operation of high-speed railways, faults in the traction power supply system are inevitable. Each protection method, through the setting of trip values and logical coordination, ensures that faults or abnormal conditions in the traction power supply system are quickly isolated or promptly signaled to minimize the extent of the fault, reduce fault losses, and ensure safe system operation.

Due to the inherent design life of protection equipment, and the potential for issues arising during component manufacturing processes, configuration settings, and maintenance, relay protection equipment also have the possibility of malfunctioning, inevitably leading to incorrect operations. Conventional protections may fail to act correctly under severe conditions, leading to serious incidents such as the burning or breaking of the traction network, destruction of power supply equipment, and prolonged power interruptions to high-speed trains. The qualitative analysis of the impact of relay protection on the reliability of the traction power supply system is divided into the following two categories:

(1) Impact of Protection Misoperation: In March 2009, at a substation during the initial no-load energization of a traction transformer, the 220 kV side low voltage startup overcurrent protection of 1 T and 2 T operated and caused tripping multiple times, affecting normal power transmission for more than 2 h. In April of the same year, while the #1 feeder was supplying power and 1 T was operating normally, a short circuit fault between the T line and F line on feeder 211# occurred. Before the feeder protection acted, the phase B differential protection of the traction transformer tripped, causing an instantaneous power outage across the substation. The line auto-reclosing equipment then switched the operation to the #2 feeder and 2 T. Upon on-site investigation, it was determined that the causes of tripping were a misoperation of the 220 kV side low voltage lockout overcurrent protection and an operational misoperation of the traction transformer differential protection during operation. It is evident that relay protection misoperation can lead to erroneous alarm signals or tripping, thus interrupting the normal power supply of the system.

(2) Impact of Protection Failure to Operate: When a fault occurs in the traction power supply system that necessitates isolation, and the main relay protection fails to operate while the backup protection activates, the duration of the fault extends compared to when the main protection operates correctly. This extended period of the fault increases the exposure of transformers, busbars, and switchgear insulation to the short-circuit impact, thereby causing wear to the traction power supply equipment and increasing the risk of faults. Additionally, the short-term surge currents can cause insulation damage to the traction power supply equipment, leading to a complete failure of the entire traction power supply equipment. Severe protection failure to operate conditions can lead to expanded power outage areas and occurrences of out-of-step and cross-regional power supply issues. For instance, after the maintenance of feeder protection at a traction substation, poor contact at the current transformer secondary wiring terminals of the feeder circuit breaker led to a ground fault on the corresponding line where the feeder circuit breaker protection failed to operate. The fault escalated to secondary tripping of the main transformer, expanding the power outage area and leading to erroneous analysis and misguided remedial actions, thus prolonging the accident handling time.

Therefore, when assessing the reliability of the traction power supply system, it is necessary to consider the impact of relay protection. To ensure the traction power supply system can supply power reliably and safely, on one hand, relay protection must meet the operational requirements of electric locomotives under normal system conditions; on the other hand, during system faults and abnormal operating conditions, relay protection should quickly and selectively isolate faults to ensure the safe operation of the power supply system and equipment.

5.3.2 Traction Power Supply System Reliability Assessment Model and Indicators

Taking a certain traction substation and its OCS within the Beijing-Shanghai High-Speed Railway as a case study, a traction power supply system reliability assessment model is established. Figure 5.14 shows the main electrical wiring diagram of the traction power supply system for this section of the Beijing-Shanghai High-Speed Railway, which includes the traction substation and the OCS sections. To enhance the continuous power supply capacity, the traction power supply system includes backup feed lines and transformers. In the diagram, two sets of 'Incoming line-line-main transformer' operate in a primary and backup configuration. For ease of subsequent fault tree model development, the traction power supply system is divided into zones (taking the Zhujianong downward OCS power supply as a case study, with other power supply directions similarly categorized): Zone 1 is the area from the Incoming line to the traction transformer, split into Zones 1.1 and 1.2, represented by solid red and dashed lines respectively; Zone 2 is from the high

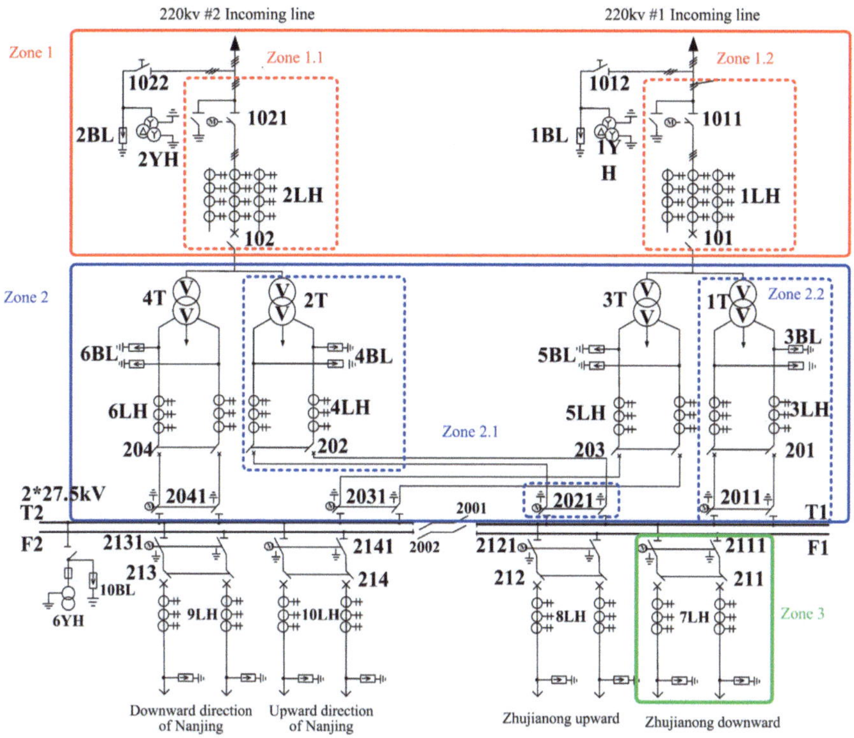

Fig. 5.14 Main electrical wiring diagram of a Beijing-Shanghai high-speed railway traction power supply system

voltage side of the transformer to the 27.5 kV busbar, divided into Zones 2.1 and 2.2, shown by solid blue and dashed lines respectively; the green solid line represents Zone 3, which includes the section from the 27.5 kV busbar to the OCS.

5.3.2.1 Traction Power Supply System FMEA

By surveying the actual site conditions and cases, combined with the structure, components, and operational mode of the traction power supply system, the FMEA method is used to qualitatively analyze the failure modes and consequences of each equipment and its protection within the traction power supply system. This analysis helps to organize the coupling relationships between relay protection and each primary equipment, facilitating the establishment of a system fault tree model.

The FMEA tables for the primary equipment and relay protection of the traction power supply system are shown in Tables 5.3 and 5.4 respectively. Since the failure modes for relay protection are all related to hardware and software failures, these are omitted in the FMEA table, and only the fault consequences are listed.

Table 5.3 FMEA for primary equipment of the traction power supply system

Traction substation			OCS		
Component	Failure mode	Consequence impact	Component	Failure mode	Consequence impact
220 kV #1 or #2 Incoming line	Wire break, power outage	Disconnect the corresponding transformer, another main transformer takes over all loads, reducing system power supply reliability	Contact wire	Break, excessive wear, electrical burn	OCS power outage and service interruption
Main transformer 1 T, 2 T, 3 T, 4 T	Shell electric shock breakdown, inter-turn short circuit, fire, flooding, etc.	Faulty transformer is disconnected, another main transformer takes over all loads, reducing system power supply reliability	OCS wire	Break, non-standard sag	Non-standard suspension, pantograph-OCS coupling faults, pantograph lifting, pantograph insertion
Circuit breaker 101, 102; isolator switch 1011, 1021	Electromechanical failure, equipment outage	Incoming line power outage, disconnects the faulty side transformer, the other side transformer takes over all traction load, reducing system power supply reliability	Dropper	Wire break electrical burn fatigue limit exceeded	Non-standard suspension, poor pantograph-OCS coupling quality
Circuit breaker 201, isolator switch 2011	Electromechanical failure, equipment outage	If powered by #1, Zhujiannong upward and downward OCS lose power, reducing system power supply reliability	Line clamp	Loosening, electrical burn	Contact wire positioning not meeting standard requirements
Circuit breaker 202, isolator switch 2021	Electromechanical failure, equipment outage	If powered by #2, Zhujiannong upward and downward OCS lose power, reducing system power supply reliability	Positioner, positioner bracket	Excessive bending, brittle fracture	Positioning equipment failure
Circuit breaker 203, isolator switch 2031	Electromechanical failure, equipment outage	If powered by #1, both upward and Downward direction of Nanjing descent lose power, reducing system power supply reliability	Positioning hook, positioning pin casing	Electrical burn, fracture	Positioning equipment failure
				Wear, fracture	Support equipment failure

Circuit breaker 204, isolator switch 2041	Electromechanical failure, equipment outage	If powered by #2, both upward and Downward direction of Nanjing descent lose power, reducing system power supply reliability	Diagonal arm and horizontal arm		
Circuit breaker 211, isolator switch 2111	Electromechanical failure, equipment outage	Downward OCS at Zhujiannong loses power	OCS wire seat, positioning ring, double-ear casing	Corrosion exceeding limits, fatigue cracking	Supporting equipment connection failure
Circuit breaker 212, isolator switch 2121	Electromechanical failure, equipment outage	Upward OCS at Zhujiannong loses power	Pole	Tilting, collapsing	Network collapse, disruption of operations
Circuit breaker 213, isolator switch 2131	Electromechanical failure, equipment outage	Downward direction of Nanjing descent loses power	Foundation	Sinking, not secure	Pole tilting, collapsing
Circuit breaker 214, isolator switch 2141	Electromechanical failure, equipment outage	Upward direction of Nanjing descent loses power	Insulator	Damage, flashover, breakdown	Reduced insulation level, short circuit fault

Table 5.4 Traction power supply system relay protection FMEA

Protection type		Consequences of failure to operate	Consequences of protection misoperation
Transformer protection	Differential current protection	Internal faults in the transformer cannot be isolated, leading to system stability disruption; the faults are cleared by subsequent backup protection activation	Isolating the transformer, causing load and line power supply interruption, leading to system stability disruption
	Low voltage startup overcurrent protection for high/low side	In case of an internal short circuit or external phase-to-phase short circuit in the transformer, failure to isolate the fault allows the transformer to carry significant fault current, expanding the accident area	(1) Overcurrent Protection Misoperation during Transformer Normal Operation: Causes unnecessary transformer cut-off, leading to system power supply interruption (2) Transformer Short Circuit: If the main protection operates due to an internal short circuit within the transformer and there is an overcurrent protection misoperation, it can expand the range of the power outage, affecting the reliable operation of the system
	Overload protection	Running the transformer under overload conditions can cause overheating, increased wear, and reduced lifespan; in severe cases, it may lead to transformer burnout	Incorrect activation of overload protection can lead to the transformer being disconnected from the system, causing power supply interruptions and damaging system stability
	Temperature protection	If the transformer's upper oil or winding temperatures exceed safe levels and no alarm is triggered, this can lead to an escalation of the fault.	Sending false alarm signals can lead to the unnecessary disconnection of the transformer, causing power supply interruptions
	Oil level abnormality protection	If the transformer experiences oil leaks or the oil level falls below or rises above set values without triggering an alarm, it can lead to transformer damage	
Transformer protection	Gas protection	Fails to detect various short circuit faults within the transformer and adverse phenomena like dropping oil levels, potentially leading to overheating and destruction	Gas protection misoperation can send incorrect signals misleading operators, and in severe cases, it can result in the transformer being unnecessarily disconnected

(continued)

Table 5.4 (continued)

Protection type		Consequences of failure to operate	Consequences of protection misoperation
		of the transformer; backup protection can still operate to isolate the fault	from the system, causing power supply interruptions and significant losses
	Pressure relief protection	If abnormal pressure inside the oil tank is not released, it can, in severe cases, lead to transformer explosions; backup protection can still operate to isolate the fault	If the pressure relief valve opens and triggers the pressure relief protection contacts to close, it can incorrectly trigger the disconnection of transformer switches on all sides, leading to an erroneous disconnection of the transformer from the system and causing power supply interruptions
Feeder protection	Distance protection (impedance protection)	(1) Action of busbar differential protection to isolate the faulted line, causing step-up tripping of the main transformer's backup protection (2) When a OCS short circuit occurs, backup protection is activated	Power loss on a unidirectional single track may lead to train service suspension
	Low voltage start-up instantaneous current break protection	(1) In the case of severe faults, the fault range further expands; (2) When there is a short circuit fault and a significant transitional resistance at the short circuit point, incremental current protection is activated; (3) This leads to the tripping of the main transformer's backup protection out of sequence; (4) Backup protection is activated for OCS short circuit faults	Normal operation with misoperation of instantaneous overcurrent protection may cause system power supply interruption
	Overcurrent protection Incremental current protection	(1) This leads to the tripping of the main transformer's backup protection out of sequence; (2) For OCS short circuit faults, there may be electrical burns in OCS components	(1) This causes power supply interruption in the system, leading to train service suspension; (2) In the event of a short circuit in the system and normal operation of the feeder's main protection, misoperation expands the scope of the power outage

5.3.2.2 Integrated Primary and Secondary Equipment FTA Model for Traction Power Supply Systems

The traction power supply system comprises both the traction substation and the OCS. Due to differences in the makeup of equipment and components, service environment characteristics, and other factors, maintenance and repair work at the site is divided between two separate departments. Therefore, this chapter will establish the FTA model for the traction power supply system and subsequent reliability assessment separately for the traction substation and the OCS. This division allows for more specialized focus and tailored strategies in handling the unique challenges each subsystem faces.

Traction Substation FTA Model

The traction substation is the heart of the high-speed railway traction power supply system. Its primary function is to convert the three-phase high voltage electricity delivered from the power system into electrical energy suitable for use by the train sets and to transmit it to the OCS. To perform an FTA modeling of the traction substation, an appropriate top event must be selected. Given that the traction substation needs to deliver the transformed electrical energy to the upward and downward OCS in two directional destinations, under the premise of consistent functionality and main wiring structure, this chapter selects "traction substation unable to normally supply power to the downward OCS at Zhujiannong" as the top event for analysis. This choice reflects actual on-site conditions and differentiates from the generic expression of "traction substation fault" used previously.

To ensure the high reliability of power supply from the traction substation and meet the requirements for continuous power supply, the substation uses mutually backup No. 1 and No. 2 feed lines, as well as main transformers 1 T and 2 T as backups for each other (this backup scenario is specifically for the top event "traction substation unable to normally supply power to the downward OCS at Zhujiannong"). Therefore, to prevent a simple equipment failure from leading to a OCS power supply interruption in any direction, the traction substation FTA is modeled by zoning as shown in Fig. 5.14. This approach models regional faults as intermediate events, considering the actual backup structure of the traction facility, and effectively connects to the basic events.

The traction substation FTA model established is shown in Fig. 5.15. The protection of the traction power supply system primarily involves transformer protection and feeder protection. Therefore, the basic events of transformer failures and feeder power failures consider the effects of relay protection. Specifically, the transformer outage is divided into three scenarios:

(1) Transformer operates normally, but due to protection misoperation, it is forced to outage;

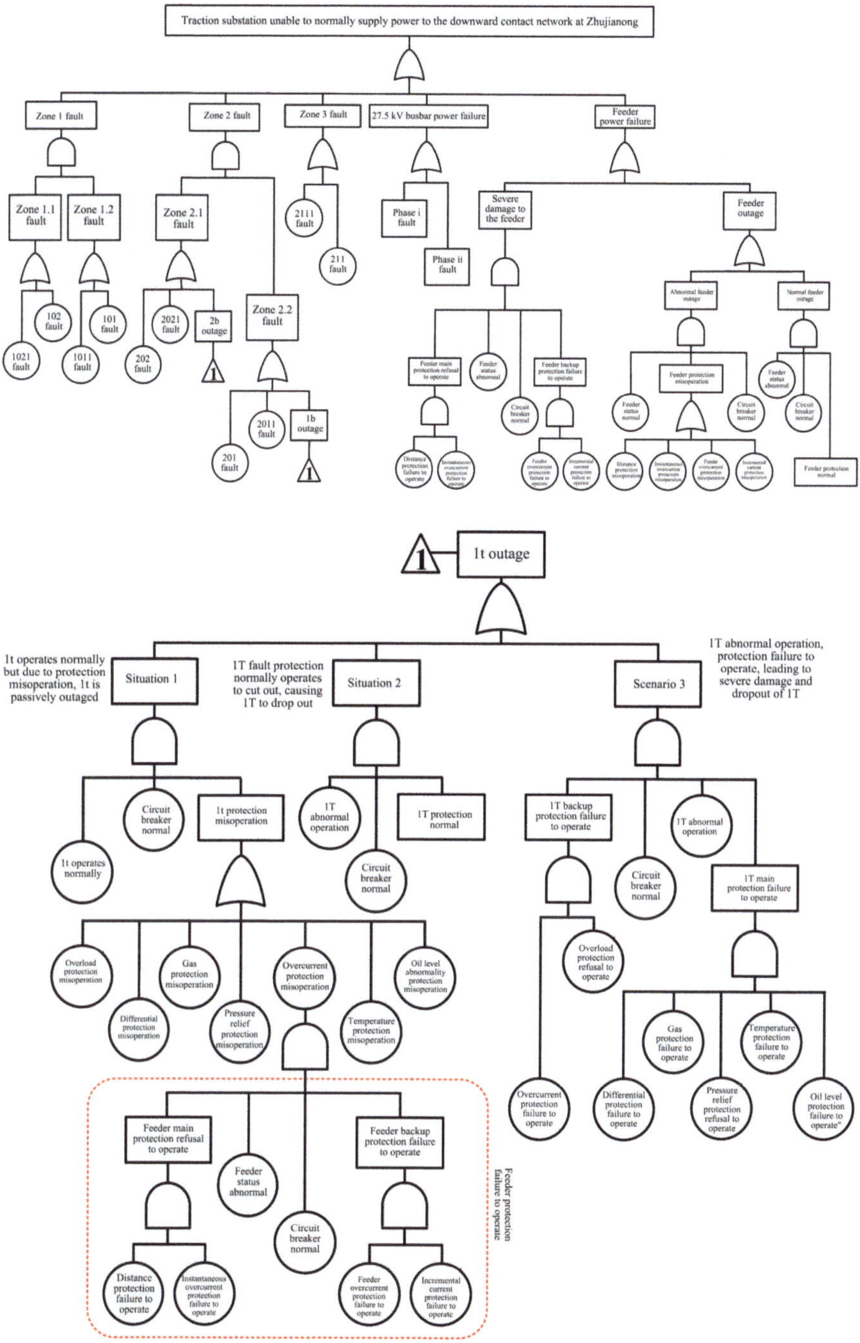

Fig. 5.15 Traction substation FTA model

(2) Transformer fault occurs, protection operates correctly leading to the transformer's outage;
(3) Transformer fault occurs, protection fails to operate, resulting in severe damage and outage of the transformer.

Scenario 1 considers the impact caused by the misoperation of the transformer protection and the situation where the feeder is abnormal and the feeder protection fails to operate, triggering the transformer protection. This reflects the logical coupling between protection, primary equipment, and different protections. Scenario 3 highlights the consequences of transformer protection refusing to operate. The refusal and misoperation of feeder protection are considered in the contexts of severe damage to the feeder and feeder outage, respectively.

FTA Model of OCS

The FTA model of the OCS is presented in Fig. 5.16. The OCS, which connects the traction substation and the electric multiple unit, consists of a series of mechanical equipment. By conducting an FMEA on the OCS, its structural components were examined, with the OCS failure designated as the top event. The fault tree model for the OCS illustrates that, due to the operational requirements and conditions of the OCS, its components do not have backups, resulting in a series of connections in the fault tree. This means any failure at the layer just below the top event can trigger the top event, indicating that the failure of one or more components can significantly increase the failure risk for the entire OCS. Additionally, a relay protection failure to operate usually results in a failure to correctly isolate a fault in the OCS, leading to problems like electrical burn or breakdown of the OCS components; whereas protection misoperation can cause abnormal OCS power supply, leading to tripping or power loss. Consequently, a deeper analysis into the causes of failures was conducted in aspects like contact wire suspension, electrical connection, and tripping/power loss events, considering the impacts of relay protection failure to operate and misoperation. For example, when the OCS is in an abnormal state, a feeder protection failure to operate could lead to a wire break in the contact wire suspension due to burning, thereby causing the contact wire to fail and trigger the failure of the entire OCS.

Reliability Assessment Indicators for the Traction Power Supply System

To comprehensively assess the reliability of the traction power supply system, four types of evaluation metrics are employed, including the conventional time metrics (MTBF, MTTR), frequency metrics (failure rate), and probability metrics (availability, unavailability). Additionally, two types of importance metrics are defined to characterize the vulnerability and criticality of equipment or components, particularly the impact of relay protection on the system.

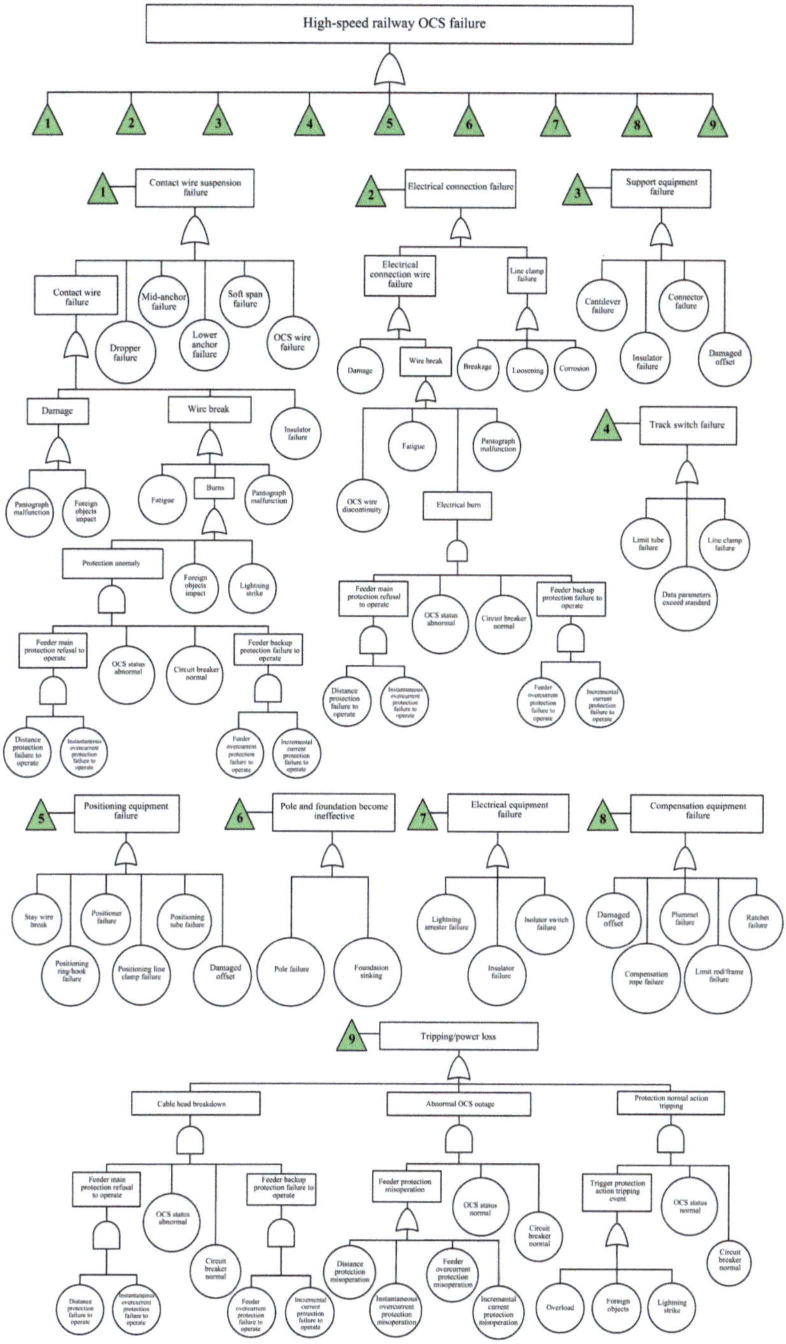

Fig. 5.16 FTA model of OCS

(1) Time Metrics

(a) MTBF

MTBF is the mathematical expectation of the operational time without failure for a system or equipment, represented as:

$$\mathrm{MTBF} = \int_0^{\infty} f(t) \cdot t\mathrm{d}t \tag{5.37}$$

where, $f(t)$ denotes the probability of the system or equipment experiencing its first failure in the interval $(t, t + \Delta t]$, also known as the failure density.

(b) MTTR

MTTR is the expected value of repair time, represented as:

$$\mathrm{MTTR} = \int_0^{\infty} g(t) \cdot t\mathrm{d}t \tag{5.38}$$

where $g(t)$ denotes the repair density of the system or equipment.

(2) Frequency Metrics

Frequency metrics use failure rate to measure. The failure rate $F_{i,f}(t)$ characterizes the extent to which system or equipment i fails before time t, with the equation:

$$F_{i,f}(t) = \int_0^t f_i(t)\mathrm{d}t = \frac{n_{i,f}(t)}{N_i(t)} \tag{5.39}$$

where $f_i(t)$ is the failure rate of equipment i at time t; $n_{i,f}(t)$) is the number of failure occurrences for equipment i before time t; $N_i(t)$ is the total number of trials for equipment i before time t.

(3) Probability Metrics:

Availability $A(t)$ refers to the probability that the system or equipment, initially working normally, will continue to function at time t. Particularly, when equipment failures and repairs follow an exponential distribution, the availability $A(\infty)$ converges to a constant independent of t, that is, steady-state availability A, calculated as:

$$A = A(\infty) = \lim_{t \to \infty} A(t) = \frac{\mathrm{MTBF}}{\mathrm{MTBF} + \mathrm{MTTR}} \tag{5.40}$$

Correspondingly, steady-state unavailability A'' is:

$$A' = 1 - A \tag{5.41}$$

Particularly, when equipment failures and repairs follow an exponential distribution, steady-state unavailability can be expressed as:

$$A' = 1 - A = 1 - \lim_{t \to \infty} A(t) = \frac{\text{MTTR}}{\text{MTBF} + \text{MTTR}} \tag{5.42}$$

(4) Importance Metrics

(a) Basic Unit Importance

Basic unit importance I_i indicates the ratio of the number of system faults z_i caused by the failure of equipment i to the number of failures of equipment i itself Z_i. It reflects the vulnerability of the equipment itself and can show the impact level of each equipment failure on the system, calculated as:

$$I_i = \frac{Z_i}{Z_i} \tag{5.43}$$

(b) Basic Unit Mode Importance

Basic unit mode importance M_i indicates the ratio of the number of system faults z_i caused by equipment i to the total number of system faults Z. Basic unit mode importance is related to the equipment's failure rate, indicating the likelihood that a system's fault is caused by that equipment, reflecting the importance of the equipment to the system. Its calculation equation is:

$$M_i = \frac{z_i}{Z} \tag{5.44}$$

5.3.2.3 Reliability Assessment Method of the Traction Power Supply System Considering Relay Protection

The framework of the reliability assessment method proposed in this chapter is shown in Fig. 5.17.

Implementation steps:

(1) Establishing a Traction Power Supply System FMEA. Based on the main wiring diagram of the traction power supply system and related protection configurations, an FMEA table is established. This involves identifying failure modes of various equipment and protection systems, analyzing the consequences and impacts of these failures, and their interrelationships, to qualitatively assess system reliability.
(2) Establishing a Traction Power Supply System Fault Tree Model. Using the FMEA, fault tree models for both primary and secondary equipment linked to traction substations and the OCS are developed. This step involves identifying

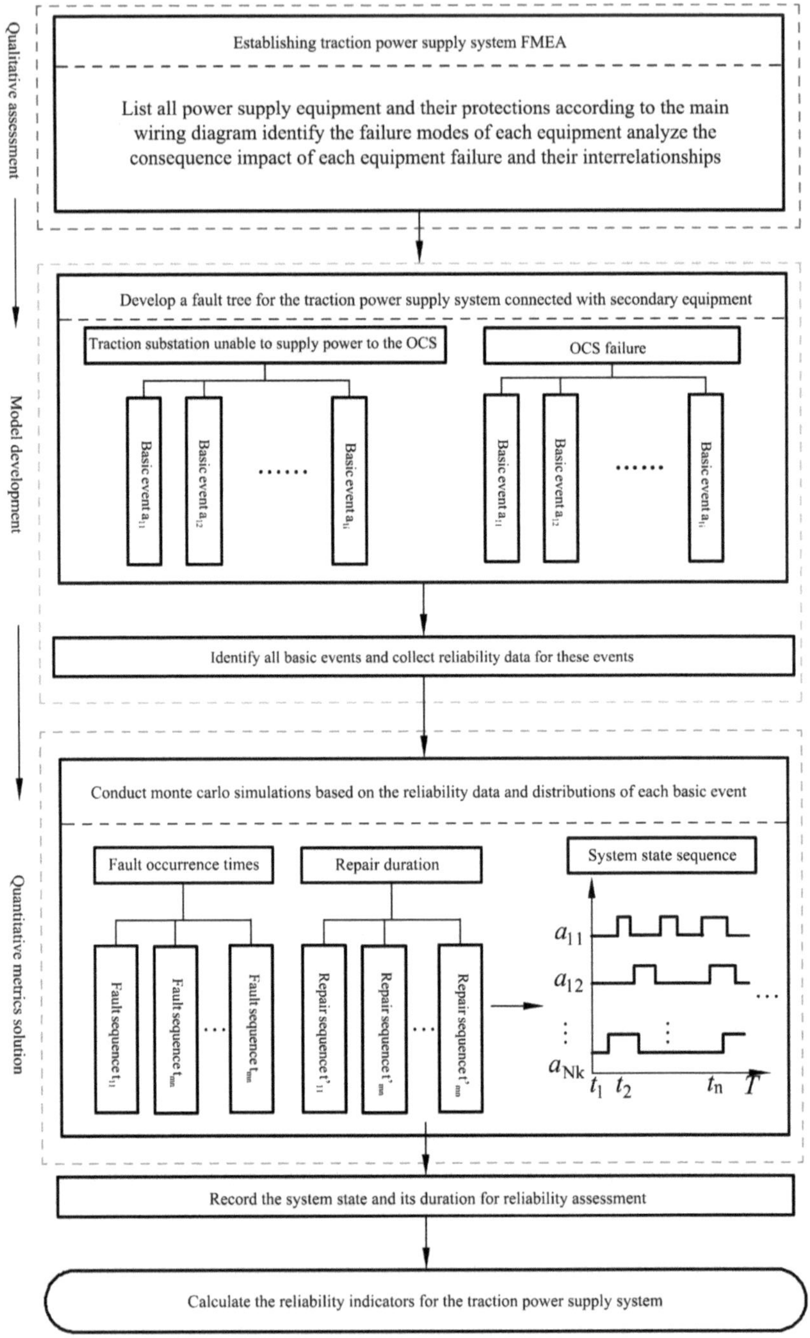

Fig. 5.17 Framework diagram of the high-speed railway traction power supply system reliability assessment method

basic events and collecting reliability data for these events, preparing inputs for quantitative assessment.

(3) Monte Carlo simulation: Based on the reliability statistical data and distribution characteristics of each basic event, generate a sequence of random numbers, apply the inverse function method to the fault distribution function and repair distribution function of basic events, simulate the sequence of events, and derive the structure function from the fault tree, thereby obtaining the system state at each moment within the period and recording simulation-related data.

(4) Reliability indicator calculation: By capturing a sufficient length of the simulation time window (or a sufficient number of samples), calculate the corresponding reliability indicators until the indicators converge, indicating that the sample size is adequate. Using the recorded sequence states from the Monte Carlo simulation, obtain and analyze the reliability assessment indicators, achieving a quantitative assessment of the traction power supply system's reliability.

5.3.3 *Case Study of Traction Power Supply System Reliability Assessment*

5.3.3.1 Basic Data Input

Currently, data on the reliability of traction power supply equipment and protection foundations is scarce, with almost no complete case samples available for validating the proposed reliability methods. Therefore, this section takes a traction substation and its OCS on the Beijing-Shanghai High-Speed Railway as a case study. The model is based on the actual main electrical wiring diagram shown in Fig. 5.14, using real statistical data from the site to test the proposed reliability assessment method. The reliability parameters for the key equipment of the traction substation, relay protection, and components of the OCS are listed in Tables 5.5, 5.6 and 5.7.

Considering operational department statistical norms and related research, the input data in Tables 5.5, 5.6 and 5.7 follow specific distribution forms. The failure rate functions of transformers, feeders/busbars, circuit breakers, and isolator switches in the traction substation adhere to a two-parameter Weibull distribution, while all other components conform to an exponential distribution, meaning both

Table 5.5 Traction substation key equipment reliability parameters

Equipment name	Failure rate	Repair rate
Transformer	$\lambda(t)=\left(\frac{16.6}{167.25^{16.6}}\right)\cdot(t)^{15.60}$	5.15×10^{-2}
Feeder/busbar	$\lambda(t)=\left(\frac{10.63}{174.48^{10.63}}\right)\cdot(t)^{9.63}$	0.20
Circuit breaker	$\lambda(t)=\left(\frac{7.75}{178.67^{7.75}}\right)\cdot(t)^{6.75}$	0.10
Isolator switch	$\lambda(t)=\left(\frac{10.88}{175.98^{10.88}}\right)\cdot(t)^{9.88}$	0.25

Table 5.6 Relay protection reliability parameters

Protection name	Status	Failure rate	Repair rate
Transformer protection	Refusal to operate	5.85×10^{-3}	0.021
	Misoperation	2.92×10^{-2}	0.021
Feeder protection	Refusal to operate	8.48×10^{-4}	0.021
	Misoperation	8.48×10^{-3}	0.021

Table 5.7 OCS component reliability parameters

Equipment names	Failure rates	Repair rates (times/h)	Equipment names	Failure rates	Repair rates (times/h)
Dropper	0.009575	0.524587	Positioning equipment	0.000946	0.603622
Mid anchor	0.001423	0.822467	Positioning hook/ring	0.000941	0.818898
Lower anchor	0.002566	0.500485	Positioning line clamp	0.001353	0.717213
Soft span	0.000830	0.859599	Positioner	0.005823	0.485165
OCS wire	0.013331	0.513773	Positioning tube	0.001703	0.553088
Cantilever	0.002179	0.379843	Limiting tube	0.000939	0.668435
Connection parts	0.000716	0.597610	Data parameters	0.001266	0.250000
Support insulator	0.009672	0.566509	Line clamp	0.000616	0.500000
Misalignment	0.002946	0.500000	Electrical connection wire	0.002248	0.442073
Pantograph failure	0.000296	0.000116	Compensation rope	0.001187	0.577164
Foreign object intrusion	0.007306	0.500000	Damage/fatigue	0.000134	0.200000
Plummet	0.001270	0.561587	Pole	0.004345	0.328438
Compensation ratchet	0.000160	0.712166	Foundation	0.000873	0.196239
Limiting rod/frame	0.000299	0.477454	Additional conductor	0.013026	0.559028
Insulator	0.002771	0.659211	Lightning	0.004566	0.250000
Lightning arrester	0.000950	0.661070	Overload	0.001826	6.000000
Network isolator switch	0.001935	0.619593	–	–	–

failure and repair rates for these parts are constant. These reliability parameters are used as input data, and reliability quantitative metrics are obtained through a Monte Carlo simulation-based model-solving approach.

5.3.3.2 Quantitative Metrics and Assessment of Traction Power Supply System Reliability

Traction Substation

Set the number of samples for state transitions of basic events to 80,000. After several rounds of simulation, the converged time metrics, frequency metrics, and probability metrics for the traction substation are shown in Table 5.8.

According to Table 5.8, it is apparent that the traction substation's MTBF, fault frequency, and steady-state unavailability are higher than those of individual zones and equipment; whereas its MTTR is 0.0487 months, which is lower than that of failure events in different zones and equipment. This is because, in the fault tree model, the logical gates connecting the traction substation to various zones and equipment are OR gates. Any failure in the lower system regions or equipment can cause the traction substation to be unable to normally supply power to the downward Zhujianong OCS. Given the consideration of dual-sided backups, the logical gates connecting Zones 1 and 2 to the lower level are AND gates. From the data in Table 5.8, it is evident that Zones 1.1 and 1.2 have longer MTBFs and lower fault frequencies, thus the reliability indicator calculations for Zone 1 show an extremely low fault probability, indicated as 0. Zone 2 contains critical traction power supply equipment—the traction transformer—and the analysis of the traction transformer considers the impact of transformer protection on it. Comparing scenarios 1, 2, and 3 and their indicators, it can be observed that the frequency of transformer outage due to misoperation of transformer protection is higher than that of transformer

Table 5.8 Traction substation reliability metrics

	Time metrics		Frequency metrics	Probability metrics
Regional type	MTBF/month	MTTR/month	Fault frequency/(times/month)	Steady-state unavailability
Traction substation	8.9198	0.0487	0.1115	5.37×10^{-3}
Zone 1	N/A	0	0	N/A
Zone 1.1	167.8780	0.0362	0.0060	4.34×10^{-5}
Zone 1.2	83.8497	0.0094	0.0119	1.15×10^{-4}
Zone 2	15.3228	0.0627	0.0650	3.96×10^{-3}
Zone 2.1	12.9677	0.0546	0.0768	4.07×10^{-3}
Zone 2.2	14.0359	0.0665	0.0709	4.45×10^{-3}
Transformer	15.3261	0.0627	0.0650	3.95×10^{-3}
Scenario 1	16.9175	0.0665	0.0589	3.79×10^{-3}
Scenario 2	162.2150	0.0260	0.0062	1.66×10^{-4}
Scenario 3	N/A	0	0	N/A
Zone 3	84.2746	0.0097	0.0119	1.16×10^{-4}
27.5 kV busbar	83.4637	0.0068	0.0120	8.16×10^{-5}
Feeder	43.3443	0.0503	0.0230	1.22×10^{-3}

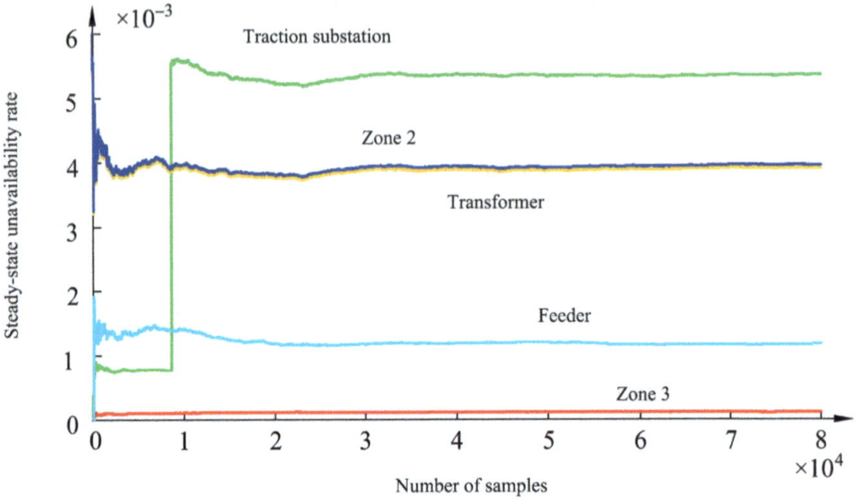

Fig. 5.18 Steady-state unavailability of the traction substation and each zone and equipment

outage due to transformer anomalies; for scenario 3, when the transformer operates abnormally, both the main and backup protections of the transformer must refuse to operate to cause the transformer to outage, and since the probability of transformer protection refusal is low, the incidence of transformer outage due to protection refusal is very rare. Zone 3 and the 27.5 kV busbar have similar reliability indicators, with an MTBF of about 84 months, and a fault frequency of approximately 0.0120; while the feeder's fault frequency is nearly twice theirs, with an MTBF of about 43 months.

Figure 5.18 illustrates the steady-state unavailability of the traction substation and each zone and equipment, derived using the simulation method described in this paper. The steady-state unavailability of the traction substation primarily reflects the extent of failures during the operational cycle due to various faults. Combining data from Table 5.8 and Fig. 5.18, it is evident that Zone 2 has the highest steady-state unavailability among all zones, mainly due to the inclusion of the traction transformer in its structure. Influenced by the higher steady-state unavailability of the transformer, the unavailability trend in Zone 2 nearly mirrors that of the transformer itself. This is followed by the feeder and Zone 3. In contrast, Zone 1, which represents the system's incoming line section, has almost zero steady-state unavailability, and the 27.5 kV busbar's unavailability is two orders of magnitude lower than that of Zone 2, Zone 3, and the feeder.

After obtaining the reliability assessment results for time, frequency, and probability metrics, importance metrics are calculated according to Eqs. (5.43) and (5.44) to reflect the vulnerability of the equipment and its potential to cause system faults. The results, as shown in Fig. 5.19, indicate that the basic unit importance for the transformer, the 27.5 kV busbar, and the feeder is 1. This suggests that a failure in any of these equipment types could lead to a system fault, indicating the highest

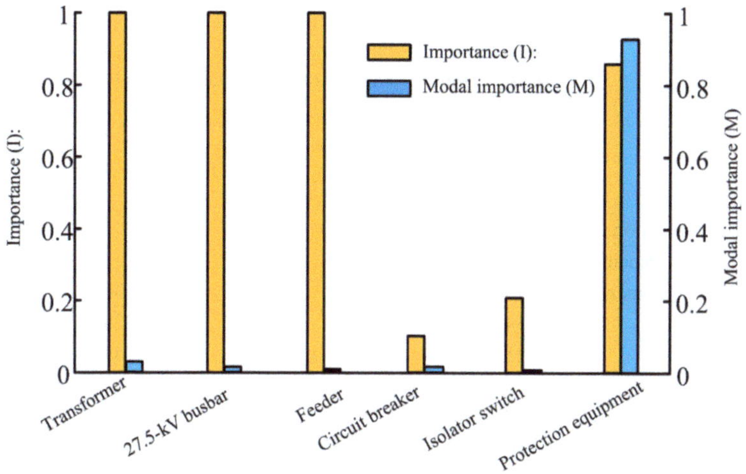

Fig. 5.19 Importance metrics for traction substation equipment

impact level on the system, which necessitates focused attention during routine inspections and scheduled maintenance. The importance of protection equipment is 0.8588, which is relatively lower for switch-type equipment. Moreover, the modal importance of protection equipment is the highest, indicating that system faults are most likely to be caused by failures in protection equipment. Given the presence of backups, the modal importance of the transformer is only 0.0295, with other equipment having even lower importance, suggesting that these equipment' failures have a lesser impact on the normal operation of the system. Thus, enhancing the reliability of protection equipment could positively affect the normal and stable operation of the system.

OCS

Like the traction substation, the OCS's simulations were set with 80,000 samples and underwent multiple rounds to achieve convergence in the time metrics, frequency metrics, and probability metrics. The results are displayed in Table 5.9.

From Table 5.9, the MTBF for the contact wire suspension and tripping power loss is notably short, at 39.51 h and 63.76 h respectively; their fault frequencies are relatively high, at 0.0241 occurrences/h and 0.0154 occurrences/h respectively. Due to the lack of backups, the OCS equipment is configured in a series structure in the system's FTA. Consequently, the reliability indicators for the OCS are primarily impacted by the contact wire suspension and tripping power loss, resulting in an MTBF of only 15.04 h and steady-state unavailability of approximately 10.72%. Comparing Table 5.8 with Table 5.9 highlights that the reliability level of the OCS is significantly lower than that of the traction substation. This difference is due to the

Table 5.9 OCS Reliability indicators

	Time metrics		Frequency metrics	Probability metrics
Regional type	MTBF/h	Mttr/h	Fault frequency (times/h)	Steady-state unavailability
OCS	15.04	1.81	0.0594	0.1072
Contact wire suspension	39.51	1.99	0.0241	0.0479
Electrical connection	137.92	1.98	0.0071	0.0141
Supporting equipment	122.16	1.87	0.0081	0.0151
Positioning equipment	131.78	1.82	0.0074	0.0136
Track switch	809.34	2.25	0.0012	0.0028
Compensation equipment	322.21	1.87	0.0031	0.0058
Pole foundation	650.92	3.25	0.0015	0.0049
Electrical equipment	273.97	1.59	0.0036	0.0058
Tripping power loss	63.76	0.98	0.0154	0.0152

traction substation mostly comprising stationary indoor power supply equipment with backups in place, whereas OCS equipment is exposed outdoors, highly susceptible to external environmental factors like foreign objects and lightning, and lacks backups. For instance, the contact wire suspension is also subject to dynamic impacts during the vehicle-network coupling process, leading to a relatively higher failure rate.

Given the FTA structure of the OCS, failures in OCS equipment will invariably cause a system fault, hence the importance metrics for each equipment are set to 1 (indicating that the system also manifests faults when individual equipment fail). Thus, Fig. 5.20 solely depicts the modal importance of the OCS components. The relatively high modal importance of contact wire suspension and tripping power loss indicates their significant impact on the OCS; conversely, the track switch and pole foundation have the lowest relative importance, suggesting that their failures have a lesser impact on normal system operations. Therefore, reducing the failure rate of contact wire suspension and enhancing the OCS's resistance to external environmental disturbances to prevent tripping power loss are crucial for improving the overall reliability of the OCS.

5.3.3.3 Impact of Relay Protection Equipment on the Unavailability of the Traction Power Supply System

To further measure the impact of relay protection on the steady-state unavailability of the traction power supply system, a comparative analysis under five different operating conditions is: conducted:

(1) Variation in transformer protection failure rate (at the traction substation).
(2) Variation in feeder protection failure rate (at the traction substation).

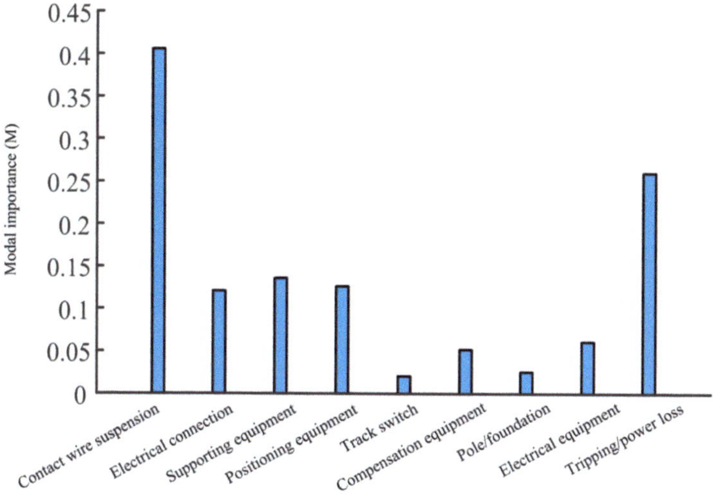

Fig. 5.20 Modal importance indicators of the OCS

(3) Variation in feeder protection failure rate (in the OCS).
(4) Simultaneous changes in primary equipment failure rate and protection misoperation rate (at the traction substation).
(5) Simultaneous changes in primary equipment failure rate and protection failure to operate rate (at the traction substation).

The misoperation and refusal rates for protections are set according to Table 5.6 and change with the same transformation ratio; considering the limitations of simulation accuracy and to demonstrate the impact of simultaneous changes in primary equipment and protection failure to operate rates, the scale parameter of the distribution function that the primary equipment adheres to is adjusted to reflect changes in their failure rates, with transformation ratios set between 0.5–1.5. The same transformation ratio settings apply to changes in the refusal rates. Simulation results showing changes in system steady-state unavailability under each operating condition are depicted in Figs. 5.21 and 5.22.

Key Insights from Figs. 5.21 and 5.22:

(1) For the Traction Substation, the impact of misoperation rates (protection misoperation) of transformer protection and feeder protection is more significant than their refusal rates, with the system's steady-state unavailability noticeably increasing as the misoperation rates rise. Similarly, the steady-state unavailability of the OCS also significantly increases with the rise in feeder protection misoperation rates.
(2) Under fixed misoperation rates, the impact of refusal rates on system steady-state unavailability is relatively minor. This is due to the high risk associated with protection failure to operate in cases of some equipment failures in the traction

Fig. 5.21 Impact of protection failure rate on system steady-state unavailability. (**a**) Impact of transformer protection failure rate on traction substation steady-state unavailability. (**b**) Impact of feeder protection failure rate on traction substation steady-state unavailability. (**c**) Impact of feeder protection failure rate on OCS steady-state unavailability

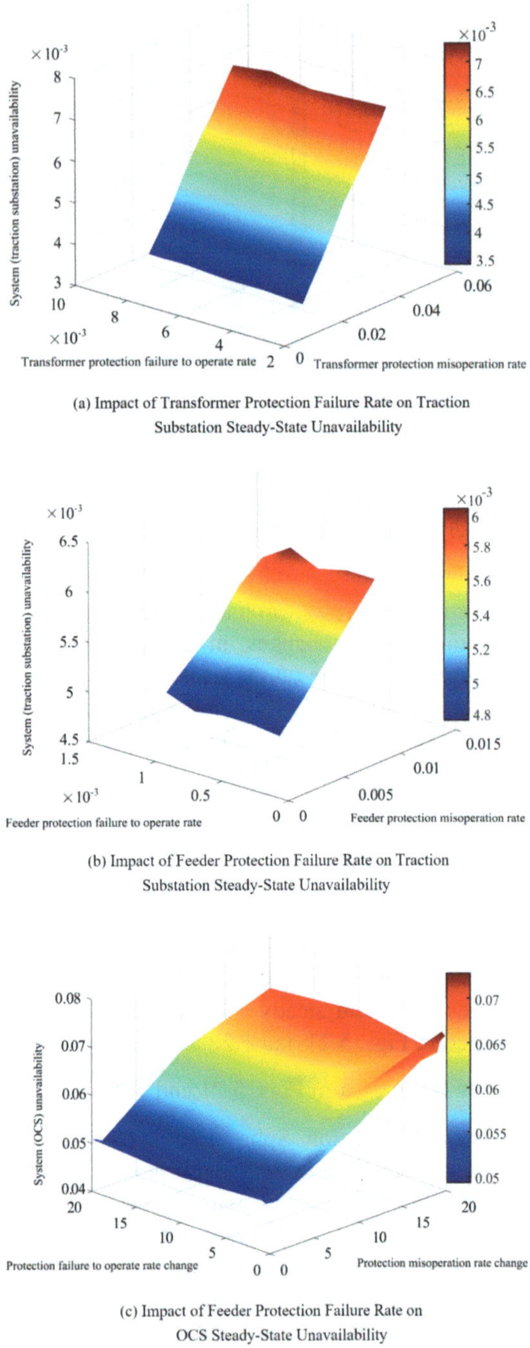

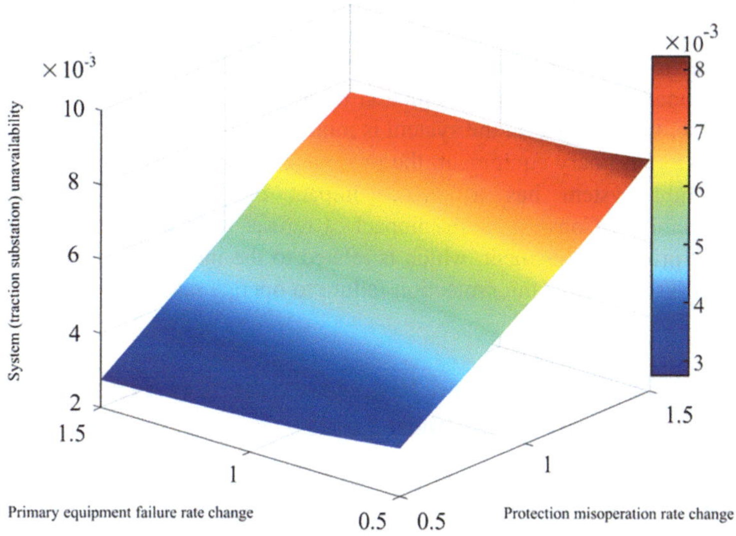

(a) Impact of Primary Equipment/Protection Misoperation Rate on System Steady-State Unavailability

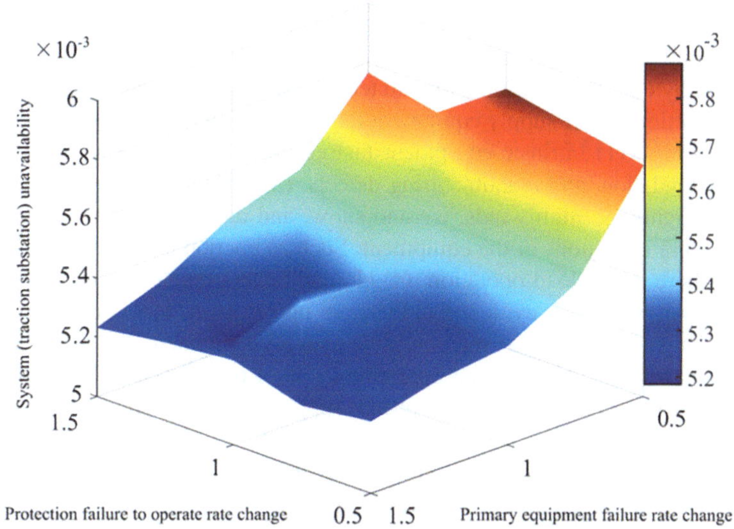

(b) Impact of Primary Equipment/Protection Failure to Operate Rate on System Steady-State Unavailability

Fig. 5.22 Impact of primary equipment/protection failure rate on system steady-state unavailability. (**a**) Impact of primary equipment/protection misoperation rate on system steady-state unavailability. (**b**) Impact of primary equipment/protection failure to operate rate on system steady-state unavailability substation. The operational mandate often prefers protection misoperation over non-operation (erring on the side of caution rather than risk under-operation), leading to a very low refusal rate configured within the protection systems. Additionally, since the transformation ratios set for these rates are not widely variable, their impact on the system is minimal.

(3) Relative to primary equipment, the misoperation rate of the traction substation protection system has a greater impact on the system's steady-state unavailability. Conversely, the impact of refusal rate changes is smaller than that of primary equipment, which is related to the inherently small values and limited variability of the protection failure to operate rates.

5.4 Multifactor Integrated Risk Assessment of Traction Power Supply Equipment Fault

In practice, China's high-speed railways are predominantly elevated, with most traction power supply equipment being exposed outdoors. This makes the equipment particularly vulnerable to external natural environmental impacts, with weather conditions like ice and snow, thunderstorms, and high winds significantly increasing the probability of incidental failures. Therefore, in addition to the aging and operating conditions of the traction power supply equipment, the harsh service environment also plays a critical role in triggering failures in high-speed railway traction power supply equipment.

Driven by on-site operation and maintenance records and statistical data, factors causing traction power supply equipment failures and outages can be classified into four categories: (1) Intrinsic equipment factors: Such as equipment aging, performance degradation, and manufacturing defects. (2) Natural environmental impact: Including lightning, high winds, ice and snow, and damage from birds or snakes. (3) Operating conditions of the equipment: Such as traction load characteristics and whether there is overload operation. (4) Human factors: Such as operational errors and negligent work. It is evident that there are numerous factors contributing to the failure of high-speed railway traction power supply equipment. On the other hand, due to the short construction and operation cycle of high-speed railways, there are few statistical samples of traction power supply equipment failures or outages. Consequently, accurately assessing the probability of equipment outage under conditions of limited samples remains a hot issue and bottleneck in the risk assessment process of the traction power supply system.

Additionally, high-speed railways primarily focuses on passenger transport, ensuring passengers arrive at their destinations on time, safely, and comfortably, which reflects the essential purpose and service mission of high-speed trains as public transportation. Therefore, the most crucial aspect people care about is whether the traction power supply system has a safe, reliable, and continuous power supply capacity to ensure the normal operation of high-speed trains. The relevant evaluation indicators should not only include the loss costs of traction power supply equipment but also consider the consequences of time loss due to equipment outages.

This section is driven by actual data related to equipment failures in the traction power supply system on-site, taking into account the influence of multiple factors and proposing a new method for assessing the risk of traction power supply equipment failure, tailored to the needs of small sample data.

5.4.1 Interval Estimation Method for Traction Power Supply Equipment Failure Probability

Given the current situation of the high-speed railway traction power supply system, there are numerous factors affecting the downtime of traction power supply equipment failures, and the number of effective samples is limited, making it difficult to satisfy the conditions for the law of large numbers to apply in reality. Therefore, precise single-value probability estimates are challenging to reflect the true possibility of equipment downtime, and the resultant risk assessment may greatly deviate from the actual situation, losing its original guiding significance and value.

In interval probability theory, the imprecise dirichlet model (IDM) is a well-established and frequently used method. IDM is suitable for interval probability statistical estimation of variables that follow a multinomial distribution and offers significant advantages in dealing with probability estimation problems under small sample conditions. This method can alleviate some of the uncertainties associated with using single-value probabilities or averages. Meanwhile, a credal network is a type of interval probability graph model used to express uncertain knowledge and perform causal reasoning. Essentially, a credal network is an extension of the classic Bayesian network, but its nodes and arcs can have interval probability measures, relaxing the strict modeling requirements of classical Bayesian networks. Thus, credal networks can integrate graph models with interval probability theory, enabling them to better express general uncertainty issues and perform uncertainty reasoning. The inference goal of the credal network model is to estimate the outage probability interval of the equipment under given operating conditions and for a specified period. To address the aforementioned issues, this chapter employs a combination of credal network and IDM to estimate the outage probability interval of traction power supply equipment failures.

5.4.1.1 Credal Network Model for Traction Power Supply Equipment

Based on the statistical results from on-site operation and maintenance records, this chapter identifies various factors affecting the operational reliability level of traction power supply equipment as node evidence variables to construct a credal network, as shown in Fig. 5.23. Given that under the same operating conditions, equipment with

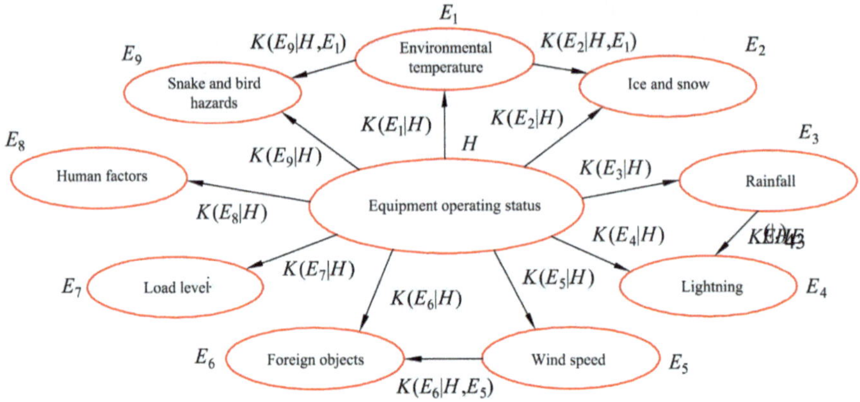

Fig. 5.23 Credal network model for traction power supply equipment

higher degrees of aging during service duration tends to have a higher outage probability, this chapter also considers the impact of service duration on the operational reliability level of the equipment, treating it as a main factor influencing the prior equipment downtime probability representing long-term average reliability levels.

As shown in Fig. 5.23, in the credal network, the parent nodes are two state random variables *H* representing the equipment operating status: H_1 for equipment failure leading to shutdown, and H_2 for no failure leading to shutdown. The child nodes represent the operating conditions of the equipment with nine random (evidence) variables: environmental temperature, ice and snow, rainfall, lightning, wind speed, foreign objects, load level, human factors, snake and bird hazards, designated as E_1, E_2, E_3, E_4, E_5, E_6, E_7, E_8, and E_9, respectively. Among these, environmental temperature E_1 is a three-state random variable: high temperature (temperature above 26 °C), low temperature (temperature below 4 °C), and normal temperature (temperature between 4 °C and 26 °C); the other eight variables are two-state random variables. The specific state divisions of these nine evidence variables are as detailed in Table 5.10.

To accommodate the interval probability expressions of the node random variables, the credal network utilizes a credal set to conduct probabilistic inference. The credal set *K(Xi)* of a random variable *Xi* is defined as a closed convex set composed of all possible probability mass functions *P(Xi)*,

$$K(X_i) = \{P(X_i) : P(X_i) = \alpha_1 P_1(X_i) + \alpha_2 P_2(X_i) + \cdots + \alpha_L P_L(X_i)\} \tag{5.45}$$

where $P_1(X_i), P_2(X_i), \ldots, P_L(X_i)$ are the probability mass functions at the *L* vertices of the credal set $K(X_i)$. The coefficients of the probability mass functions in Eq. (5.45) must satisfy the following condition:

Table 5.10 Division of evidence variable states

Evidence variable	State 1	State 2	State 3
E_1	$E_{1,1}$ Low temperature	$E_{1,2}$ Normal temperature	$E_{1,3}$ High temperature
E_2	$E_{2,1}$ Ice and snow	$E_{2,2}$ No ice and snow	N/A
E_3	$E_{3,1}$ Rainfall	$E_{3,2}$ No rainfall	N/A
E_4	$E_{4,1}$ Lightning	$E_{4,2}$ No lightning	N/A
E_5	$E_{5,1}$ Strong wind	$E_{5,2}$ No strong wind	N/A
E_6	$E_{6,1}$ Foreign objects	$E_{6,2}$ No foreign objects	N/A
E_7	$E_{7,1}$ Normal load	$E_{7,2}$ Overload	N/A
E_8	$E_{8,1}$ Human factors	$E_{8,2}$ No human factors	N/A
E_9	$E_{9,1}$ snake and bird hazards	$E_{9,2}$ No snake and bird damage	N/A

$$\sum_{k=1}^{L} \alpha_k = 1, 0 \leqslant \alpha_k \leqslant 1 \tag{5.46}$$

In practical applications, the vertices of the credal set can be directly determined by identifying combinations of endpoints that satisfy normative constraints and appear in the probability intervals for each state of the random variables.

At the same time, to achieve an imprecise conditional estimation of the outage probability of equipment under given operating conditions, it is necessary to know the prior probability of equipment operating status and the probabilistic association between equipment operating status and operating conditions. The prior probability of equipment outage can be substituted with statistically derived equipment outage probabilities. For the probabilistic associations between equipment operating status and its operating conditions, conditional credal sets $K(E_a|H_z)$ for $a \in \{1, 3, 5, 7, 8\}$ and $z \in \{1, 2\}$, $K(E_2|H_z,E_{1,b1})$, $K(E_4|H_z,E_{3,b3})$, $K(E_6|H_z,E_{5,b5})$, $K(E_9|H_z,E_{1,b1})$ with $b_1 \in \{1, 2, 3\}$, $b_3 \in \{1, 2\}$, $b_5 \in \{1, 2\}$, and $z \in \{1, 2\}$ are used to represent these relationships. Specifically, for $K(E_a|H_2)$, since the equipment spends much more time in a non-failure state than in a failure state, the sample size for non-failure conditions is: abundant, allowing the use of precise probability forms to simplify calculations. Thus, the probability of each operating condition occurring during non-failure times can be approximated by the ratio of the cumulative duration of each operating condition during non-failure periods to the total non-failure duration of the equipment. In this case, $K(E_a|H_2)$ simplifies to $P(E_a|H_2)$. On the other hand, for $K(E_a|H_1)$ in the credal network model, the IDM method is used for estimation.

5.4.1.2 Credal Set Derivation

To leverage the advantages of credal networks over traditional Bayesian networks, the IDM is employed to derive the credal sets, allowing the nodes and arcs of the credal network to have interval probability measures. This facilitates the subsequent credal network reasoning and resolution process.

IDM is an extension of the deterministic Dirichlet model. Suppose a random variable following a multinomial distribution has n possible states, with the probability of occurrence of each state represented by $\boldsymbol{\theta} = (\theta_1, \theta_2, \mathrm{L}, \theta_n)$. The deterministic Dirichlet model, using Bayesian statistical principles, treats the unknown parameters as random variables $\boldsymbol{\theta}$ and employs the Dirichlet distribution to obtain their prior probability density function. With the observed sample values $\boldsymbol{M} = \{m_1, m_2, \mathrm{L}, m_n\}$, the Bayesian process updates this to form the posterior Dirichlet probability density function, thus obtaining the expected values of the posterior distribution to estimate the parameters θ_i:

$$E(\theta_i) = (m_i + s \cdot r_i)/(s + M) \tag{5.47}$$

where, m_i represents the number of occurrences of state i for the random variables; M is the total sample size, i.e., $M = m_1 + m_2 + \ldots + m_{\mathrm{n}}$. r_i is the hyperparameter representing the mean of θ_i, which satisfies $0 \leqslant r_i \leqslant 1$ and $\sum_{i=1}^{n} r_i = 1$; s is a set parameter reflecting the extent of the influence of prior information on the statistical results, typically within the range of [1, 2]. $s \cdot r_i$ represents the weight of the prior information for the parameter θ_i, satisfying $\sum_{i=1}^{n} (s \cdot r_i) = s$.

Compared to the deterministic Dirichlet model, the IDM does not use a single probability density function to estimate parameter $\boldsymbol{\theta}$, but rather employs a collection of prior probability density functions under a given s parameter. This collection consists of all possible Dirichlet distributions under the given s. Consequently, IDM can avoid the drawback of being overly influenced by the prior distribution in small sample conditions. Using Bayesian principles, IDM updates the collection of prior probability density functions to a collection of posterior probability density functions, thereby obtaining an interval for the probabilities of $\boldsymbol{\theta}$ (the boundaries of this interval are obtained where the hyperparameter r_i is set at 0 and 1):

$$\theta_i \in \left[\underline{E}(\theta_i), \overline{E}(\theta_i)\right] = \left[\frac{m_i}{M + s}, \frac{m_i + s}{M + s}\right] \tag{5.48}$$

Thus, IDM can utilize small sample data to estimate the interval probability of the occurrence of states for random variables under given conditions, eliminating the adverse impact of improper prior settings on the probability estimation of events under small sample conditions.

5.4.1.3 Credal Network Reasoning and Resolution Process

Based on the IDM method for interval probability estimation, this section illustrates the credal network reasoning process using the credal set $K(E_5|H_1)$ for the condition of strong wind during equipment failure outages as a case study. According to the IDM method (Eq. 5.48) and using historical record data, we obtain the interval probabilities $P_{in}(E_{5,1}|H_1)$ and $P_{in}(E_{5,2}|H_1)$ for occurrences of strong wind and no strong wind during equipment failure outages, respectively:

$$\begin{gathered} P_{in}(E_{5,1}\,|\,H_1)=\left[\underline{P}(E_{5,1}\,|\,H_1),\overline{P}(E_{5,1}\,|\,H_1)\right] \\ \underline{P}(E_{5,1}\,|\,H_1)=\frac{m_{5,1}}{M+s},\overline{P}(E_{5,1}\,|\,H_1)=\frac{m_{5,1}+s}{M+s} \\ P_{in}(E_{5,2}\,|\,H_1)=\left[\underline{P}(E_{5,2}\,|\,H_1),\overline{P}(E_{5,2}\,|\,H_1)\right] \\ \underline{P}(E_{5,2}\,|\,H_1)=\frac{m_{5,2}}{M+s},\overline{P}(E_{5,2}\,|\,H_1)=\frac{m_{5,2}+s}{M+s} \\ M=m_{5,1}+m_{5,2} \end{gathered} \tag{5.49}$$

Where, M represents the total number of occurrences of equipment failure outages; $m_{5,1}$ and $m_{5,2}$ represent the number of occurrences of strong wind and no strong wind weather conditions during equipment failure outages, respectively; s is the parameter set in IDM, with values typically between [1, 2].

Further, according to Eqs. (5.45) and (5.49), the credal set $K(E_5|H_1)$ can be derived:

$$\begin{aligned} K(E_5\,|\,H_1)=\{&P(E_5\,|\,H_1):P(E_{5,1}\,|\,H_1)\in\left[P(E_{5,1}\,|\,H_1),\overline{P}(E_{5,1}\,|\,H_1)\right] \\ &\times P(E_{5,2}\,|\,H_1)\in\left[\underline{P}(E_{5,2}\,|\,H_1),\overline{P}(E_{5,2}\,|\,H_1)\right],P(E_{5,1}\,|\,H_1)+P(E_{5,2}\,|\,H_1)=1\} \end{aligned} \tag{5.50}$$

where the probability mass functions $P(E_{5,1}|H_1)$ and $P(E_{5,2}|H_1)$ within the credal set $K(E_5|H_1)$ take values within their respective interval probability boundaries and must satisfy the constraint that their probabilities sum to one.

At this point, by using the combinations of the upper and lower boundaries of the imprecise probabilities that satisfy the sum-to-one constraint in the credal set $K(E_5|H_1)$, we can obtain the vertices of the credal set, ext.[$K(E5|H1)$].

$$\text{ext}[K(E_5\,|\,H_1)]=\left\{\left(\underline{P}(E_{5,1}\,|\,H_1),\overline{P}(E_{5,2}\,|\,H_1)\right),\left(\overline{P}(E_{5,1}\,|\,H_1),\underline{P}(E_{5,2}\,|\,H_1)\right)\right\} \tag{5.51}$$

For evidence variables E2, E4, E6, and E9 in Fig. 5.23, which are influenced by multiple parent nodes, the conditional credal sets for the states of these evidence variables under different operating conditions of the equipment must consider the occurrence of each state of all their parent nodes.

By traversing the combinations of vertices of each credal set in the credal network to form a joint probability mass function, and using conventional Bayesian network inference methods, we calculate and solve for the maximum and minimum boundary values of the imprecise conditional probability $P_{in}(H_1|\boldsymbol{E})$. The inference equation is:

$$\overline{P}(H_1,\boldsymbol{E}) = max\left\{\frac{\prod\limits_{a=1,3,5,7,8} P(H_1)P(E_{a,b_a}\,|\,H_1)P(E_{2,b_2}\,|\,H_1,E_{1,b_1})}{\sum\limits_{z=1}^{2}\prod\limits_{a=1,3,5,7,7,8} P(H_z)P(E_{a,b_a}\,|\,H_z)P(E_{2,b_2}\,|\,H_z,E_{1,b_1})}\right.$$
$$\left.\frac{P(E_{4,b_4}\,|\,H_1,E_{3,b_1})P(E_{6,b_6}\,|\,H_1,E_{5,b_5})P(E_{9,b_9}\,|\,H_1,E_{1,b_1})}{P(E_{4,b_4}\,|\,H_z,E_{3,b_3})P(E_{6,b_6}\,|\,H_{z,},E_{5,b_5})P(E_{9,b_9}\,|\,H_z,E_{1,b})}\right\}$$
$$\underline{P}(H_1,\boldsymbol{E}) = min\left\{\frac{\prod\limits_{a=1,3,5,7,8} P(H_1)P(E_{a,b_a}\,|\,H_1)P(E_{2,b_2}\,|\,H_1,E_{1,b_1})}{\sum\limits_{z=1}^{2}\prod\limits_{a=1,3,5,7,8} P(H_z)P(E_{a,b_a}\,|\,H_z)P(E_{2,b_2}\,|\,H_z,E_{1,b_1})}\right.$$
$$\left.\frac{P(E_{4,b_4}\,|\,H_1,E_{3,b_3})P(E_{6,b_6}\,|\,H_1,E_{5,b_5})P(E_{9,b_0}\,|\,H_1,E_{1,b_1})}{P(E_{4,b_4}\,|\,H_z,E_{3,b_3})P(E_{6,b_6}\,|\,H_z,E_{5,b_5})P(E_{9,b_9}\,|\,H_z,E_{1,b})}\right\} \tag{5.52}$$

$$\begin{cases} P(E_{a,b_a}\,|\,H_1) \in \text{ext}[K(E_a\,|\,H_1)] \\ P(E_{2,b_2}\,|\,H_1,E_{1,b_1}) \in \text{ext}[K(E_2\,|\,H_1,E_{1,b_1})] \\ P(E_{4,b_4}\,|\,H_1,E_{3,b_3}) \in \text{ext}[K(E_4\,|\,H_1,E_{3,b_3})] \\ P(E_{6,b_6}\,|\,H_1,E_{5,b_5}) \in \text{ext}[K(E_6\,|\,H_1,E_{5,b_5})] \\ P(E_{9,b_9}\,|\,H_1,E_{1,b_1}) \in \text{ext}[K(E_9\,|\,H_1,E_{1,9})] \\ \text{for } a=1, b_a \in \{1,2,3\} \\ \text{for } a=2,3,4,5,6,7,8,9, b_a \in \{1,2\} \end{cases} \tag{5.53}$$

In Eq. (5.52), $P(E_{a,b_a}\,|\,H_1)$ should be selected from the vertices of the credal set ext.[$K(Ea|H1)$], and similarly for $P(E_{2,b_2}\,|\,H_1,E_{1,b_1})$, $P(E_{4,b_4}\,|\,H_1,E_{3,b_9})$, $P(E_{6,b_6}\,|\,H_1,E_{5,b_5})$, $P(E_{9,b_9}\,|\,H_1,E_{1,b_1})$. However, $P(E_{a,b_a}\,|\,H_2)$, $P(E_{2,b_2}\,|\,H_2,E_{1,b_1})$, $P(E_{4,b_4}\,|\,H_2,E_{3,b_9})$, $P(E_{6,b_6}\,|\,H_2,E_{5,b_5})$, $P(E_{9,b_9}\,|\,H_2,E_{1,b_1})$ are taken as precise probabilities statistically obtained.

Summarizing the above, the specific steps for the interval estimation of traction power supply equipment failure probability are as follows:

(1) Based on relevant operating conditions from actual failure records, construct a credal network for estimating the probability of traction power supply equipment failure.
(2) Using historical data, calculate the prior probability $P(H_z)$ for different operating states of the equipment and the conditional credal sets $K(E_a|H_z)$ corresponding to each operating condition evidence node.
(3) Utilize Eqs. (5.52) and (5.53) in the credal network based on the given operating conditions for the estimated target period to conduct probabilistic reasoning, ultimately obtaining the interval probability estimation of traction power supply equipment failure under the given operating conditions for the target period.

5.4.2 Analysis and Quantification of Consequences of Traction Power Supply Equipment Failures

5.4.2.1 Risk-Based Failure Mode and Effect Analysis

Another important part of risk assessment is the analysis and quantification of the consequences of risks. Failures of traction power supply equipment not only result in equipment loss and increased maintenance costs but can also lead to power supply interruptions, potentially causing train delays or even suspensions. Thus, the time loss caused by equipment failures is a concern not only for the maintenance departments of the traction power supply system but also for society at large. This is because the primary task of the traction power supply system is to ensure the timely, safe, and reliable operation of high-speed railways, making the time loss due to equipment failures a significant issue.

As illustrated in Fig. 5.24, the typical high-speed railway traction power supply system's main electrical wiring diagram includes two major components: the traction substation and the OCS, together fulfilling the functions of power transformation, distribution, and delivery within the traction power supply system. The traction substation serves as the heart of the high-speed railway traction power supply system. Its primary task is to convert the three-phase high voltage electricity delivered from the power system into electrical energy suitable for use by electric multiple unit, which it then transmits to the OCS. The traction substation involves numerous electrical components such as the traction transformer, circuit breaker, isolator switch, current transformer, lightning arrester, and busbar. To meet the requirements for continuous power supply, the traction substation employs a backup system consisting of Line 1 and Line 2 for the incoming feed, and two main transformers, 1 T and 2 T, which are also backups for each other. The low-voltage side busbar uses a single busbar with a sectional wiring configuration. During normal operations, one set of "incoming line—line—main transformer" functions as the main while the other serves as the backup. Circuit breakers (such as circuit breaker 103) and isolator switches (such as isolator switches 1001 and 1002) are typically in the off position, as are isolator switches 2122 and 2123. For instance, if one feeder line loses power, isolator switches 1001 and 1002 are first closed, followed by the closing of circuit breaker 103, thus allowing the alternate feeder line to handle the full load of the station. The OCS acts as the interface between the traction substation and the electric multiple unit. Unlike the stationary loads of a typical power transmission line, the OCS's load is the dynamically moving electric multiple unit. The main components of the OCS include the contact wire, OCS wire, dropper, supporting equipment, positioning equipment, pole, and foundation. These components ensure the continuous and reliable transmission of electrical power to the electric multiple unit as they move at high speeds.

To further analyze the fault modes, impact consequences, and risk conditions of each key equipment in the traction power supply system, traction power supply equipment failures are classified into three levels, as specifically described in

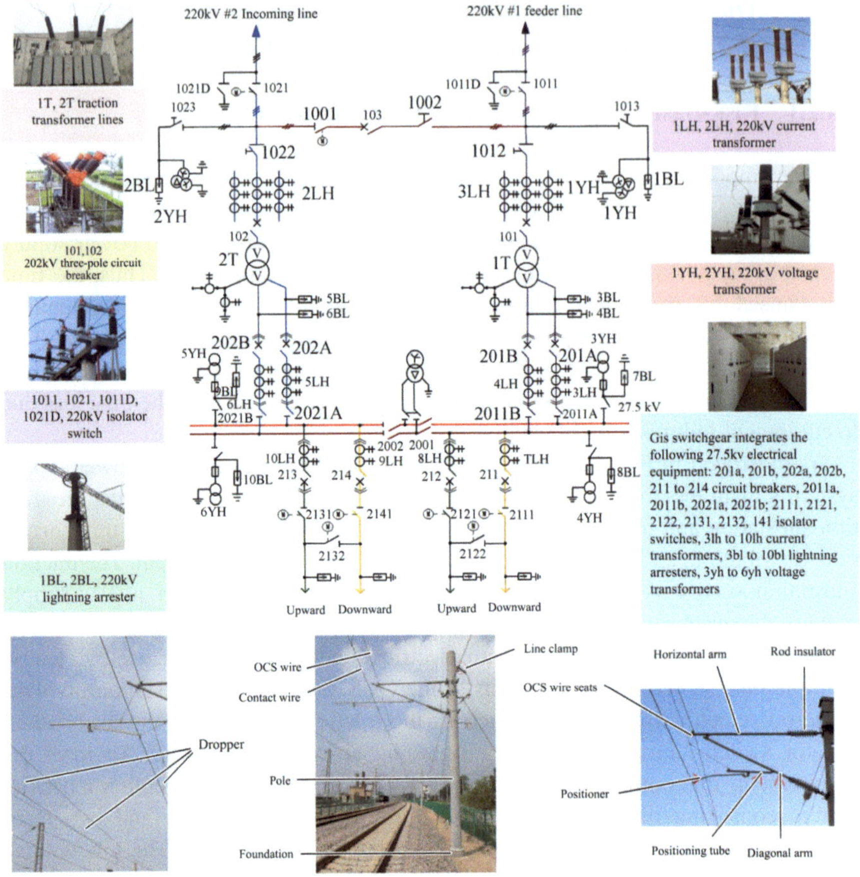

Fig. 5.24 High-speed railway traction power supply system electrical main connection and key equipment

Table 5.11 Traction power supply equipment fault level classification

Fault level	Fault characteristic description
A	Equipment failure causes system power outages and shutdown
B	Equipment failure does not cause a power outage but reduces system power supply reliability
C	Equipment failure does not affect power supply capability and reliability

Table 5.11 for each fault level. Furthermore, a risk-based failure mode and effect analysis (RB-FMEA) table is established, based on the analysis of potential fault modes of the equipment and their impact on system functions, considering the risk conditions of equipment failure, with results as shown in Table 5.12.

Table 5.12 Traction power supply equipment RB-FMEA

Component	Failure mode	Consequence impact	Inspection method	Countermeasures	Equipment fault level
220 kV feeder line	Wire break, power outage	Disconnect the corresponding transformer, another main transformer responsible for power supply, system power supply reliability decreases	Current/voltage measurement	Switch to another feed line for power supply; timely repair or replacement	B
Main transformer	Inter-turn short circuit, core fault, casing electrical breakdown, fire, water flooding, etc.	Disconnect the corresponding transformer, another main transformer responsible for power supply, system power supply reliability decreases	Current/voltage/sound/oil level/oil temperature measurement	Switch to another feed line for power supply; timely repair or replacement	B
220 kV circuit breaker, isolator switch	Electromechanical failure, equipment outage	Corresponding feeder line power outage, switch main transformer to undertake all traction load, system power supply reliability decreases	Current/voltage measurement	Switch power supply from another feeder line; timely repair or replacement	B
201A/2011A	Electromechanical failure, equipment outage	When powered by #2, loss of power in the OCS of the left power supply zone's up and Downwards, system power supply reliability decreases	Current/voltage measurement	Switch the power supply to the left power supply zone from 2T; timely repair or replacement	B
201B/2011B	Electromechanical failure, equipment outage	When powered by #2, loss of power in the OCS of the right power supply zone's up and Downwards, system power supply reliability decreases	Current/voltage measurement	Switch the power supply to the right power supply zone from 1T; timely repair or replacement	B
202A/2021A	Electromechanical failure, equipment outage	When powered by #1, loss of power in the OCS of the left power supply zone's up and Downwards, system power supply reliability decreases	Current/voltage measurement	Switch the power supply to the left power supply zone from 1T; timely repair or replacement	B

(continued)

Table 5.12 (continued)

Component	Failure mode	Consequence impact	Inspection method	Countermeasures	Equipment fault level
202B/2021B	Electromechanical failure, equipment outage	When powered by #1, loss of power in the OCS of the right power supply zone's up and Downwards, system power supply reliability decreases	Current/voltage measurement	Switch the power supply to the right power supply zone from 1T; timely repair or replacement	B
211/2111	Electromechanical failure, equipment outage	Loss of power in the Downward of the right power supply zone	Current/voltage measurement	Switch the power supply by shifting to 2122 from an adjacent feeder	B
212/2121	Electromechanical failure, equipment outage	Loss of power in the Upward of the right power supply zone	Current/voltage measurement	Switch the power supply by shifting to 2122 from an adjacent feeder	B
214/2141	Electromechanical failure, equipment outage	Loss of power in the Downward of the left power supply zone	Current/voltage measurement	Switch the power supply by shifting to 2132 from an adjacent feeder	B
213/2131	Electromechanical failure, equipment outage	Loss of power in the Upward of the left power supply zone	Current/voltage measurement	Switch the power supply by shifting to 2132 from an adjacent feeder	B
Contact wire	Break, excessive wear, electrical burn	OCS power outage and service interruption	Detection vehicle, manual inspection	Timely repair and replacement	A
OCS wire	Break, non-standard sag	Non-standard OCS hanging, pantograph-OCS coupling failure, pantograph bowing, pantograph sticking	Detection vehicle, manual inspection	Timely repair and replacement	A
Insulator	Damage, flashover, breakdown	Insulation level reduced, short circuit fault	Manual patrol inspection	Timely repair and replacement	A
Dropper	Wire break, electrical burn, fatigue limit exceeded	OCS not hanging to standard, poor pantograph-OCS coupling quality	Detection vehicle, manual inspection	Timely repair and replacement	C

Line clamp	Loose, electrical burn	Contact wire not positioned to standard requirements	Detection vehicle, manual inspection	Timely repair and replacement	C
Positioner and support	Excessive bending, brittle fracture	Positioning equipment failure	Detection vehicle, manual inspection	Timely repair and replacement, adjust positioner and positioning tube angle	B
Positioning hook, positioning sleeve pin	Electrical burn, fracture	Positioning equipment failure	Detection vehicle, manual inspection	Timely repair and replacement, adjust positioner and positioning tube angle	B
Diagonal arm and horizontal arm	Wear, fracture	Support equipment failure	Detection vehicle, manual inspection	Timely repair and replacement	B
OCS wire seat, positioning ring, sleeve double ears	Corrosion beyond limits, fatigue cracking	Supporting equipment connection failure	Detection vehicle, manual inspection	Timely repair and replacement	B
Pole	Tilting, collapse	Net collapse, traffic interruption	Manual patrol inspection	Timely repair and replacement	A
Foundation	Settling, instability	Pole tilting, collapse	Manual patrol inspection	Timely repair and replacement	A

5.4.2.2 Quantification of Traction Power Supply Equipment Failure Consequences

The risk of the traction power supply system stems from its inherent performance degradation, environmental impacts, and changes in operational conditions, which are characterized by their probability features. Widely used deterministic criteria in the power system, such as the $N - 1$ or $N - k$ principle, do not reflect the probabilistic attributes of traction power supply system behavior. Although the outage probability of the traction power supply system is low, the consequences of an outage are very severe; therefore, a theoretical method that can estimate the outage probability and consider the severity of the consequences is needed.

Considering that risk theory meets these requirements, it is incorporated into the risk assessment of the traction power supply system and its equipment. According to IEEE standards, the most widely used classic risk assessment equation can be expressed as:

$$R = P(\mathrm{A}) \cdot P(\mathrm{C} \mid \mathrm{A}) \cdot L(\mathrm{C}) \tag{5.54}$$

where R represents the risk indicator for various components of the system; $P(\mathrm{A})$ is the probability of occurrence of event or condition A; $P(\mathrm{C} \mid \mathrm{A})$ represents the probability of the system experiencing consequence C triggered by the event or condition A; $L(\mathrm{C})$ represents the severity or loss caused by the occurrence of consequence C.

In the risk assessment process of the traction power supply system, R represents the risk indicator for each equipment in the traction power supply system. The probability of affecting the traction power supply system under environmental weather factors and operating conditions A is denoted by $P(\mathrm{A})$. $P(\mathrm{C} \mid \mathrm{A})$ represents the probability of consequence C occurring under environmental weather factors and operating conditions A; $L(\mathrm{C})$ represents the severity or loss brought about by the occurrence of consequence C. Given that the actual data from the traction power supply system site consists of records of traction power supply equipment failures or system outages, and to facilitate the quantification of the impact of environmental weather factors and operating conditions on traction power supply equipment, this study directly uses the product of $P(\mathrm{A})$ and $P(\mathrm{C} \mid \mathrm{A})$ (i.e., $P(\mathrm{A}) \cdot P(\mathrm{C} \mid \mathrm{A})$) as the probability result of the traction power supply system experiencing a fault outage under various factors. In other words, it is considered that $P(\mathrm{A})$ represents the probability of the traction power supply system experiencing a fault outage (consequence C) under environmental weather factors and operating conditions A, where $P(\mathrm{C} \mid \mathrm{A}) = 1$ at this time. Therefore, the risk assessment equation focused on in this chapter is:

$$R = P(\mathrm{A}) \cdot L(\mathrm{C}) \tag{5.55}$$

where $P(\mathrm{A})$ and $L(\mathrm{C})$ correspond respectively to the first and second parts of the risk assessment method framework; R is the result of the equipment outage risk assessment under given operating conditions.

For the managers and users of the traction power supply system, the impact and consequences of traction power supply equipment failures on the system are made clear through a risk level-based failure mode and effect analysis. Since the traction substation employs a dual-feed power supply structure, equipment failures within it generally do not directly cause a power outage shutdown but result in reduced power supply reliability, so fault levels are classified as Category B. In contrast, OCS equipment almost has no backup setup, hence equipment failures are likely to lead to power outages, such as with contact wires and insulators, which are categorized as fault level A. However, the risk loss and its extent caused by traction power supply equipment failures have always been a concern and trouble for the managers and users of the traction power supply system. To further quantify the impact and consequences of traction power supply equipment failures on the traction power supply system, this chapter, starting from actual situations and circumstances, combines RB-FMEA for qualitative analysis and fault grading studies on the risk loss caused by traction power supply equipment failures.

Category A Fault Level

A Category A fault level indicates that a traction power supply equipment failure will cause a power outage. Therefore, for such failures, both economic loss and time loss are considered.

(1) Economic loss. This part of the loss primarily measures the economic consequences brought about by a traction power supply equipment failure. Economic loss consists of direct equipment loss and maintenance and repair costs, as follows:

$$C_{LA} = C_{\text{equipment loss}} + C_{\text{maintenance costs}} \tag{5.56}$$

Direct equipment loss is divided into two scenarios: one is where the equipment cannot be repaired after failure, necessitating the replacement of equipment or parts, incurring significant costs; the other is where the equipment can be repaired without needing new equipment or components, and this part of the loss is calculated as the depreciation cost of the equipment. Thus, direct equipment loss is expressed as:

$$C_{\text{equipment loss}} = \begin{cases} C_{\text{replacement}}, & \text{if replacement of equipment or parts is needed} \\ C_{\text{depreciation}}, & \text{if no replacement of equipment or parts is needed} \end{cases} \quad (5.57)$$

Maintenance costs include the cost of preparing and utilizing spare parts for traction power supply equipment and labor costs of maintenance work, expressed as:

$$C_{\text{maintenance costs}} = C_{\text{spare parts}} + C_{\text{labor costs}} \quad (5.58)$$

(2) Time loss. Time lost due to power outages caused by traction power supply equipment failure, and the potential impact on train operation times. The time loss considered for Category A faults includes two parts: one is the power outage time in the power supply section caused by equipment failure; the other is the total time high-speed trains spend traveling and stopping in this power supply section within a day, specifically expressed as:

$$\begin{aligned} T_L &= T_{\text{power outage shutdown}} + T_{\text{time spent traveling and stopping in the section}} \\ &= T_{\text{power outage shutdown}} + \sum_{i=1}^{n} T_i \end{aligned} \quad (5.59)$$

where $T_{\text{power outage shutdown}}$ represents the downtime due to traction power supply downtime caused by equipment failure, numerically equivalent to the average repair time of that equipment; $\sum_{i=1}^{n} T_i$ represents the total time high-speed trains spend traveling and stopping in the affected range of the traction power supply equipment failure within a day, with T_i being the time the i-th train spends traveling and stopping in the affected section, and n being the number of trains passing through this power supply section within a day.

Category B fault level

Unlike Category A faults, when a Category B fault occurs, it generally does not directly cause a power outage (substations generally achieve uninterrupted power supply through switch rerouting; equipment failures on the OCS affect support and positioning performance but do not directly cause power loss). However, the overall power supply reliability of the traction power supply system will be reduced. If another equipment failure occurs on the power supply side at this time, the severity of the consequences will greatly increase. Therefore, economic loss not only considers the cost loss of the traction power supply equipment itself and the maintenance costs incurred by maintenance activities but also takes into account the risk cost caused by the reduced system power supply reliability following an equipment failure. This impact is quantified through risk cost as follows:

$$C_{LB} = C_{\text{equipment loss}} + C_{\text{maintenance costs}} + C_{\text{risk cost}} \tag{5.60}$$

where direct equipment loss and maintenance and repair costs are the same as for Category A faults and are quantified using Eqs. (5.57) and (5.58). The defined risk cost is equivalent to the cost of direct equipment loss, representing the impact of reduced system power supply reliability caused by equipment failure.

Category C Fault Level

Since it does not cause a power outage nor does it decrease the system's power supply reliability, Category C fault level only needs to consider the economic loss associated with traction power supply equipment, which is represented the same as the economic loss for Category A fault level:

$$C_{LC} = C_{\text{equipment loss}} + C_{\text{maintenance costs}} \tag{5.61}$$

5.4.2.3 Traction Power Supply Equipment Fault Risk Indicator

According to the risk assessment framework Eq. (5.55), the final risk indicator is expressed as the fault probability multiplied by the loss, yielding both the economic risk indicator and the time risk indicator for traction power supply equipment failure:

$$R_e = P_{\text{in}}(H_1 \mid \boldsymbol{E}) \cdot C_L \tag{5.62}$$

$$R_t = P_{\text{in}}(H_1 \mid \boldsymbol{E}) \cdot T_L \tag{5.63}$$

where $P_{\text{in}}(H_1 \mid \boldsymbol{E})$ is the probability interval obtained through the credal network and IDM; T_L is the time loss during a Category A fault level event; C_L is the economic loss, corresponding to different equipment fault levels, namely:

$$C_L = \begin{cases} C_{LA}, \text{Category A fault level} \\ C_{LB}, \text{Category B fault level} \\ C_{LC}, \text{Category C fault level} \end{cases} \tag{5.64}$$

5.4.3 Case Study of Traction Power Supply Equipment Fault Risk Interval Estimation

The method for risk assessment of traction power supply equipment outages consists of reliability estimation and risk loss estimation. The method framework is shown in Fig. 5.25.

(1) Interval Estimation of Failure Probability of Traction Power Supply Equipment: This part involves constructing a credal network based on historical outage data and predicted operating conditions for the target period. It uses the IDM with multi-state random variables to obtain the imprecise conditional probability of nodes under small sample conditions, thus estimating the range of equipment outage probabilities under specific operating conditions.
(2) Failure Consequence Analysis and Quantification of Traction Power Supply Equipment: This section starts with the traction power supply system's main wiring structure to establish an RB-FMEA table. It qualitatively analyzes the impact of traction power supply equipment outages, evaluates and categorizes equipment fault levels, and then quantitatively measures the risk consequences of equipment failure, focusing on economic loss and time lost during outages.

Testing the Proposed Risk Assessment Method with OCS Equipment Fault Records from a Railway Bureau's Power Supply Section, 2011-2016 Over these six years, there were 43 failures (power outage) in the contact wire on this line. The prior fault probability is taken from the average failure rate over the last 3 years, 0.0012

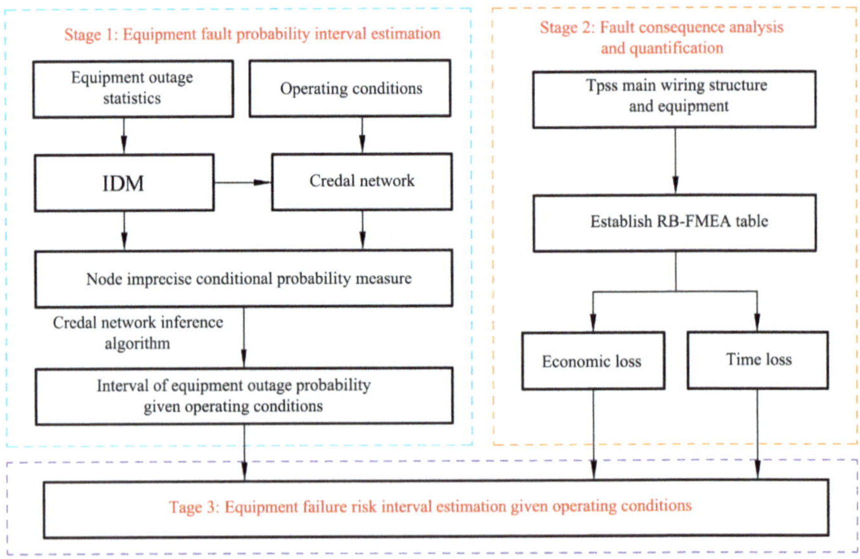

Fig. 5.25 Multifactor integrated traction power supply equipment fault risk assessment framework

Table 5.13 Fault occurrence statistics under various operating conditions

Operating conditions	Number of occurrences	Operating conditions	Number of occurrences
Low temperature	10	No strong wind	15
Normal temperature	9	With foreign objects and strong wind	18
High temperature	24	With foreign objects and no strong wind	4
Ice and snow and low temperature	5	No foreign objects and with strong wind	10
Ice and snow and normal temperature	0	No foreign objects and no strong wind	11
Ice and snow and high temperature	0	Overload	4
No ice and snow and low temperature	5	Normal load	39
No ice and snow and normal temperature	9	Human factors	3
No ice and snow and high temperature	24	No human factors	40
With rainfall	26	With snakes and birds damage and low temperature	5
No rainfall	17	With snakes and birds damage and normal temperature	7
Thunderstorm and with rainfall	17	With snakes and birds damage and high temperature	6
Thunderstorm and no rainfall	3	No snakes and birds damage and low temperature	5
No thunderstorm and with rainfall	9	No snakes and birds damage and normal temperature	2
No thunderstorm and no rainfall	14	No snakes and birds damage and high temperature	18
With strong wind	28		

occurrences per hour. The fault occurrences under various operating conditions, based on the established credal network, are shown in Table 5.13.

Using the historical statistical data from Table 5.13 and according to Eqs. (5.52) and (5.53), precise results and interval probability results for the contact line fault probability are obtained, as shown in Table 5.14.

Based on these results, simulate the changes in the operating conditions of the contact wire on this OCS line for the next 8 h, as shown in Table 5.15. Using the aforementioned reliability estimation method, the variations in the fault outage probability of the contact wire equipment on this OCS line for the upcoming period are illustrated in Fig. 5.26.

Table 5.14 Comparison of interval probability distribution and precise probability for contact wire fault state under operating conditions

Operating condition probability expression	Interval probability	Precise probability	Operating condition probability expression	Interval probability	Precise probability
$P_{in}(E_{1,1}\|H_1)$	[0.227 3,0.250 0]	0.233 1	$P_{in}(E_{5,2}\|H_1)$	[0.340 9,0.363 6]	0.349 0
$P_{in}(E_{1,2}\|H_1)$	[0.204 5,0.227 3]	0.209 7	$P_{in}(E_{6,1}\|H_1\ E_{4,1})$	[0.620 7,0.655 2]	0.644 8
$P_{in}(E_{1,3}\|H_1)$	[0.545 5,0.568 2]	0.557 2	$P_{in}(E_{6,1}\|H_1\ E_{4,2})$	[0.250 0,0.312 5]	0.268 7
$P_{in}(E_{2,1}\|H_1\ E_{1,1})$	[0.454 5,0.545 5]	0.488 6	$P_{in}(E_{6,2}\|H_1\ E_{4,1})$	[0.344 8,0.379 3]	0.355 2
$P_{in}(E_{2,1}\|H_1\ E_{1,2})$	[0,0.100 0]	0	$P_{in}(E_{6,2}\|H_1\ E_{4,2})$	[0.687 5,0.750 0]	0.731 2
$P_{in}(E_{2,1}\|H_1\ E_{1,3})$	[0,0.040 0]	0	$P_{in}(E_{7,1}\|H_1)$	[0.090 9,0.113 6]	0.092 4
$P_{in}(E_{2,2}\|H_1\ E_{1,1})$	[0.454 5,0.545 5]	0.511 4	$P_{in}(E_{7,2}\|H_1)$	[0.886 4,0.909 1]	0.907 6
$P_{in}(E_{2,2}\|H_1\ E_{1,2})$	[0.900 0,1.000 0]	1.000 0	$P_{in}(E_{8,1}\|H_1)$	[0.068 2,0.090 9]	0.069 6
$P_{in}(E_{2,2}\|H_1\ E_{1,3})$	[0.960 0,1.000 0]	1.000 0	$P_{in}(E_{8,2}\|H_1)$	[0.909 1,0.931 8]	0.930 4
$P_{in}(E_{3,1}\|H_1)$	[0.590 9,0.613 6]	0.604 1	$P_{in}(E_{9,1}\|H_1\ E_{1,1})$	[0.454 5,0.545 5]	0.488 6
$P_{in}(E_{3,2}\|H_1)$	[0.386 4,0.409 1]	0.395 9	$P_{in}(E_{9,1}\|H_1\ E_{1,2})$	[0.700 0,0.800 0]	0.785 7
$P_{in}(E_{4,1}\|H_1\ E_{3,1})$	[0.629 6,0.666 7]	0.654 3	$P_{in}(E_{9,1}\|H_1\ E_{1,3})$	[0.240 0,0.280 0]	0.252 5
$P_{in}(E_{4,1}\|H_1\ E_{3,2})$	[0.166 7,0.222 2]	0.175 2	$P_{in}(E_{9,2}\|H_1\ E_{1,1})$	[0.454 5,0.545 5]	0.511 4
$P_{in}(E_{4,2}\|H_1\ E_{3,1})$	[0.333 3,0.370 4]	0.345 7	$P_{in}(E_{9,2}\|H_1\ E_{1,2})$	[0.200 0,0.300 0]	0.214 3
$P_{in}(E_{4,2}\|H_1\ E_{3,2})$	[0.777 8,0.833 3]	0.824 8	$P_{in}(E_{9,2}\|H_1\ E_{1,3})$	[0.720 0,0.760 0]	0.747 5
$P_{in}(E_{5,1}\|H_1)$	[0.636 4,0.659 1]	0.651 0	–	–	–

It is clear from Fig. 5.26 that due to varying operating conditions each hour, the fault probability of the contact wire changes, with different probabilities for each estimated period. Moreover, the impact of various factors on contact wire faults, as derived from historical data on the number of occurrences and prior probability, not only reflects the influence of the equipment's aging but also highlights the characteristics of these impacts. For example, the second hour is affected by low temperatures, ice and snow, and snake and bird hazards, while the seventh hour is disrupted by high temperatures, rainfall, and strong winds, leading to relatively higher fault

Table 5.15 Contact wire continuous 8-period operating conditions

Period	E_1	E_2	E_3	E_4	E_5	E_6	E_7	E_8	E_9
1	$E_{1,1}$	$E_{2,2}$	$E_{3,2}$	$E_{4,2}$	$E_{5,2}$	$E_{6,2}$	$E_{7,2}$	$E_{8,2}$	$E_{9,2}$
2	$E_{1,1}$	$E_{2,1}$	$E_{3,2}$	$E_{4,2}$	$E_{5,2}$	$E_{6,2}$	$E_{7,2}$	$E_{8,2}$	$E_{9,1}$
3	$E_{1,2}$	$E_{2,2}$	$E_{3,2}$	$E_{4,2}$	$E_{5,2}$	$E_{6,2}$	$E_{7,2}$	$E_{8,2}$	$E_{9,1}$
4	$E_{1,2}$	$E_{2,2}$	$E_{3,1}$	$E_{4,1}$	$E_{5,2}$	$E_{6,2}$	$E_{7,2}$	$E_{8,2}$	$E_{9,2}$
5	$E_{1,2}$	$E_{2,2}$	$E_{3,2}$	$E_{4,2}$	$E_{5,1}$	$E_{6,1}$	$E_{7,2}$	$E_{8,2}$	$E_{9,2}$
6	$E_{1,2}$	$E_{2,2}$	$E_{3,2}$	$E_{4,2}$	$E_{5,2}$	$E_{6,2}$	$E_{7,1}$	$E_{8,2}$	$E_{9,1}$
7	$E_{1,3}$	$E_{2,2}$	$E_{3,1}$	$E_{4,2}$	$E_{5,1}$	$E_{6,2}$	$E_{7,2}$	$E_{8,2}$	$E_{9,2}$
8	$E_{1,3}$	$E_{2,2}$	$E_{3,2}$	$E_{4,2}$	$E_{5,2}$	$E_{6,2}$	$E_{7,2}$	$E_{8,1}$	$E_{9,2}$

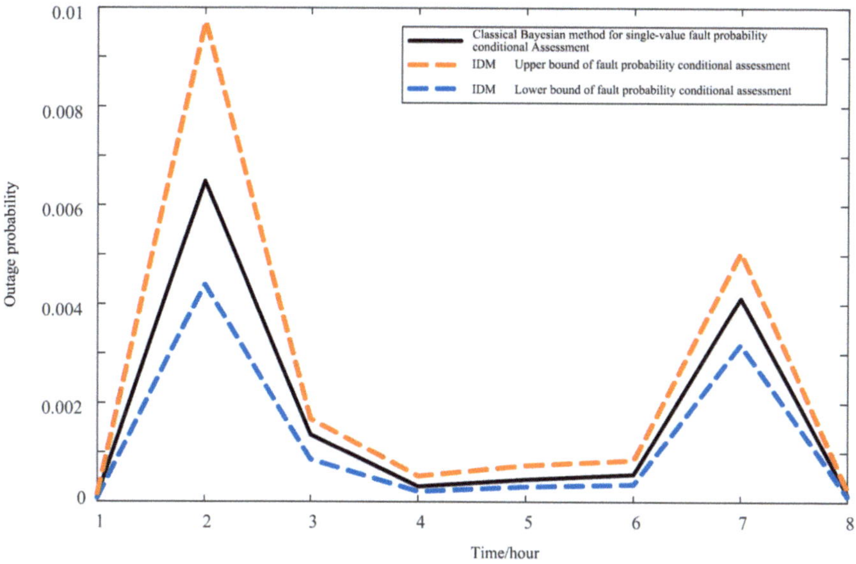

Fig. 5.26 Interval probability/single-value probability estimation of contact wire during continuous periods

probabilities under harsher operating conditions. Therefore, the method proposed in this paper demonstrates the time-varying nature of the operational reliability level of traction power supply equipment as its operating conditions change. It also reflects the combined influence of multiple factors such as equipment aging, natural environmental and weather conditions, load state, and human threats on the probability of traction power supply equipment failure.

From Table 5.14 and Fig. 5.26, it can be observed that the precise probabilities estimated using the classical deterministic Bayesian network method fall within the interval probability range. This implies that the interval probability results derived from credal network inference encompass the precise single-value conditional probability estimates provided by the classical Bayesian network. Interval probabilities provide a more comprehensive description of equipment operational reliability

Table 5.16 OCS Equipment risk loss foundation data

Equipment	Risk level	Economic loss			Time loss	
		Equipment loss	Maintenance costs	Risk cost	Power outage	Travel or stay within the interval
Contact wire	A	1 000	1 000	N/A	35.7 min	450 min
Cantilever	B	500	500	500	N/A	N/A
Dropper	C	300	500	N/A	N/A	N/A

Table 5.17 Three types of OCS equipment continuous 18-period operating conditions

Period	E_1	E_2	E_3	E_4	E_5	E_6	E_7	E_8	E_9
1	$E_{1,2}$	$E_{2,2}$	$E_{3,2}$	$E_{4,2}$	$E_{5,2}$	$E_{6,2}$	$E_{7,2}$	$E_{8,2}$	$E_{9,2}$
2	$E_{1,2}$	$E_{2,2}$	$E_{3,1}$	$E_{4,2}$	$E_{5,2}$	$E_{6,2}$	$E_{7,2}$	$E_{8,2}$	$E_{9,2}$
3	$E_{1,2}$	$E_{2,2}$	$E_{3,1}$	$E_{4,1}$	$E_{5,2}$	$E_{6,2}$	$E_{7,2}$	$E_{8,2}$	$E_{9,2}$
4	$E_{1,2}$	$E_{2,2}$	$E_{3,1}$	$E_{4,1}$	$E_{5,1}$	$E_{6,2}$	$E_{7,2}$	$E_{8,2}$	$E_{9,2}$
5	$E_{1,2}$	$E_{2,2}$	$E_{3,1}$	$E_{4,1}$	$E_{5,1}$	$E_{6,1}$	$E_{7,2}$	$E_{8,2}$	$E_{9,2}$
6	$E_{1,2}$	$E_{2,2}$	$E_{3,2}$	$E_{4,2}$	$E_{5,2}$	$E_{6,1}$	$E_{7,2}$	$E_{8,2}$	$E_{9,1}$
7	$E_{1,3}$	$E_{2,2}$	$E_{3,2}$	$E_{4,2}$	$E_{5,2}$	$E_{6,2}$	$E_{7,2}$	$E_{8,2}$	$E_{9,2}$
8	$E_{1,3}$	$E_{2,2}$	$E_{3,1}$	$E_{4,2}$	$E_{5,2}$	$E_{6,2}$	$E_{7,2}$	$E_{8,2}$	$E_{9,2}$
9	$E_{1,3}$	$E_{2,2}$	$E_{3,1}$	$E_{4,1}$	$E_{5,2}$	$E_{6,2}$	$E_{7,2}$	$E_{8,2}$	$E_{9,2}$
10	$E_{1,3}$	$E_{2,2}$	$E_{3,1}$	$E_{4,1}$	$E_{5,1}$	$E_{6,2}$	$E_{7,2}$	$E_{8,2}$	$E_{9,2}$
11	$E_{1,3}$	$E_{2,2}$	$E_{3,1}$	$E_{4,1}$	$E_{5,2}$	$E_{6,1}$	$E_{7,2}$	$E_{8,2}$	$E_{9,2}$
12	$E_{1,3}$	$E_{2,2}$	$E_{3,2}$	$E_{4,2}$	$E_{5,2}$	$E_{6,1}$	$E_{7,2}$	$E_{8,2}$	$E_{9,1}$
13	$E_{1,1}$	$E_{2,2}$	$E_{3,2}$	$E_{4,2}$	$E_{5,2}$	$E_{6,2}$	$E_{7,2}$	$E_{8,2}$	$E_{9,2}$
14	$E_{1,1}$	$E_{2,1}$	$E_{3,2}$	$E_{4,2}$	$E_{5,2}$	$E_{6,2}$	$E_{7,2}$	$E_{8,2}$	$E_{9,2}$
15	$E_{1,1}$	$E_{2,1}$	$E_{3,2}$	$E_{4,2}$	$E_{5,1}$	$E_{6,2}$	$E_{7,2}$	$E_{8,2}$	$E_{9,2}$
16	$E_{1,1}$	$E_{2,1}$	$E_{3,2}$	$E_{4,2}$	$E_{5,2}$	$E_{6,2}$	$E_{7,2}$	$E_{8,2}$	$E_{9,1}$
17	$E_{1,1}$	$E_{2,2}$	$E_{3,2}$	$E_{4,2}$	$E_{5,2}$	$E_{6,2}$	$E_{7,2}$	$E_{8,2}$	$E_{9,1}$
18	$E_{1,1}$	$E_{2,2}$	$E_{3,2}$	$E_{4,2}$	$E_{5,2}$	$E_{6,1}$	$E_{7,2}$	$E_{8,2}$	$E_{9,1}$

levels, demonstrating that the interval probabilities derived in this chapter can reflect the estimation errors and uncertainties caused by insufficient sample data.

To further quantify risk loss, this chapter combines actual site data to statistically organize three types of OCS equipment: contact wire, cantilever, and dropper. The fault occurrences for the contact wire remain the same as previously mentioned. Over the six years of statistics, the cantilever and dropper on this line experienced 30 and 26 faults respectively. The prior fault probabilities are taken from the average failure rate over the last three years, which are 0.000617 occurrences per hour for the cantilever and 0.000540 occurrences per hour for the dropper. The foundational data for the risk loss of these three types of OCS equipment are shown in Tables 5.16 and 5.17 presents the risk conditions of each equipment failure under the simulated continuous 18 periods of operating conditions.

The continuous 18-period operating conditions set in Table 5.17 are divided into segments of 6 h each (1–6 h, 7–12 h, 13–18 h), progressively deteriorating under

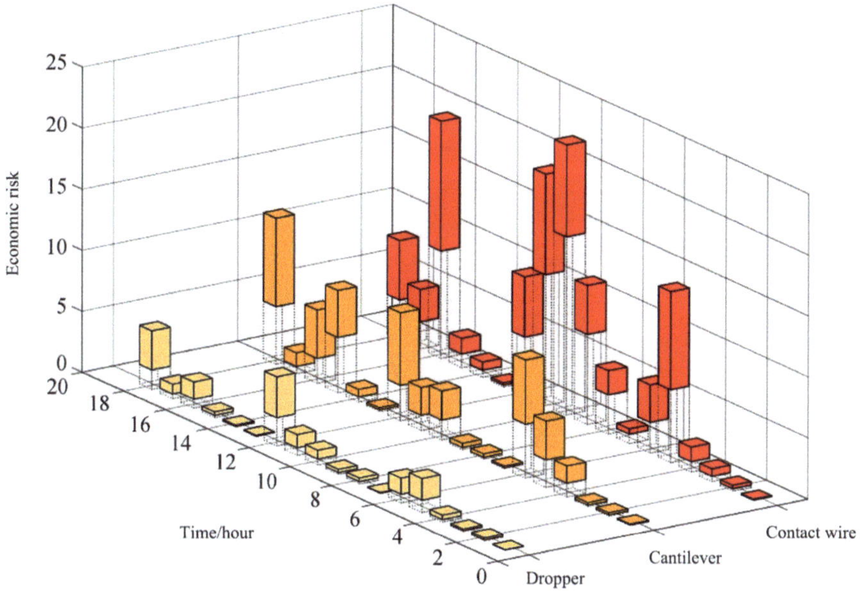

Fig. 5.27 Economic risk interval estimation for OCS equipment faults over continuous periods

conditions of normal temperature, high temperature, and low temperature (with gradually increasing influencing factors). Using Eq. (5.62), the economic risk indicators for faults in contact wire, cantilever, and dropper can be determined, as illustrated in Fig. 5.27. Figure 5.27 reflects the expected economic losses caused by faults in the contact wire, cantilever, and dropper under various period operating conditions. By conducting a horizontal comparison of the three types of OCS equipment, the contact wire, which belongs to Category A fault risk level due to relatively higher equipment loss and maintenance costs, also shows a relatively higher economic risk expectation. Cantilever faults, categorized under Category B fault level, consider not only direct equipment loss and maintenance and repair costs but also the risk cost that may cause a power outage, making the overall risk of cantilever faults higher than that of droppers.

Although the foundational data on economic losses differ among the three types of OCS equipment, a similar trend in economic risk is observed over the 6-h segmented intervals. In each segment, economic risk gradually increases during the first 4 h. However, in the subsequent two hours, as operating conditions and influencing factors change, the magnitude and trend of risk reflected by the three types of OCS equipment also differ. For example, comparing the fifth and sixth hours under consistent operating conditions, the contact wire and dropper are more significantly affected by rainfall, thunderstorms, and strong winds than by snake and bird hazards, hence the economic risk expectation is higher in the fifth hour. In contrast, the cantilever is more affected by snake and bird hazards, making its

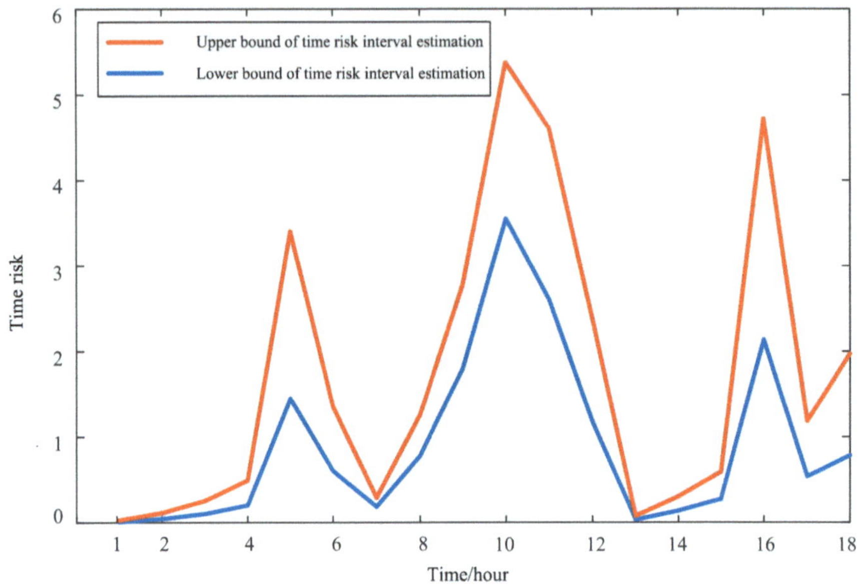

Fig. 5.28 Time risk interval estimation for contact wire faults over continuous periods

economic risk expectation for the sixth hour higher than in the fifth. These trends align with the historical statistical data patterns of OCS equipment faults.

For the contact line with a Category A fault level, faults not only cause economic loss but also lead to a power outage shutdown, resulting in time loss. Figure 5.28 illustrates the time risk indicator for contact wire faults, derived using Eq. (5.63). This figure shows the trend of the time risk interval for contact wire faults over continuous 18 periods as influenced by changing operating conditions. Affected by a variety of factors, the time risk for contact wire faults is higher during the fifth, 10th, 11th, and 16th hours, comprehensively quantifying the likelihood of contact wire faults and the consequences of outages.

This section reveals that the proposed method can combine multiple factors under traction power supply equipment reliability interval estimation results, quantifying fault risks and considering the consequences of economic and time losses to provide a comprehensive risk indicator. Figures 5.26, 5.27, and 5.28 demonstrate that the interval estimates include classical Bayesian network precise single-value conditional estimation results, showcasing the ability of interval estimation to handle small sample data and more comprehensively describe the operational risk level of traction power supply equipment. This provides a complete and reliable reference for operational and maintenance decision-making in traction power supply systems.

5.5 Conclusion

This section first outlines the reliability and risk assessment indicators, reliability assessment models, system risk sources, and research challenges of the high-speed railway traction power supply system, discussing the significance of such assessments in the context of modern high-speed railway systems. The chapter analyzes load characteristics of the high-speed railway traction power supply system, establishes a traction transformer reliability assessment model that considers load characteristics, and conducts a reliability assessment of the traction transformer at the Shangyu North traction substation on the Hangzhou-Ningbo high-speed railway. It also examines the impact of relay protection failures on the reliability of the traction power supply system and analyzes the reliability assessment and case studies of traction power supply systems with relay protection. Finally, the chapter introduces an interval estimation method for traction power supply equipment failure probability, analyzes and quantifies the consequences of equipment failures, and conducts a multifactor integrated traction power supply equipment fault risk assessment, exploring practical case studys of risk assessment methods applied to high-speed railway traction power supply equipment.

Chapter 6
Maintenance Decision-Making of High-Speed Railway Traction Power Supply System

Over the years, China's maintenance and repair strategies for electrified railway traction power supply equipment and systems have typically borrowed models from power systems or drawn on experiences from conventional railway operations. This approach has resulted in outdated maintenance philosophies, a lack of sophisticated repair theories, and stagnant maintenance technologies. Consequently, on-site maintenance work has failed to adapt effectively to the operational characteristics of high-speed railways. It has been challenging to tailor maintenance activities to the specific service environments of traction power supply equipment or to address the considerable challenges presented by the transition to full operational maintenance phases for large-scale, network-based high-speed railway lines globally. The deficiencies in determining operating statuses, handling faults, or the modes of maintenance and repair within traction power supply systems and equipment contribute to these issues.

The most apparent issue is the outdated nature of the maintenance and repair models currently used in traction power supply systems, which still rely on periodic maintenance, manual inspections, and emergency repairs. This passive approach often results in either insufficient or excessive maintenance. A major flaw is the lack of specificity in developing maintenance strategies; some of these strategies are based on operating procedures, maintenance guidelines, and technical standards from the power system sector, while others simply apply existing protocols from conventional railways without adequately considering the structural characteristics and service conditions unique to high-speed railway traction power supply systems. This often leads to inefficient and even misguided maintenance activities. Furthermore, the implementation of maintenance activities tends to be blind and speculative, with maintenance personnel relying on experience or even luck to bring equipment out of a fault state. Although this may temporarily eliminate the fault on the surface, such maintenance, lacking scientific basis, poses hidden risks for subsequent operation of the equipment. Therefore, this passive maintenance and repair model results in low maintenance efficiency, high economic costs, and increased workload and work intensity.

Z. He et al., *Prognostics and Health Management of High-Speed Railway Traction Power Substations*, https://doi.org/10.1007/978-981-95-4538-4_6

As technology related to high-speed railways continues to advance, including design, manufacturing, and construction techniques, the deficiencies in operational maintenance capabilities for traction power supply systems have become a significant constraint. These shortcomings hinder the full operational maintenance and improvement of availability, security, and economic efficiency in high-speed railways. Thus, effectively addressing these deficiencies and establishing a robust maintenance and repair theory and technology would significantly enhance the service capability of equipment and the efficiency of system operations. This would not only save on operational maintenance costs but also significantly reduce the workload and pressure on maintenance personnel, supporting the transition to a comprehensive operational maintenance phase for high-speed railways.

Based on the reasons discussed, this chapter explores the decision-making and optimization of maintenance strategies for traction power supply equipment and systems from various perspectives, including degradation modeling, fault prognostics, risk assessment, and multi-attribute decision-making.

6.1 Overview of Maintenance Strategy for High-Speed Railway Traction Power Supply System

6.1.1 Concept of Maintenance for High-Speed Railway Traction Power Supply Equipment

Maintenance of high-speed railway traction power supply equipment refers to the technical activities carried out to restore the function of the equipment after technical condition deterioration or fault occurrence. Maintenance encompasses both the activities of maintenance and repair.

Maintenance involves a series of tasks such as cleaning, inspection, washing, lubrication, tightening, adjustment, and anti-corrosion, collectively referred to as upkeep. It is performed according to a predefined plan or specified technical conditions to timely detect and address abnormal phenomena that occur during operation, reducing the rate of performance degradation and lowering the likelihood of faults. This is crucial for ensuring normal operation and extending the lifespan of the equipment.

Maintenance tasks vary in depth and scope and can be classified into routine maintenance, primary maintenance, and secondary maintenance. Most enterprises in China adopt the secondary maintenance mode.

Repair refers to all technical work performed to restore the specified technical performance and proper working status of equipment when a fault occurs or when the technical condition degrades to a certain critical level. Since repairs often rely on inspection results and are combined with inspections, this process is also referred to as maintenance. Repair is one of the primary means to restore equipment performance, ensure normal operation, and extend the material life of the equipment.

Repairs can be categorized by function into corrective repair and improvement repair. Typically, the term *repair* refers to restorative repair, aimed at bringing equipment back to its proper working condition. Improvement repair, however, involves making enhancements or modifications to areas of the equipment with a high failure rate, through changes in structure, parameters, materials, and manufacturing processes, to reduce or eliminate the occurrence of faults.

The objective of maintenance is to keep the equipment in good condition with minimal consumption, cost, time, and resources, achieving the highest repair rate possible. This enhances equipment availability and availability, slows down degradation, extends service life, ensures security, and meets environmental protection standards.

6.1.2 Development of Maintenance Techniques for High-Speed Railway Traction Power Supply Equipment

Over the past half-century, concepts, theories, techniques, and management models related to maintenance and repair have continuously evolved. From the initial stage of breakdown maintenance (BM), where repairs were only conducted after equipment damage or faults, to the subsequent stage of preventive maintenance (PM), involving regular maintenance activities based on set intervals and schedules. More recently, the last few decades have seen the development of condition-based maintenance (CBM), where maintenance activities are planned based on the actual operational status of the equipment. This shift reflects a gradual change from post-fault repairs to pre-fault state monitoring and maintenance.

Breakdown maintenance refers to a maintenance method where repairs are conducted after equipment faults occur. This was the earliest form of maintenance regime. Due to the unpredictability of when and where faults may occur, this approach often lacks pre-repair preparation, leading to lengthy downtimes for equipment repairs. Additionally, these unplanned repairs can disrupt production schedules, potentially affecting product quality and delivery times, and resulting in significant losses for businesses. While advanced maintenance modes have largely replaced it, breakdown maintenance remains applicable for handling unexpected faults.

Preventive maintenance is aimed at preventing faults before they occur, based on the principle of prevention and averting potential issues. This method requires maintaining upkeep during equipment operation, enhancing routine inspections, and performing regular checks. Based on the patterns of equipment faults, wear and tear of parts, and inspection results, preventive maintenance involves planning repairs or replacements in advance for parts that are prone to wear or fault. This proactive approach helps prevent equipment failures or avoids sudden faults.

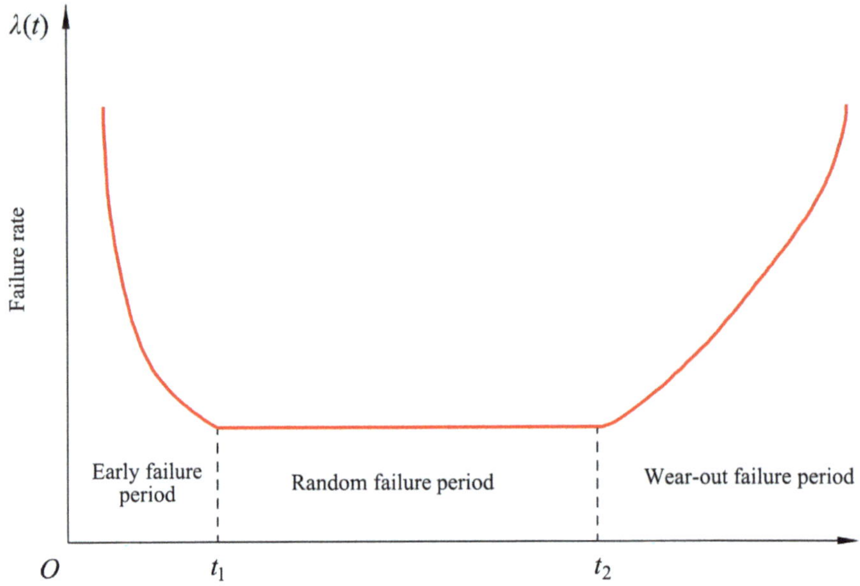

Fig. 6.1 Common maintenance mode classification

Preventive maintenance focuses on calculating the operational cycles and monitoring the usage frequency of equipment. Measurements of data must be precise, and this type of maintenance is typically employed for handling wear-prone parts. Establishing a preventive maintenance regime is essential for consumables prone to damage; otherwise, sudden shutdowns or accidents could occur, disrupting safe and continuous production. Preventive maintenance offers advanced methods for managing equipment, ensuring safe operation, and enhancing the overall economic benefits of an enterprise.

State-based maintenance relies on advanced state monitoring and diagnostic technologies to provide information about equipment conditions. This approach involves detecting abnormalities and predicting potential faults in equipment, allowing for the rational scheduling of maintenance projects and intervals based on anticipated fault information. Thus, maintenance plans are organized according to the health condition of the equipment, making maintenance tasks more scientific, reducing unnecessary work, and saving on labor and costs (Fig. 6.1).

6.1.3 Life Distribution Law of High-Speed Railway Traction Power Supply Equipment

A complex system or piece of equipment is composed of various components, and faults in any of these can lead to system-wide failures. The increasing complexity of

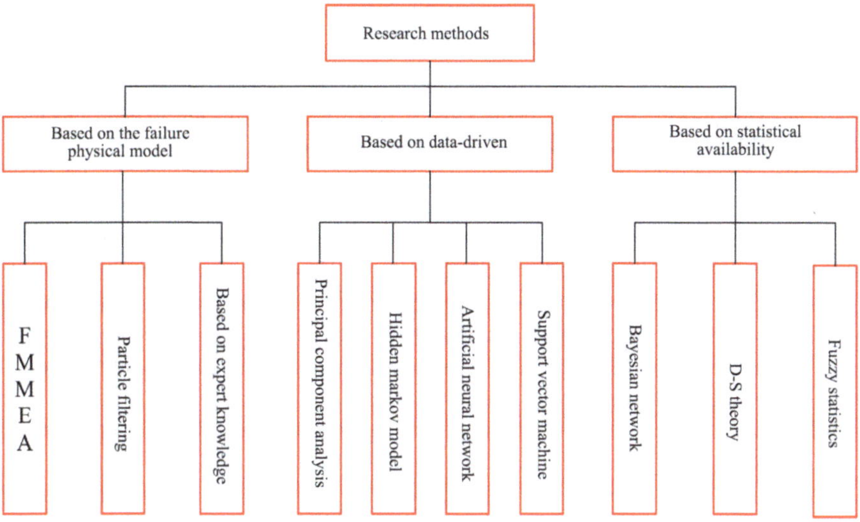

Fig. 6.2 Bathtub curve

modern equipment has diversified failure modes, leading to a failure rate that evolves over time and typically follows a pattern resembling a bathtub curve. This pattern is known in availability research as the *bathtub curve*, as illustrated in Fig. 6.2.

Generally, the bathtub curve consists of three parts, as shown in Fig. 6.2. At the left edge, the failure rate decreases; in the middle section, the failure rate approaches a constant; and at the right edge, the failure rate increases. During early usage, due to internal defects left bymanufacturing and assembly processes, the product's failure rate is often higher. An early failure elimination test is usually conducted before the product leaves the factory to eliminate early faults. After the product goes through the early failure period, the failure rate stabilizes at a lower level, approaching a constant. This period, characterized by failures often caused by random reasons, is called the random failure period or normal operation period. The corresponding usage time during this period is called the effective service life. After experiencing a relatively stable failure rate period, the failure rate begins to increase due to reasons such as aging and wear of components. This period is referred to as the wear-out failure period (Fig. 6.2).

6.1.4 Main Research Methods for Maintenance Strategy of High-Speed Railway Traction Power Supply System

State-based maintenance represents the future trend in maintenance methodologies. System health management builds on the foundation of traditional fault diagnosis and availability research, incorporating techniques from human physiological monitoring, pathological diagnostics, immune system functions, and disease treatment and recovery to address complex systems. This approach involves comprehensive research in fault prognostics, risk early warning, availability analysis, health assessment, and maintenance management. PHM is currently the hottest and most developed theory and technology in this research area.

PHM demands the ability to predict system faults and manage the health status systematically. Its core principle involves integrating advanced sensors and employing various intelligent algorithms and inference models to monitor, assess, predict, and manage the health status of systems. This proactive approach provides timely predictive warnings before system failures occur, integrating data, information, and resources from various channels to formulate comprehensive maintenance and repair measures. PHM represents a paradigm shift from traditional fault diagnosis, moving from diagnostics based on single sensors to prognostics based on intelligent systems and extending in-situ testing capabilities and condition monitoring technologies. Currently, there are three relatively mature and commonly used methods in PHM research: methods based on the physics of failure, data-driven methods, and statistical availability-based methods (Fig. 6.3).

6.1.4.1 Maintenance Decision-Making Methods Based on Failure Physical Models

Uses the knowledge of lifecycle loads and failure mechanisms for prognostics or diagnostics. This method requires in-depth knowledge specific to the domain of the object being analyzed, focusing on creating physical models of failure. These models are supported by mathematical modeling where possible, offering high prognostic accuracy. However, their application is limited when dealing with complex dynamic systems where precise mathematical models are hard to establish. Typical methods include Failure Mode, Mechanisms and Effects Analysis (FMMEA), Kalman filtering/Particle filtering, and Expert Knowledge-based methods.

FMMEA refers to a novel technology in availability engineering where potential failure modes and mechanisms of each component of a product are analyzed during the design and production process. This analysis focuses on how these failures could affect the product's functionality, performance, and long-term degradation, and suggests possible preventive and improvement measures.

The concept of Kalman filtering/Particle filtering is based on Monte Carlo methods, which use a particle set to represent probabilities that can be applied in any form of state space model. Its core idea involves using random state particles drawn from the posterior probability to represent the distribution characteristics, employing a sequential importance sampling method.

6.1.4.2 Data-Driven Maintenance Decision Method

In many practical situations, it is impossible or difficult to obtain or solve the mathematical model of complex systems. In such cases, installing various sensors

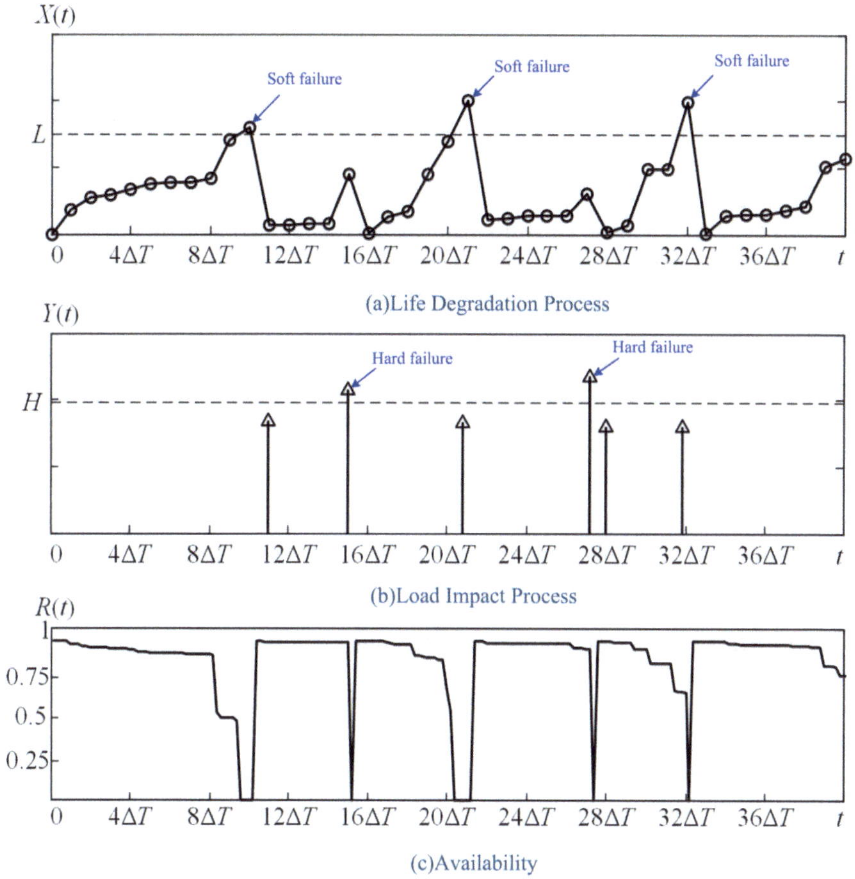

Fig. 6.3 Main research methods in maintenance strategy

at critical locations on the subject of study or conducting multiple experiments becomes an effective way to understand the system's service patterns. Data-driven methods utilize data collected from these sensors or experiments to analyze and predict system health. These methods are broadly applicable and can achieve significant accuracy. Typical data-driven methods include Principal Component Analysis (PCA), Hidden Markov Models (HMM), Support Vector Machines (SVM), and Artificial Neural Networks (ANN).

PCA serves to reduce dimensionality by combining multiple indicators into a few independent composite indicators (the principal components). Each principal component reflects most of the information of the original variables without redundancy. This method simplifies complex issues by reducing them to a few principal components, facilitating a more scientifically effective analysis.

HMM is a statistical model that describes a Markov process with hidden unknown parameters. The challenge lies in identifying these hidden parameters from observable data, which are then used for further analysis.

SVM is a classification algorithm that aims to minimize structural risk to enhance the generalization capability of learning machines, achieving a balance between empirical risk and confidence interval minimization. It is effective even with a small sample size. Essentially, SVM is a type of binary classification model. Its basic model is defined as a linear classifier with the maximum margin in the feature space. The learning strategy of the support vector machine is to maximize the margin, which can ultimately be transformed into solving a convex quadratic programming problem.

ANN consists of numerous neurons connected by tunable connection weights, characterized by large-scale parallel processing, distributed information storage, and robust self-organizing learning capabilities. Parameters such as the number of middle layers, the number of units in each layer, and the learning rate of the network can be flexibly set based on specific needs. ANNs are widely used in various fields such as optimization, signal processing, pattern recognition, intelligent control, and fault diagnosis due to their flexibility and broad application prospects.

6.1.4.3 Statistical Availability-Based Maintenance Decision-Making Method

Due to various constraints, it is often difficult to solve the system's mathematical model or obtain sufficient status data. In such cases, as long as enough historical failure data has been accumulated, statistical availability-based methods can be used to diagnose and predict faults. Statistical availability-based methods involve fitting an object's life distribution from an availability perspective and estimating the average remaining life. Due to a lack of detailed information, the accuracy of these methods can vary significantly, often necessitating the inclusion of a confidence level to indicate the precision of the prognostics. Common methods in this category include the Bayesian network method, D-S theory, and fuzzy statistical method.

A Bayesian network is one of the most effective theoretical models for expressing and inferring uncertainty. It consists of a directed acyclic graph with nodes representing random variables and directed edges connecting these nodes, indicating inter-variable relationships (from parent nodes to subnodes) expressed through conditional probabilities. Nodes without parents express information through prior probabilities. This framework is suited for expressing and analyzing events characterized by uncertainty and probabilistic dependencies, and it is applicable to decision-making that depends conditionally on multiple control factors. Bayesian networks facilitate inference from incomplete, imprecise, or uncertain knowledge or information.

D-S theory extends the Bayesian inference method by using Bayesian conditional probabilities from probability theory to provide known conditions. Unlike Bayesian methods, D-S theory does not require prior probabilities to represent *uncertainty* effectively, making it widely used for handling uncertain data. It is particularly applicable in scenarios involving information fusion, expert systems, intelligence analysis, legal case analysis, and multi-attribute decision analysis.

The fuzzy statistical method is a comprehensive evaluation approach based on fuzzy mathematics. Utilizing the theory of membership degrees from fuzzy mathematics, this method converts qualitative assessments into quantitative evaluations. It provides clear results and is systematic, effectively addressing problems that are fuzzy and difficult to quantify. This method is well-suited for resolving various types of uncertainties, offering a holistic assessment of entities or phenomena influenced by multiple factors.

6.2 Maintenance Strategy for Traction Power Supply Equipment Based on Degradation Modeling

The service performance degradation and failure processes of traction power supply equipment exhibit typical randomness, manifested in both temporal and spatial dimensions. For instance, the occurrence time of load impacts shows temporal randomness, while the magnitude of load impacts and the degradation amount per unit time display spatial randomness. Therefore, based on the foundation laid in Sect. 6.2, this section employs the random process method to model the service performance degradation and failure processes of traction power supply equipment, aiming to describe the impacts of temporal and spatial randomness. Furthermore, based on competitive failure modes of soft and hard failures, the availability indicators for the full lifecycle service period of traction power supply equipment are derived. Additionally, a proactive maintenance strategy is based on the *combined monitoring, inspection, and maintenance strategy*. This strategy, combined with economic considerations over the full lifecycle service period, defines performance evaluation indicators. The proactive maintenance strategy is then optimized to achieve the best performance under the optimal configuration of decision variables. Finally, a sensitivity analysis method is employed to study how the optimal proactive maintenance decision results vary with changes in the actual model parameters used on site.

6.2.1 Basic Assumptions

The life degradation of traction power supply equipment is largely concealed, with many early faults also being non-obvious. Due to constraints related to cost and availability, the number and variety of online monitoring equipment currently deployed in operational settings are quite limited, typically only covering basic functions such as temperature, gas pressure, and oil chromatography monitoring. This significantly restricts monitoring the extent of service performance degradation (health condition) of the equipment and hinders the detection of early internal faults. Therefore, in addition to effectively implementing proactive maintenance for traction power supply equipment, periodic inspections should also be employed. Periodic inspection is the most commonly used maintenance approach in the traction power supply system, involving regular online or offline manual inspections, preventive tests, handheld equipment measurements, and non-contact flaw detection to determine the service status of equipment, identify hidden faults, and guide maintenance activities. Based on this, the proposed proactive maintenance strategy for traction power supply equipment in practice is a *combined monitoring, inspection, and maintenance strategy*: continuous health indicators (such as the temperature of traction transformers and leak rates of GIS high-voltage cabinets) and traction load data are collected by online monitoring equipment and SCADA systems, while intermittent health indicators (such as the dielectric loss factor of traction transformers and the main circuit resistance of GIS high-voltage cabinets) are obtained through periodic inspection methods. After assessing the equipment's health condition and service performance degradation, maintenance is performed. Importantly, the critical thresholds within this proactive maintenance strategy are not fixed but need to be dynamically adjusted based on the specific service conditions and performance requirements of the equipment to achieve the best overall performance of the maintenance strategy. Maintaining high availability throughout the full lifecycle service period and reducing maintenance costs are two major factors in measuring the performance of a maintenance strategy. The optimal performance of the proactive maintenance strategy seeks to balance these factors.

Under the premise of comprehensive maintenance, the time spent on inspection and maintenance activities is almost negligible compared to the full lifecycle of traction power supply equipment. Therefore, it is assumed that inspections are instantaneous, complete, and non-destructive:

(1) Instantaneity of inspection: Inspections are completed instantaneously without occupying the normal service time of the equipment.
(2) Completeness of inspection: The results of inspections can accurately determine the health condition and the degree of life degradation of the equipment.
(3) Non-destructive nature of inspection: Inspections do not affect the health condition or the degree of life degradation of the equipment, nor do they impact its normal operation.

Similarly, it is assumed that maintenance actions can be implemented immediately after inspections and possess qualities of instantaneity and completeness:

(1) Instantaneity of maintenance action: Maintenance actions are completed instantaneously without occupying the normal service time of the equipment.
(2) Completeness of maintenance action: Each maintenance action restores the equipment to a *like new* condition.

6.2.2 Availability Indicators Describing Equipment Service Performance Degradation

The service performance of traction power supply equipment gradually degrades over its entire lifecycle, undergoing several faults and repairs. availability indices are used to assess the service performance of traction power supply equipment during this process. Core indices like availability and fault probability density not only reflect the current service performance status of the equipment but also predict its future service performance trends. Therefore, availability and fault probability density are used to describe the service performance of traction power supply equipment throughout its full lifecycle in a real-time and accurate manner.

Under the assumption of a complete repair model, it is assumed that the life degradation process (soft failure) and load impact process (hard failure) are independent of each other. Since the failure of traction power supply equipment is a competitive failure mode of soft and hard failures, the availability of the equipment at time t is equal to the product of the probabilities of two events: "the equipment survives all S(t) load impacts until time t (i.e., $Y_i < H$, $i = 1, 2, \cdots, S$)" and "the life degradation amount of the equipment at time t is less than the threshold (i.e., $X(t) < L$)." Therefore, the availability of the traction power supply equipment at time t is given by:

$$
\begin{aligned}
R(t) &= Pr\{X(t) < L, Y_1 < H, Y_2 < H, \cdots, Y_{S(t)} < H\} = Pr\{X(t) < L, S(t) = 0\} \\
&\quad + \sum_{i=1}^{S} Pr\{X(t) < L, Y_1 < H, Y_2 < H, \cdots, Y_{S(t)} < H, S(t) = i\} \\
&= Pr\{X(t) < L \mid S(t) = 0\} \cdot Pr\{S(t) = 0\} \\
&\quad + \sum_{i=1}^{S} F_Y^i(H) \cdot Pr\{X(t) < L \mid S(t) = 0\} \cdot \Pr\{S(t) = i\}
\end{aligned}
\tag{6.1}
$$

Because the time of soft failure τ_L is related to the cumulative distribution function of degradation $X(t)$ as $F_{X(t)}(L) = 1 - F_{\tau_L}(t)$, substituting this into Eq. (6.1) yields:

$$
R(t) = \frac{\Gamma(\alpha) - \Gamma(\alpha, L\beta)}{\Gamma(\alpha)} e^{-\lambda t} + \sum_{i=1}^{S} \Phi^i\left(\frac{H - \mu}{\sigma}\right) \cdot \frac{\Gamma(\alpha) - \Gamma(\alpha, L\beta)}{\Gamma(\alpha)} \cdot \frac{(\lambda t)^{i-1} e^{-\lambda t}}{i!}
\tag{6.2}
$$

Thus, the fault probability density function $f_\tau(t)$, which is the probability density function of the random failure time τ, can be derived.

$$f_{\tau}(t) = -\frac{\mathrm{d}R(t)}{\mathrm{d}t} = \lambda\frac{\Gamma(\alpha)-\Gamma(\alpha,L\beta)}{\Gamma(\alpha)}\mathrm{e}^{-\lambda t} - \sum\nolimits_{i=1}^{S}\Phi^{i}\left(\frac{H-\mu}{\sigma}\right)\cdot\frac{\Gamma(\alpha)-\Gamma(\alpha,L\beta)}{\Gamma(\alpha)}\cdot\frac{\lambda(-\lambda t+i)(\lambda t)^{i-1}\mathrm{e}^{-\lambda t}}{i!} \tag{6.3}$$

In this way, as a function of time t, the availability function in Eq. (6.2) and the fault probability density function in Eq. (6.3) can reflect the service performance of the equipment at any time, ensuring real-time and accurate depiction of the equipment's availability throughout its full lifecycle service period.

Under the *combined monitoring, inspection, and maintenance strategy* for proactive maintenance, traction power supply equipment undergoes periodic inspections with an interval ΔT. Simultaneously, the SCADA system, serving as an online monitoring equipment, continuously collects real-time data on traction load. At a specific inspection time $T_j = j \cdot \Delta T$ (where $j = 1, 2, \cdots$), if the degradation amount exceeds the life degradation threshold L, or if a load impact amplitude within the monitoring period $(T_{j-1}, T_j]$ exceeds the threshold H, then a maintenance action must be immediately performed at the inspection completion time T_j. If neither of these conditions occurs, no maintenance action is taken at the inspection completion time, and the maintenance decision process is deferred to the next inspection time T_{j+1}.

This proactive maintenance strategy based on the *combined monitoring, inspection, and maintenance strategy* ensures that the traction power supply equipment undergoes several cycles of *service performance degradation* → *failure* (*soft failure* + *hard failure*) → *restored as new* throughout its full service life. This cycle represents the service performance evolution law of the traction power supply equipment over its entire lifecycle. Figure 6.4 illustrates the service performance evolution law of typical traction power supply equipment over a specific period and its corresponding availability.

6.2.3 Economic Indicators Describing Equipment Service Performance Degradation

In the context of a full lifecycle service period for a traction power supply system, maintenance costs accumulate over the equipment's service life, eventually becoming a significant expense in high-speed railway operations and severely impacting economic efficiency. When considering the maintenance of traction power supply equipment, it is essential to evaluate both the availability of the equipment and the economic aspects of the maintenance activities. Therefore, the performance of a proactive maintenance strategy for traction power supply equipment should be assessed from both availability and economic perspectives, where economics are primarily reflected through maintenance costs. An effective proactive maintenance

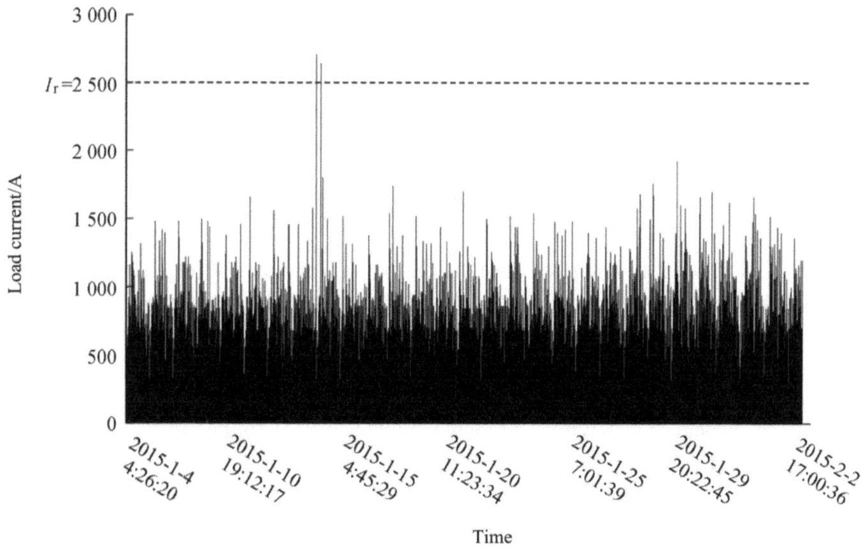

Fig. 6.4 Service performance evolution law and availability of typical traction power supply equipment

strategy should balance maintaining as high availability as possible with minimizing maintenance costs.

Due to the long service life of traction power supply equipment, maintenance work spans a long time scale, and the occurrence of faults and the implementation of maintenance actions are random in time. Therefore, it is difficult to describe the economic aspects by accurately calculating the total cost over the service lifecycle. Therefore, a scale transformation method based on renewal reward theory is used. This method divides the long service period into finite renewal cycles and introduces the expected maintenance costs per unit time to statistically describe the economics of the proactive maintenance strategy. A renewal cycle refers to the interval between two consecutive maintenance actions, encompassing the process from the equipment being restored as new, through gradual life degradation, the occurrence of soft or hard failures, the implementation of maintenance actions, to being fully restored as new. According to the renewal reward theorem, the expected maintenance cost per unit time is given by:

$$C_{ex} = \lim_{t \to \infty} \frac{C(t)}{t} = \frac{E[C(G)]}{E[G]} \tag{6.4}$$

where $C(t)$ represents the cumulative total maintenance cost up to time t, G is the length of a renewal cycle, $C(G)$ is the total maintenance cost within a renewal cycle, and E[ ] denotes the expectation of the random variables inside the brackets.

Through Eq. (6.4), the time scale for calculating maintenance costs is shortened from the equipment's full lifecycle service period to an expected renewal cycle.

The total maintenance cost within an expected renewal cycle equals the sum of inspection costs, downtime losses, and maintenance action costs. Thus, the expected total maintenance cost within a renewal cycle can be expressed as:

$$E[C(G)] = C_I E[N_I] + C_D E[T_D] + C_M \tag{6.5}$$

where C_I is the cost of one inspection, N_I is the total number of inspections within a renewal cycle, C_D is the cost rate of downtime loss (i.e., the loss cost per unit of downtime), and T_D is the downtime caused by equipment failure within a renewal cycle, C_M is the cost of a single maintenance action

Since equipment failures can only be detected through periodic inspections, the number of inspections N_I within a renewal cycle is related to the random failure time τ. In practice, the event {k inspections occur within a renewal cycle} is equivalent to the event {a failure occurs between the $(k-1)$th and kth inspections}. Therefore, the expected total number of inspections within a renewal cycle can be obtained as:

$$\begin{aligned} E[N_I] &= \sum\nolimits_{k=1}^{m} kPr\{N_I = k\} \\ &= \sum\nolimits_{k=1}^{m} kPr\{(k-1)\Delta T < \tau \leqslant k\Delta T\} = \sum\nolimits_{k=1}^{m} k \cdot \int_{(k-1)\Delta T}^{k\Delta T} f_\tau(t)\mathrm{d}t \end{aligned} \tag{6.6}$$

where m represents the total number of renewal cycles within the equipment's full lifecycle service period; $f_\tau(t)$ is the fault probability density function, obtained from Eq. (6.3).

Within the interval $((N_I - 1)\Delta T, N_I\Delta T]$ of the random failure time τ, the equipment downtime T_D spans from the moment of failure τ to the next inspection moment $N_I\Delta T$, hence $T_D = N_I\Delta T - \tau$ holds. Therefore, the expected downtime of equipment within a renewal cycle is:

$$\begin{aligned} E[T_D] &= \sum\nolimits_{k=1}^{m} E[T_D \mid N_I = k]Pr\{N_I = k\} \\ &= \sum\nolimits_{k=1}^{m} E[N_I\Delta T - \tau \mid N_I = k]Pr\{N_I = k\} \\ &= \sum\nolimits_{k=1}^{m} \int_0^{\Delta T} (k\Delta T - t) f_\tau(t)\mathrm{d}t \cdot \int_{(k-1)\Delta T}^{k\Delta T} f_\tau(t)\mathrm{d}t \end{aligned} \tag{6.7}$$

The length of a renewal cycle depends on the number of inspections within that cycle, and a renewal cycle always ends at the moment of the last inspection $N_I\Delta T$. Thus, the expected length of a renewal cycle is:

$$E[G] = \sum_{k=1}^{m} E[G \mid N_I = k] Pr\{N_I = k\} = \sum_{k=1}^{m} k\Delta T \cdot \int_{(k-1)\Delta T}^{k\Delta T} f_\tau(t)\mathrm{d}t \tag{6.8}$$

By combining Eqs. (6.4)–(6.8), the expected maintenance cost per unit time can be obtained as:

$$C_{ex} = \frac{C_I \cdot \sum_{k=1}^{m} k \cdot \int_{(k-1)\Delta T}^{k\Delta T} f_\tau(t)\mathrm{d}t \quad C_D \cdot \sum_{k=1}^{m} \int_0^{\Delta T} (k\Delta T - t) f_\tau(t)\mathrm{d}t \cdot \int_{(k-1)\Delta T}^{k\Delta T} f_\tau(t)\mathrm{d}t + C_M}{\sum_{k=1}^{m} k\Delta T \cdot \int_{(k-1)\Delta T}^{k\Delta T} f_\tau(t)\mathrm{d}t} \tag{6.9}$$

This expected maintenance cost per unit time provides a measure of the economics of the proactive maintenance strategy, shortening the long time scale of a full lifecycle service period to an expected renewal cycle, unaffected by time randomness, and more consistent with statistical laws.

6.2.4 Evaluation Indicators for Maintenance Strategy Performance

The performance of a proactive maintenance strategy for traction power supply equipment is measured by its ability to maintain high availability while minimizing maintenance costs. Therefore, the effectiveness of a proactive maintenance strategy should be assessed based on both availability and economic factors. The ideal proactive maintenance strategy maintains a high level of availability and controls maintenance costs to achieve the best balance between availability and economics.

To quantitatively evaluate the performance of the proactive maintenance strategy over the full lifecycle service period of traction power supply equipment, an evaluation metric called the *availability cost ratio* is defined. The formula for RC is:

$$RC = \frac{R_{av}}{C_{\text{total}}} = \frac{\int_0^r R(t)\mathrm{d}t / r}{C_{ex} \cdot r} \tag{6.10}$$

where R_{av} represents the average availability during the full service life r of the equipment, and C_{total} is the total maintenance cost over r and across all m renewal cycles. Thus, the physical significance of the availability cost ratio RC is the level of availability maintained per unit of maintenance cost over the equipment's full lifecycle service period, reflecting the performance of the proactive maintenance strategy from both availability and economic perspectives.

Since substituting equations (6.2), (6.3), and (6.9) into Eq, (6.10) results in the expression for *RC* containing a mixed form of multiple summations and integrations, it becomes difficult to obtain *RC* through analytical calculation or conventional numerical methods. Therefore, the Monte Carlo simulation method should be used to solve for *RC*.

The decision-making process for the proactive maintenance strategy of traction power supply equipment is essentially an optimization process. The performance of the proactive maintenance strategy serves as the objective function. By adjusting the decision variables within the proactive maintenance strategy, the goal is to achieve the optimal value of the objective function. Prior to the actual deployment of traction power supply equipment, the threshold for life degradation, denoted as L, is typically preset by the manufacturer. Thus, in this optimization process, the load impact threshold H and the periodic inspection cycle ΔT are selected as decision variables. By optimizing the values of H and ΔT, the objective is to maximize the *RC*.

6.2.5 *Maintenance Decision Case Study Based on Degradation Modeling of Traction Power Supply Equipment*

6.2.5.1 Maintenance Decision for a GIS High-Voltage Cabinet

The GIS high-voltage cabinet on the load side of the 4# traction transformer at the Xincheng traction substation of the Wuhan-Guangzhou railway has a rated voltage of 31.5 kV. Table 6.1 provides its basic nameplate information, including the manufacturer, model, and rated values.

Figure 6.5 shows a segment of load current data for traction transformer #4 at the traction substation, collected over one month by the substation's SCADA system. This load current data exhibits the characteristic features of high-speed railway traction load impacts, with frequent and sharp increases in current that subject the traction power supply equipment to a harsh and demanding load impact environment. According to the international standard *IEC 62271-200 High-voltage switchgear and controlgear - Part 200: AC metal-enclosed switchgear and controlgear for rated voltages above 1 kV and up to and including 52 kV* and its corresponding national standard *GB 3906-2006 Certification specification of*

Table 6.1 Basic information of the GIS high-voltage cabinet used in the case study

Model	Manufacturer	Rated voltage	Rated frequency	Rated current
XGN-2×27.5/2500-31.5	Changzhou Pacific Electric Power Equipment (Group) Co., Ltd.	31.5 kV	50 Hz	2500 A

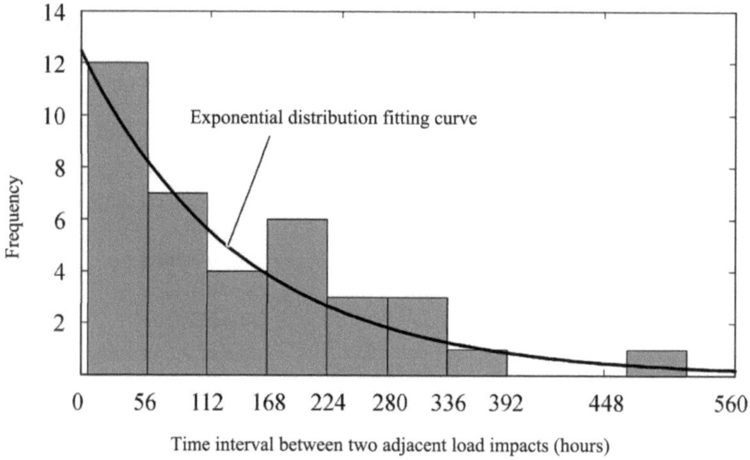

Fig. 6.5 Load data (within one month) used in the case study

alternating-current metal-enclosed switchgear and controlgear for rated voltages above 3.6 kV and up to and including 40.5 kV, a load impact is considered to occur when the load current exceeds the rated current value $I_r = 2500$ A. From Fig. 6.5, it can be observed that there were two instances where the load current exceeded the dashed line in the graph, indicating two load impacts occurred from January 4 to February 3, 2015.

To reflect the load environment condition of the GIS high-voltage cabinet during its service time, load data for one year (12 months) of traction transformer #4 at the substation was selected to establish a model for the load impact process. The model parameters for the load impact process were obtained using the maximum likelihood estimate (MLE) method and are presented with a 95% confidence interval. The actual values of the parameters obtained are shown in Table 6.2. In addition to the model parameters of the load impact process, the model parameters of the life degradation process are obtained from the GIS high-voltage cabinet equipment manufacturer. The basic parameters related to maintenance cost are jointly determined by the equipment manufacturer and the railway operating unit. These actual parameter values are also recorded in Table 6.2.

After determining the parameters of the load impact process via the MLE method, it is necessary to perform a distribution fit test on the load impact process to ensure the fitting quality of the compound Poisson process used meets the requirements. The Kolmogorov-Smirnov (K-S) test is employed for the distribution fit test, and the test results, as shown in Figs. 6.6 and 6.7, indicate that both the load impact intervals and the load impact amplitudes pass the K-S test, meeting the distribution fitting requirements.

The optimization of the proactive maintenance strategy for the GIS high-voltage cabinet is illustrated in Fig. 6.8. It can be observed that the availability cost ratio *RC* is a two-dimensional convex function with respect to the inspection cycle ΔT and the

Table 6.2 Parameter values and sources in the case study

Parameter type	Parameter	Value	Source
Load impact process parameters	λ	6.1033	MLE method (95% confidence interval)
	μ	2703.8019	MLE method (95% confidence interval)
	σ	88.2643	MLE method (95% confidence interval)
Life degradation process parameters	α	1/12	Manufacturer
	β	1/6	Manufacturer
	L	15	Manufacturer
Maintenance cost parameters	C_I	\$10	Manufacturer and railway bureau
	C_D	\$200	Manufacturer and railway bureau
	C_M	\$50	Manufacturer and railway bureau

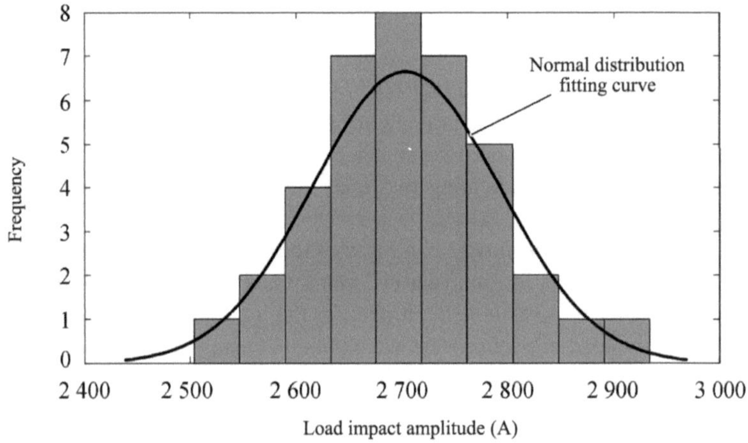

Fig. 6.6 Exponential distribution fit test result for load impact intervals

load impact threshold *H*. Therefore, the optimal performance of the GIS high-voltage cabinet proactive maintenance strategy, i.e., the maximum value of *RC*, can be achieved by selecting the optimal ΔT and *H*.

In Fig. 6.8, two orthogonal curves that pass through the maximum value point of the *RC* and are parallel to the ΔT and *H* axes, respectively, are selected and plotted separately in Figs. 6.9 and 6.10. These figures provide the optimization results for the decision variables ΔT and *H* of the proactive maintenance strategy. As shown in Fig. 6.9, the availability cost ratio *RC* reaches its maximum value of 4.775×10^{-5} when the inspection cycle ΔT is set to 394.2 hours. Figure 6.10 shows that the availability cost ratio *RC* increases with the increase in the load impact threshold *H*, reaching a stable maximum value of 4.775×10^{-5} when *H* is increased to 2980 A. Beyond this point, further increases in *H* would lead to a decrease in the average

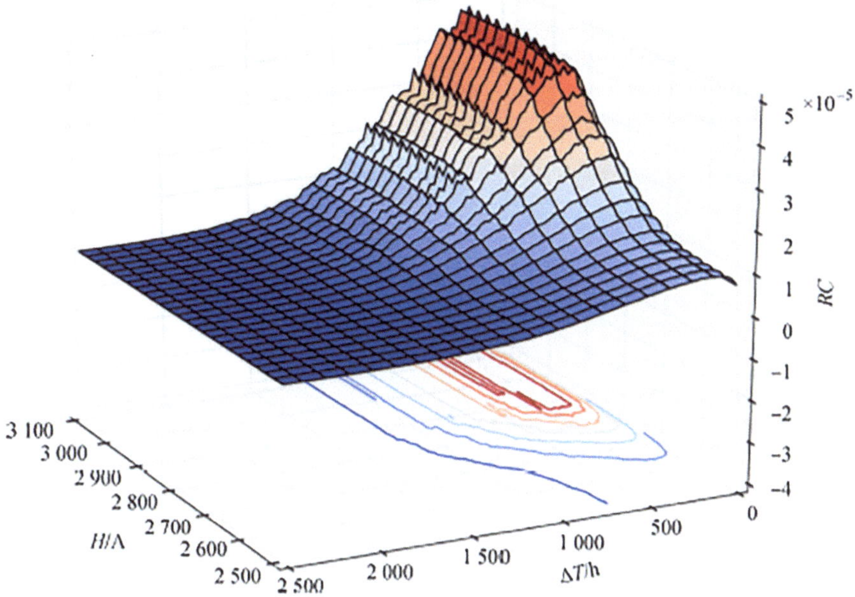

Fig. 6.7 Normal distribution fit test result for load impact amplitude

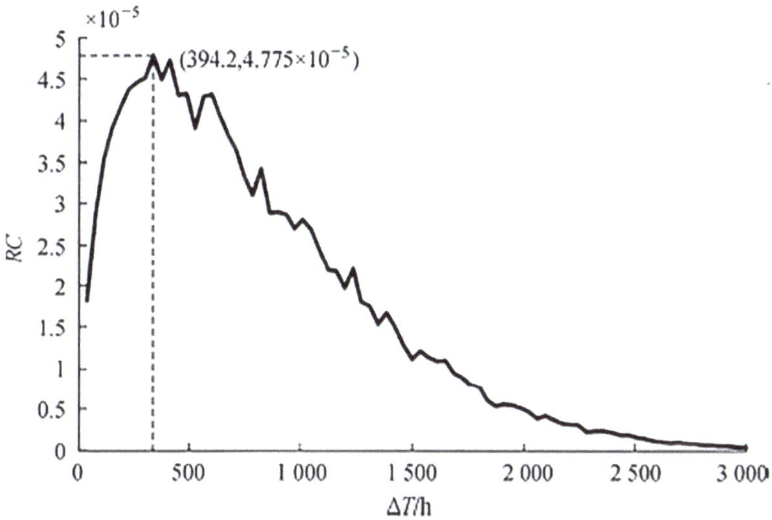

Fig. 6.8 Overall optimization result of the GIS high-voltage cabinet proactive maintenance strategy

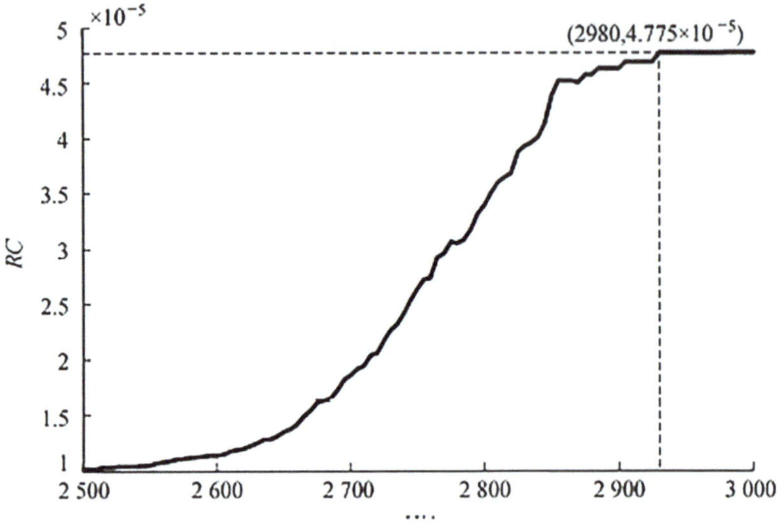

Fig. 6.9 Optimization result for inspection cycle

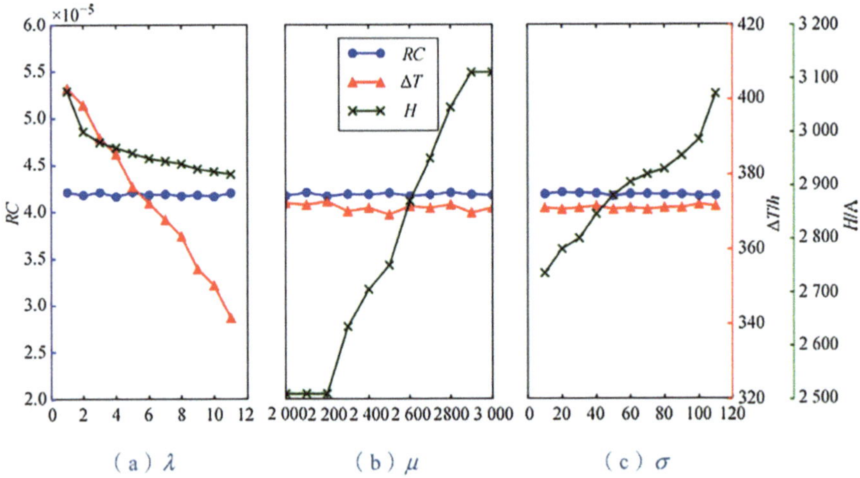

Fig. 6.10 Optimization result for load impact threshold

availability R_{av}, making 2980 A the optimal value for H. Through this optimization decision process, the GIS high-voltage cabinet's proactive maintenance strategy achieves optimal performance with the best configuration of inspection cycle and load impact threshold.

The above optimization decision results were obtained from a Monte Carlo simulation experiment for the proactive maintenance strategy of the GIS high-

Table 6.3 Final optimization decision results for the case study

Decision variables and optimization objective	Number of simulation experiments	Expected value
ΔT	1×10^5	371.1 h
H	1×10^5	2950 A
RC	1×10^5	4.180×10^{-5}

voltage cabinet. To eliminate the influence of randomness present in the random process on the decision results, the same parameters were used in 10^5 Monte Carlo simulation experiments. The expected values of the proactive maintenance strategy decision variables and the optimization objective were obtained as the final optimization decision results, as listed in Table 6.3. This outcome indicates that for the GIS high-voltage cabinet on the load side of the traction transformer4# at the Xinchang traction substation on the Wuhan-Guangzhou railway, a proactive maintenance strategy should be adopted with an inspection cycle of 371.1 hours and a load impact threshold of 2950 A. This strategy maintains a availability cost ratio of 4.180×10^{-5} during its full lifecycle service period to achieve optimal performance.

6.2.5.2 Sensitivity Analysis

In practical engineering applications, different traction power supply systems have various topological structures, load capacities, and maintenance requirements. The types, nameplate information, service environments, and performance degradation situations of the traction power supply equipment serving in these systems also differ, leading to service performance degradation and failure models built using different load data and degradation parameters that are not identical. The optimized proactive maintenance strategies obtained through decision-making using different cost parameters also vary significantly. Therefore, the optimization decision results of proactive maintenance strategies are significantly influenced by model parameters. To quantitatively study the impact of model parameters on the final optimization decision results, a sensitivity analysis is conducted on the optimization decision results of the GIS high-voltage cabinet's proactive maintenance strategy used in the case study. This case is used as a typical case study to illustrate how to adjust decision variables to achieve the best performance of a proactive maintenance strategy for traction power supply equipment with different model parameters.

In the above case study of proactive maintenance decision-making for the GIS high-voltage cabinet, all nine parameters are divided into three groups according to load impact process parameters, life degradation process parameters, and maintenance cost parameters. Each group undergoes sensitivity analysis for three independent parameters. The results of the sensitivity analysis are shown in Figs .6.11, 6.12 and 6.13.

As illustrated in Fig. 6.11, the sensitivity analysis for load impact process parameters shows that as the three parameters increase, it is necessary to adjust the

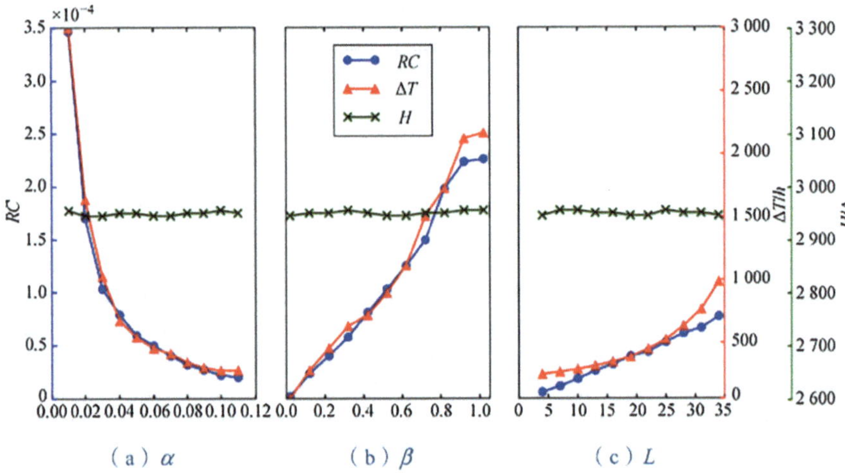

Fig. 6.11 Sensitivity analysis results for load impact process parameters

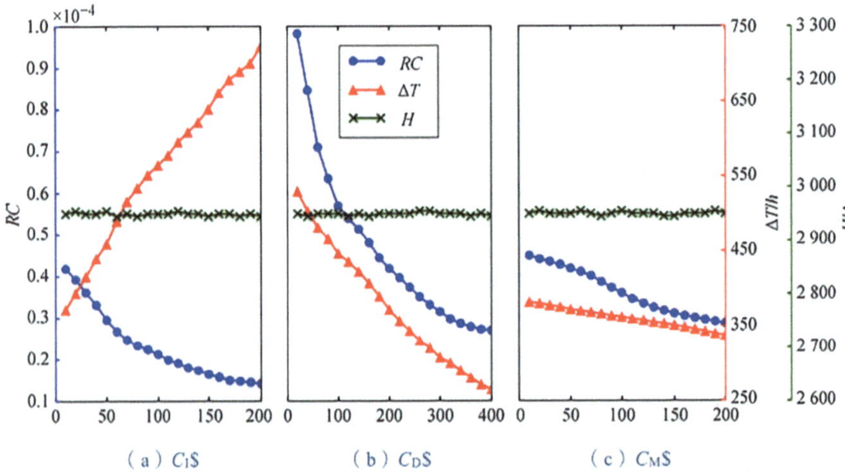

Fig. 6.12 Sensitivity analysis results for life degradation process parameters

decision variables ΔT and H synchronously with the load impact frequency and intensity to maintain the availability cost ratio RC consistently near its maximum value of 4.180×10^{-5}. Specifically, as the load impacts become more frequent (higher λ), the proactive maintenance strategy requires more frequent periodic inspections (smaller ΔT) and a lower load impact threshold (smaller H) to preserve the availability cost ratio. With an increase in the load impact amplitude (larger μ), it becomes necessary to raise the load impact threshold (larger H), though this does not affect the inspection cycle (ΔT remains stable at 371.1 h) to maintain the availability

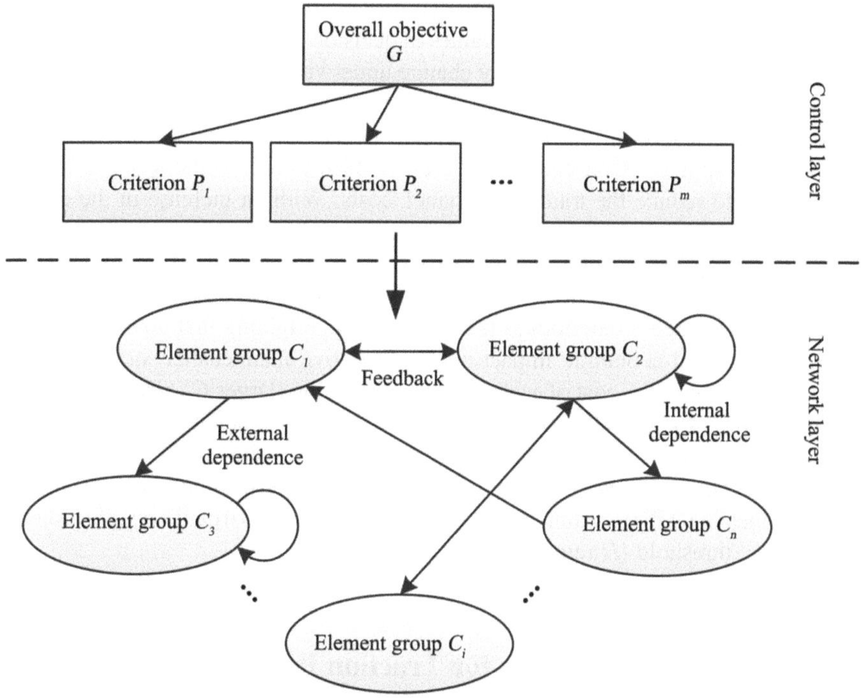

Fig. 6.13 Sensitivity analysis results for maintenance cost parameters

cost ratio. Additionally, load impacts with greater amplitude uncertainty or volatility (larger σ) require a higher load impact threshold (larger H) to mitigate the risks associated with potential excessive load impacts.

From the sensitivity analysis results for life degradation process parameters shown in Fig. 6.12, it is evident that the availability cost ratio and one of the decision variables—the inspection cycle—vary with the characteristics of the life degradation process of the GIS high-voltage cabinet. Faster life degradation (larger α and smaller β) leads to a lower availability cost ratio (smaller *RC*), necessitating more frequent periodic inspections (smaller ΔT). Conversely, a smoother life degradation process (smaller fluctuations in degradation amount, i.e., larger β) results in a higher availability cost ratio (larger *RC*) and allows for a more relaxed periodic inspection schedule (larger ΔT). Additionally, GIS high-voltage cabinets with higher life degradation thresholds (larger L) achieve higher optimal values of the availability cost ratio (larger *RC*) and require fewer periodic inspections (larger ΔT). It is also clear from Fig. 6.12 that the other decision variable—the load impact threshold (H)—is not affected by changes in life degradation process parameters, remaining stable at 2950 A. This observation confirms the assumption in the full repair model that the life degradation process and the load impact process are independent of each other.

The sensitivity analysis results for maintenance cost parameters, as shown in Fig. 6.13, illustrate how the optimal values of decision variables and the performance of the proactive maintenance strategy change under varying costs. As the inspection cost per occurrence increases (larger C_I), the availability cost ratio progressively decreases (smaller *RC*), indicating a decline in the performance of the proactive maintenance strategy. Consequently, there is a need to extend the inspection cycle (larger ΔT) to reduce the total maintenance costs. With an increase in the cost of downtime losses per unit time (larger C_D), there is a significant need to increase the frequency of periodic inspections (smaller ΔT) to minimize the GIS high-voltage cabinet's fault downtime. However, the performance of the proactive maintenance strategy still markedly deteriorates (smaller *RC*), highlighting that downtime losses have a substantial economic impact on the proactive maintenance strategy. Similarly, an increase in the cost of each maintenance action (larger C_M) leads to a rise in the frequency of periodic inspections (smaller ΔT) and a decline in the strategy's performance (smaller *RC*), though to a lesser extent. This suggests that maintenance action costs do not impact the economy as severely as downtime losses. Additionally, changes in all three maintenance cost parameters have virtually no effect on the load impact threshold (*H* remains stable at 2950 A).

6.3 Maintenance Strategy for Traction Power Supply System Based on Multi-Attribute Decision Making

For the high-speed railway traction power supply system, the maintenance strategy represents a comprehensive system engineering approach that balances system health and availability, efficacy, cost expenditure, and risk loss from a whole-system perspective. Unlike the maintenance strategies for individual traction power supply equipment, which focus on decisions over time, the maintenance strategy for the traction power supply system emphasizes spatial decisions. This involves optimizing the combination of different maintenance methods, which is essentially an optimal sorting combination problem in the realm of system engineering.

In practical traction power supply systems, the decision-making process for maintenance methods involves numerous influencing factors. These factors are categorized into several hierarchical levels, yet interlock and depend on each other within these levels. Lower-level elements may exert dominance and have feedback effects on higher-level elements, resulting in a complex network of relationships that resembles an interwoven, multi-layer network structure. Additionally, the decision-making process is complicated by the randomness in perception, subjective differences, and the ambiguity of the decision object, which can constrain the objectification of subjective weights and affect the accuracy of decision results. Therefore, it is necessary to employ suitable multi-attribute decision-making methods that adequately consider the dependencies and mutual influences between elements across layers, integrating the strengths of both subjective and objective weights, to accurately and profoundly describe the maintenance decision-making issues in the traction power supply system.

The maintenance methods for traction power supply systems are continuously evolving alongside advancements in maintenance philosophy and technology. Currently, these methods primarily consist of traditional combinations of breakdown maintenance and scheduled maintenance, supplemented to some extent by basic-level condition-based maintenance. However, as the theory and technology of state-based maintenance mature, there is a strong likelihood that advanced state-based maintenance will replace existing methods. Moreover, due to the uneven regional development in China, some areas with rapid technological growth have adopted advanced state-based maintenance and hidden danger detection techniques, while regions with slower technological development still employ more outdated methods. These factors cause the maintenance requirements of traction power supply systems to vary over time and by region, leading to temporal and spatial differences in decision-making influencing factors. Therefore, when studying the maintenance decision-making issues for traction power supply systems, it is also necessary to consider how decision outcomes vary with model parameter changes. Adjusting the priority order of maintenance methods appropriately to adapt to these differences in decision-making influencing factors across time and space is essential.

This section classifies and summarizes the current and potential future maintenance methods for traction power supply systems, highlighting their advantages and disadvantages. It establishes a set of maintenance methods and analyzes various attributes of the system during in-service operation and maintenance activities. By mapping out the complex dependency and feedback relationships among factors, a set of influencing factors is constructed. The application of network analysis methods is proposed for maintenance decision-making. On the foundation of establishing a maintenance decision model, composite weights of each maintenance method are calculated and ranked to quantitatively determine the priority order of maintenance methods for the traction power supply system. Additionally, sensitivity analysis of the decision results is conducted to flexibly adjust the maintenance strategy for traction power supply systems under different development stages and operational requirements.

6.3.1 Set of Maintenance Methods for Traction Power Supply System Maintenance Decisions

In the maintenance strategy for traction power supply systems, maintenance methods refer to the different means and categories of managing the timing, intensity, and types of maintenance work and activities to ensure the performance of the object maintained remains at a prescribed level throughout its service. The existing and potentially future applicable maintenance methods for traction power supply systems can generally be divided into four types: breakdown maintenance, scheduled maintenance, hidden danger detection, and state-based maintenance.

(1) Breakdown maintenance refers to the maintenance method that involves repairing traction power supply equipment after a functional failure or a serious fault occurs, restoring it to its initial or specified condition. Breakdown maintenance simply uses the occurrence of a functional failure as the criterion for maintenance. It is typically chosen when the consequences of the fault are minor, the repair process is straightforward, equipment is not easily replaced or scrapped, and the economic cost is manageable.
(2) Scheduled maintenance, also known as routine maintenance, is the maintenance method that involves performing varying degrees of repairs on traction power supply equipment after it has been in use for a certain period, according to a predefined maintenance cycle, to restore it to its specified working condition. Scheduled maintenance uses time as the criterion, arranging maintenance cycles based on the principle that the equipment failure rate is proportional to operating time. This method can mitigate the risks associated with breakdown maintenance by periodically restoring the condition of the traction power supply equipment and reducing the number of functional failures. Depending on the frequency and intensity of maintenance, scheduled maintenance can be divided into short-cycle, medium-cycle, and long-cycle types. As the maintenance cycle increases, there is a potential risk of insufficient maintenance for equipment in poor service conditions; conversely, as the maintenance cycle decreases, there is a potential for excessive maintenance of equipment in good service conditions.
(3) Hidden danger detection refers to a maintenance method that involves inspecting, testing, and monitoring equipment at certain time intervals or during continuous operating statuses. This method enables a thorough understanding of the equipment's service condition, provides early warnings for, adjusts, or corrects any indicators not meeting specifications, and decides on further maintenance plans. Hidden danger detection is suited for traction power supply equipment when faults are still in their early or latent stages and have not yet manifested as clear functional failures. Through further detailed inspections, this method precisely assesses the service status of the equipment to prevent the deterioration of fault conditions and the broadening impact of faults. Hidden danger detection can be divided into preventive testing and online monitoring based on the means and intervals of detection: preventive testing involves planned, periodic offline testing and inspections, while online monitoring entails real-time, continuous surveillance and data collection without disrupting normal equipment operations.
(4) State-based maintenance is an advanced method involving stringent monitoring and diagnostic approaches throughout the traction power supply equipment's entire service lifecycle. It aims to promptly detect concealed and potential faults, assess the equipment's health status and availability in real-time, and proactively carry out maintenance based on the actual service performance and health level of the equipment. Also known as condition-based maintenance, this approach operates based on actual equipment conditions, allowing for maintenance on demand. It maximizes the potential of individual units and effectively reduces system risks caused by performance degradation. Additionally, state-based

maintenance does not require equipment to be taken out of operation, facilitating maintenance without power outages. Current methods of state-based maintenance for traction power supply equipment include life cycle cost (LCC)-based state-based maintenance, availability-centered maintenance (RCM), and risk-based maintenance (RBM). LCC aims to minimize maintenance costs and total expenses during the repair process; RCM focuses on analyzing and enhancing the availability indices of equipment and systems; RBM strives to significantly reduce the risks and losses that equipment failures and maintenance activities pose to the system.

Overall, the pros and cons of various maintenance methods are summarized in Table 6.4.

6.3.2 Set of Influencing Factors for Maintenance Decisions of Traction Power Supply System

For the traction power supply system, determining the set of influencing factors for maintenance decisions is foundational for system maintenance method decision-making. It reflects the performance merits and demerits of the selected maintenance method and the importance of its impact on various system attributes. By referencing the principles of establishing systems for multi-attribute decision-making, assessment, and early warning in power systems, we distilled five major principles: scientificity, dynamics, advancement, comprehensiveness, and feasibility. These principles guide the selection of maintenance method decision influencing factors and the establishment of the set of influencing factors. This ensures that each factor has clear meaning, simple calculation, and ease of comparison, guaranteeing the integrity of the factor set and the interrelationships among factors.

Based on the operating characteristics, fault mechanisms, and maintenance activities of the high-speed railway traction power supply system, through research involving relevant railway design and operational units and listening to feedback and various demands and suggestions from onsite maintenance personnel, the set of maintenance decision influencing factors is categorized into five attributes: healthiness, availability, security risk factors, maintainability, and economic factors. Within each attribute, several influencing factors are identified, and these factors undergo screening, comparison, classification, and sorting to ensure the comprehensiveness of the set of influencing factors.

6.3.2.1 Healthiness

Healthiness primarily refers to the level at which each equipment and the entire system of the high-speed railway traction power supply system operate at their optimal working status during service. This optimal working status is defined as a

Table 6.4 Advantages and disadvantages of several maintenance methods for the traction power supply system

Maintenance method		Advantages	Disadvantages
Breakdown maintenance		Fully utilizes the remaining life of the equipment No additional monitoring costs No maintenance planning expenses	Significant security consequences and operational losses Higher maintenance costs and repair difficulty Immense pressure on maintenance staff Urgent maintenance time, compromising repair quality High system risk, low availability and health level
Scheduled maintenance	Short cycle	Better avoids latent faults Lower system risk, higher availability and health levels	Very high maintenance costs Very high maintenance planning expenses Longer downtime Lower equipment lifetime utilization
	Medium cycle	Basically avoids the occurrence of functional failures Appropriately reduces system risk, improves availability and health levels Appropriately shortens maintenance downtime and reduces maintenance costs	Higher maintenance planning expenses Lower equipment lifetime utilization Gradual growth of latent faults
	Long cycle	Shorter downtime Lower maintenance planning expenses and maintenance costs Higher equipment lifetime utilization	Higher system risk, lower availability and health levels Latent faults are more likely to develop into functional faults
Hidden danger detection	Preventive testing	Effectively avoids latent faults Appropriately reduces system risk, improves availability and health levels	Higher maintenance costs and testing expenses Longer downtime
	Online monitoring	Timely detection and effective avoidance of latent faults Significantly reduces system risk, improves availability and health levels No need for equipment shutdown	Very high costs for additional equipment Higher maintenance difficulty Availability challenges of the additional equipment themselves

(continued)

Table 6.4 (continued)

Maintenance method		Advantages		Disadvantages	
State-based maintenance	LCC	Very low total maintenance costs	Effectively avoids latent faults Very short downtime Very high equipment lifetime utilization and maintenance quality	Relatively higher system risk Relatively lower availability and health levels	Higher maintenance difficulty Higher monitoring and testing expenses
	RCM	Very high availability and health levels		Relatively higher total maintenance costs Relatively higher system risk	
	RBM	Very low system risk		Relatively higher total maintenance costs Relatively lower availability and health levels	

health condition; the presence of potential faults or latent faults can lead to a performance decline, transitioning into a sub-health condition. As service performance degrades, faults gradually evolve into functional failures. Throughout this process, multiple health grades are passed, eventually deteriorating into a severe abnormal state, forcing the equipment or system to cease operation and potentially causing severe security accidents. Thus, on one hand, the health status reflects the service performance of the traction power supply system; on the other hand, as the energy source of the high-speed railway, providing stable, reliable, high-quality electrical energy represents its optimal working status and is a functional reflection of excellent service performance. Therefore, the influencing factors for healthiness include the health value describing the equipment's health condition, service age, etc., and factors describing the system's electrical energy quality such as voltage qualification rate, frequency qualification rate, harmonic content, three-phase imbalance, as well as factors describing the quality of electric power transmission and distribution, like the average operational load rate and flow distribution indicators.

6.3.2.2 Availability

Availability and its related indicators describe the ability of traction power supply equipment and systems to complete normal power supply tasks without faults over an extended timescale. availability reflects various statistical rules of the traction power supply equipment and systems throughout their full lifecycle service period, which can be characterized from three aspects: probability, degree, and time. The representation of probability includes factors such as availability and failure rate, the representation of degree includes factors such as availability and importance, and the representation of time includes factors such as mean time to failure, mean time between failures, and remaining useful life. Among these factors, availability is an important degree characterization index. There are three different definitions of availability when describing availability: inherent availability, achievable availability, and operational availability. Inherent availability is the intrinsic representation of equipment availability, solely related to equipment operating time and repair time; achievable availability is a type of availability related to operating time, repair time, and system risk, which considers risk factors and the impact of risk mitigation measures (such as preventive testing) based on the foundation of inherent availability, hence its value is less than inherent availability; operational availability is a type of availability that reflects the real usage of equipment, considering practical factors such as technician skill level and repair tool preparation time based on the foundation of achievable availability, resulting in a further reduction in its value.

6.3.2.3 Security Risk

Security risk factors primarily characterize the hazard level of faults or accidents, such as equipment damage, line tripping, and eventual loss of power supply capacity that may occur in the traction power supply system under severe weather conditions like high winds, ice and snow, thunder, and lightning. These factors describe the possibility and probability of failure accidents that impact the safe operation of the system, as well as the severity of the consequences of these events. While healthiness and availability consider the impact of internal factors within the traction power supply system, security risk factors account for the influence of external environmental factors on the system. Severe weather and other external environmental factors significantly affect the external power supply and the catenary's power delivery to the traction power supply system. Thus, security risk factors encompass influencing factors related to external power supply risks, such as traction load reduction probability risk, traction load reduction frequency risk, insufficient traction load power risk, insufficient traction load energy risk, and railway throughput capability loss probability risk. They also include power supply risk influencing factors for the catenary exposed to harsh environments like high winds, ice and snow, and thunder and lightning, such as catenary wind deviation risk, catenary ice coverage risk, and catenary lightning strike risk.

6.3.2.4 Maintainability

Maintainability describes the performance level of the maintenance strategy across three dimensions: time, space, and technology. It also reflects the ability of the traction power supply equipment and system to maintain or restore the specified working status under the implementation of the maintenance strategy. From the temporal perspective, the influencing factors of maintenance strategy performance include maintenance time consumption, historical maintenance frequency, and inspection frequency. From the spatial perspective, the influencing factors of maintenance strategy performance include the supply level of spare parts and the proportion of online monitoring equipment. From the technological perspective, the influencing factors of maintenance strategy performance include ease of maintenance, applicability of maintenance technology, and technician skill level. The implementation of a maintenance strategy and the ability of equipment and systems to maintain or restore the specified working status include influencing factors such as maintainability and repair rate. It is evident that from the perspective of maintenance strategy performance level, maintainability has a direct dependency relationship with economic factors; from the perspective of the ability of equipment and systems to maintain or restore the specified working status, maintainability has both direct and indirect dependency relationships with healthiness, availability, and security (risk). Therefore, maintainability is one of the key attributes that needs to be considered in the maintenance decision-making of the traction power supply system.

6.3.2.5 Economic

Economic factors broadly reflect the total efficacy, cost, losses, and benefits generated during the normal operation of the traction power supply system and during maintenance activities. Guided by the principle of maximizing overall efficacy and benefits while minimizing costs and losses, economic factors encompass those influencing the productive efficacy of the traction power supply system during normal operation, such as equipment initial cost, network losses, equipment capacity utilization rate, and reactive power deficit. They also cover factors influencing the consumption incurred by maintenance activities, such as maintenance action cost, maintenance downtime loss, human resources allocation cost, spare parts storage cost, additional equipment cost, and maintenance planning expenses. Economic factors have a direct and strong dependency relationship with the previous four-factor sets. However, this dependency manifests as a positive dependency for some factors and a negative dependency for others. For instance, optimizing the influencing factors of healthiness, availability, security risk factors, and maintainability could lead to the optimization of factors representing operational benefits in economics (such as equipment capacity utilization rate), but at the same time, it could lead to the deterioration of factors representing the consumption generated by maintenance activities (such as maintenance costs). Economic factors also have a significant impact on the selection of maintenance methods in the maintenance decision-making process for the traction power supply system.

6.3.3 Maintenance Decision Method for Traction Power Supply System Based on ANP

6.3.3.1 ANP Decision Model

With the established sets of maintenance methods and influencing factors for the traction power supply system, it is necessary to construct a scientific decision model to make maintenance decisions and quantitatively determine the priority order of various maintenance methods. In the traction power supply system, numerous factors affect the decision-making on maintenance methods. These factors are interlocking and interdependent within each level, with lower-level elements having dominance and a feedback effect on higher-level elements. The relationships between system elements are complex and the system structure resembles a network structure. Decision-making requires clarifying the interactions and dependencies among influencing factors and appropriately distributing weights. The ANP, proposed by Professor T. L. Saaty from the University of Pittsburgh, effectively handles complex decision-making problems under such a network structure. It thoroughly considers the dependencies between elements and mutual influence between adjacent levels, accurately and deeply describing the decision-making problems of complex systems. Additionally, ANP is a method that gradually transforms subjective weights into objective weights, combining the advantages of subjectivity and objectivity to provide more scientifically sound decision outcomes. Thus, the maintenance decision-making for the traction power supply system is enhanced.

The ANP decision model divides all elements in the system into two levels. The first level, the control layer, includes the overall objective and various decision criteria. All decision criteria are considered independent of each other and are solely dominated by the overall objective. The second level, the network layer, consists of element groups dominated by the control layer. These groups are interconnected, mutually influencing, and interdependent, forming a network structure. A typical ANP decision model is illustrated in Fig. 6.14.

In the network layer of the ANP decision model, there are three types of connectivity between element groups: external dependence, internal dependence, and feedback. External dependence reflects the influence of one element group on the elements of another. In the provided figure, elements in the element group C_1 influence those in the element group C_3, forming an external dependence relationship. Internal dependence reflects the mutual influence among elements within the same element group. In Fig. 6.14, the mutual influence among elements within element group C_2 forms an internal dependence relationship. The feedback relationship reflects the mutual influence between elements of two element groups. In Fig. 6.14, the elements in the element group C_1 influence the elements in the element

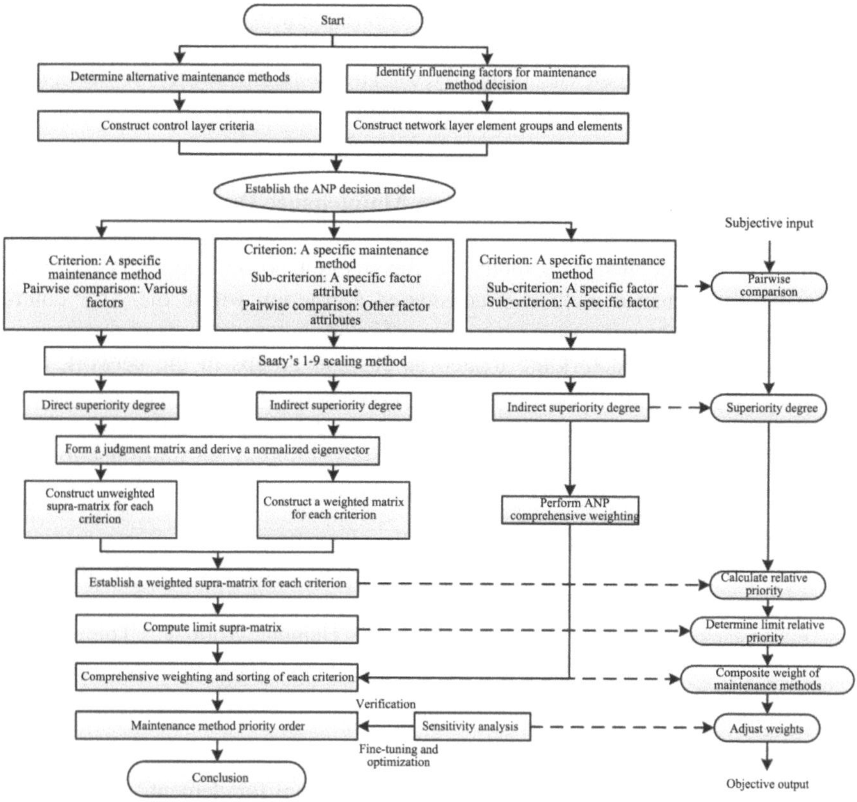

Fig. 6.14 Typical ANP decision model

group C_2, and conversely, the elements in C_2 also influence the elements in C_1. Therefore, C_1 and C_2 form a feedback relationship.

After establishing the various criteria in the control layer and identifying all elements within each element group of the network layer, it is necessary to perform pairwise comparisons of the elements dominated by a specific criterion. This process determines their priority order in terms of importance, superiority, preference for selection, and degree of mutual influence. The quantitative measure of this priority order is called the element's superiority degree. In ANP, the compared elements are rarely independent but are interdependent in complex ways. Therefore, two types of superiority degrees are defined for comparing elements:

(1) Direct Superiority Degree: This involves comparing two elements in terms of their importance relative to a given criterion.
(2) Indirect Superiority Degree: This involves comparing the extent to which two elements influence a third element (referred to as a sub-criterion) under a given criterion.

Pairwise comparison of elements to determine the superiority degree is a crucial foundation. The 1–9 scaling method proposed by Saaty is used to quantitatively express the results of pairwise comparisons. The result can reflect the multiple by which the superiority degree between the two elements differs.

6.3.3.2 Traction Power Supply System Maintenance Decision Process

In the set of maintenance methods for the traction power supply system, each selectable maintenance method is considered a criterion within the ANP control layer, denoted as the criterion $p_1, p_2, \cdots, p_m$. The attributes of the set of maintenance decision influencing factors are treated as element groups in the network layer, labeled as element group C_1, C_2, $\cdots$, C_N. Within each element group C_i, there are elements $e_{i1}, e_{i2}, \cdots, e_{in_i} (i = 1, 2, \cdots, N)$, each representing an influencing factor. Initially, for each control layer criterion $p_s(s = 1, 2, \cdots, m)$, an unweighted supra-matrix $\boldsymbol{W}_\mathrm{s}$ is constructed. The process is as follows:

(1) Using criterion p_s as the criterion and the elements $e_{jk}(k = 1, 2, \cdots, n_j)$ within element group $C_j(j = 1, 2, \cdots, N)$ as sub-criteria, conduct indirect superiority degree comparisons based on the impact of elements in the element group C_j on e_{jk} or e_{jk}'s impact on elements within the element group C_j. This forms a judgment matrix from which a normalized eigenvector is derived, and feature vectors are sorted as $\left(w_{i1}^{(jk)}, w_{i2}^{(jk)}, \cdots, w_{in}^{(jk)}\right)^\mathrm{T}$.

(2) Following this method (1), use the elements $e_{j1}, e_{j2}, \cdots, e_{jn_j}$ in C_j as sub-criteria and perform indirect superiority degree comparisons for elements $e_{i1}, e_{i2}, \cdots, e_{in_i}$ in element group C_i. Form respective judgment matrices and derive sorted normalized feature vectors. These feature vectors are then compiled column-wise into the matrix $\boldsymbol{W}_{ij}$.

$$\boldsymbol{W}_{ij} = \begin{bmatrix} w_{i1}^{(j1)} & w_{i1}^{(j2)} & \cdots & w_{i1}^{(jn_j)} \\ w_{i2}^{(j1)} & w_{i2}^{(j2)} & \cdots & w_{i2}^{(jn_j)} \\ \vdots & \vdots & & \vdots \\ w_{in_i}^{(j1)} & w_{in_i}^{(j2)} & \cdots & w_{in_i}^{(jn_j)} \end{bmatrix} \tag{6.11}$$

The matrix $\boldsymbol{W}_{ij}$ represents the influence relationship of element group C_i on element group C_j. Notably, if elements in C_j are not influenced by those in C_i, then $\boldsymbol{W}_{ij} = 0$.

(3) Using the methods outlined in steps (1) and (2), an unweighted supra-matrix $\boldsymbol{W}_\mathrm{s}$ under criterion p_s is constructed.

$$
W_s = \begin{bmatrix} W_{11} & W_{12} & \cdots & W_{1j} & \cdots & W_{1N} \\ W_{21} & W_{22} & \cdots & W_{2j} & \cdots & W_{2N} \\ \vdots & \vdots & & \vdots & & \vdots \\ W_{i1} & W_{i2} & \cdots & W_{ij} & \cdots & W_{iN} \\ \vdots & \vdots & & \vdots & & \vdots \\ W_{N1} & W_{N2} & \cdots & W_{Nj} & \cdots & W_{NN} \end{bmatrix} \tag{6.12}
$$

Each element within this supra-matrix is derived from the matrices $W_{ij}(i, j = 1, 2, \cdots, N)$ obtained in step (2). This supra-matrix stores the results of indirect superiority degree comparisons, reflecting the relative preferences and importance among the maintenance method decision influencing factors.

Similarly, by using the criterion p_1, p_2, $\cdots$, p_m as criteria, such unweighted supra-matrices can be constructed, totaling m matrices, all of which are non-negative. The supra-matrix is not column-normalized; only the sub-blocks W_{ij} are column-normalized. Therefore, the supra-matrix does not reflect the relative priority of each element. It is necessary to perform a superiority degree comparison among the element groups to transform the unweighted supra-matrix into a weighted supra-matrix.

Using criterion p_s as the criterion and element group C_j as the sub-criterion, perform indirect superiority degree comparisons based on the impact of element groups C_1, C_2, $\cdots$, C_N on C_j or the impact of C_j on element groups C_1, C_2, $\cdots$, C_N. This forms a judgment matrix from which a sorted normalized eigenvector $\left(a_1^{(j)}, a_2^{(j)}, \cdots, a_N^{(j)}\right)^{\mathrm{T}}$ is obtained. Subsequently, using each element group as a sub-criterion, perform indirect superiority degree comparisons among element groups C_1, C_2, $\cdots$, C_N, and compile the sorted normalized eigenvectors column-wise to form the weighted matrix.

$$
A_s = \begin{bmatrix} a_1^{(1)} & a_1^{(2)} & \cdots & a_1^{(N)} \\ a_2^{(1)} & a_2^{(2)} & \cdots & a_2^{(N)} \\ \vdots & \vdots & & \vdots \\ a_N^{(1)} & a_N^{(2)} & \cdots & a_N^{(N)} \end{bmatrix} \tag{6.13}
$$

This weighted matrix A_s stores the results of indirect superiority degree comparisons between element groups under criterion p_s. The sorted vector components corresponding to element groups that are unrelated to sub-criterion C_j are zero. There are m such weighted matrices under m criteria.

The weighted matrix A_s is used to weight the elements in the unweighted supra-matrix W_s, resulting in the weighted supra-matrix $\overline{W_s}$,

$$
(\overline{W_s}) = (A_s) \cdot (W_s), s = 1, 2, \cdots, m \tag{6.14}
$$

where ( ) denotes taking each element of the matrix. The weighted supra-matrix $\overline{W_s}$ is column-normalized and known as a column stochastic matrix, reflecting the relative priority of each element.

Due to the presence of dependencies and feedback relationships within the ANP decision model, determining element priorities is complex. The priority between two elements, e_i and e_j, can be determined using the element w_{ij} in the weighted supra-matrix $\overline{\boldsymbol{W}}_s$ (referred to as a one-step priority), or by using the element $\sum_{l=1}^{N} w_{il} w_{lj}$ in $\overline{\boldsymbol{W}}_s^2$ (referred to as a two-step priority), and even using elements in $\overline{\boldsymbol{W}}_s^n$ (referred to as an n-step priority). Therefore, to accurately reflect the dependencies and feedback relationships among elements, it is necessary to stabilize the weighted supra-matrix through a stabilization process, namely by calculating the limit supra-matrix to determine stable element priorities:

$$\overline{\boldsymbol{W}}_s^{\infty} = \lim_{t \to \infty} \overline{\boldsymbol{W}}_s^t \tag{6.15}$$

When this limit exists, the jth column of $\overline{\boldsymbol{W}}_s^{\infty}$ is the limit relative priority of the elements in the network layer of criterion p_s towards element e_j.

For each control layer criterion, determine the limit relative priority among network layer elements following the steps described above, and then apply ANP synthesis weighting according to the weight of each criterion to determine the composite weight of each selectable maintenance method. Based on these composite weights, the priority order of the maintenance methods is established.

Finally, it is necessary to perform sensitivity analysis on the decision results to examine how the composite weight of each maintenance method changes with variations in the limit relative weights of various influencing factors. Sensitivity analysis involves selecting an influencing factor and gradually adjusting its weight, proportionally redistributing the weight of the remaining influencing factors to cause changes in the composite criterion weight, ultimately affecting the priority order of maintenance methods. In practice, the weights of influencing factors can be adjusted according to changes in on-site operating requirements, thus using sensitivity analysis to verify, fine-tune, and optimize the maintenance decision outcomes.

In summary, the process and weight handling for maintenance decision-making in the high-speed railway traction power supply system based on ANP are illustrated in Fig. 6.15. It is evident that the ANP method is capable of mitigating the impact of subjective differences on decision results. Although the results of pairwise comparisons are somewhat subjective, the concepts of superiority degree, priority, and composite weight set by ANP gradually reduce subjectivity, ultimately yielding a more objective composite weight. This allows the maintenance decision method based on ANP to combine subjective advantages (such as intuitiveness and simplicity) with objective advantages (such as practicality), thereby providing the most scientifically sound decision results.

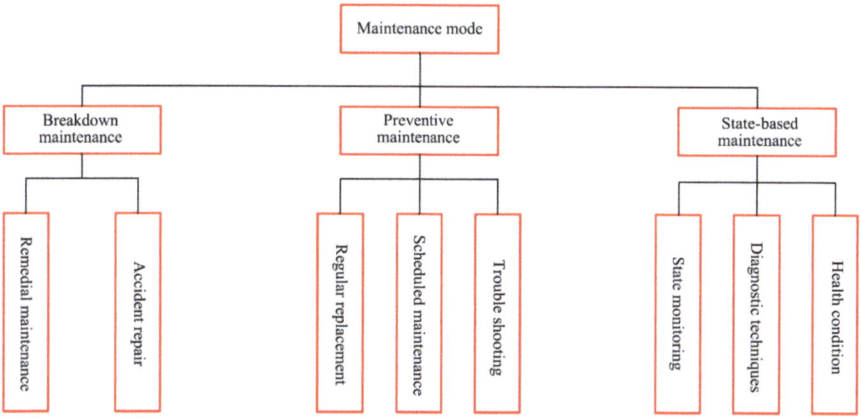

Fig. 6.15 ANP-Based High-Speed Railway Traction Power Supply System Maintenance Decision Process

6.3.4 Maintenance Decision Case Study for Traction Power Supply System Based on Multi-Attribute Decision Making

6.3.4.1 Establishing an ANP Decision Model

Using a segment of the Beijing-Shanghai railway traction power supply system as a case study, the ANP method is applied to make maintenance decisions, providing a comprehensive priority order of maintenance methods. Initially, control layer criteria are determined based on nine feasible maintenance methods, and network layer elements and element groups are identified according to a set of influencing factors, totaling 45. The ANP decision model for the traction power supply system maintenance method, as illustrated in Fig. 6.16, is established along with external dependence, internal dependence, and feedback relationships among elements within each element group.

6.3.4.2 Determining Limit Relative Priority

Selecting nine maintenance methods as criteria and five attribute types, or 45 influencing factors as sub-criteria, the pairwise comparison method is performed as depicted in Fig. 6.15. Although this process is inherently subjective, it is substantiated by extensive consultation of relevant literature, standards, and procedures, and by seeking advice from numerous designers, onsite maintenance personnel, and operational managers to ensure the results of the pairwise comparisons are as scientifically reasonable and closely aligned with onsite realities as possible. After calculating the superiority degrees through the ANP method and establishing

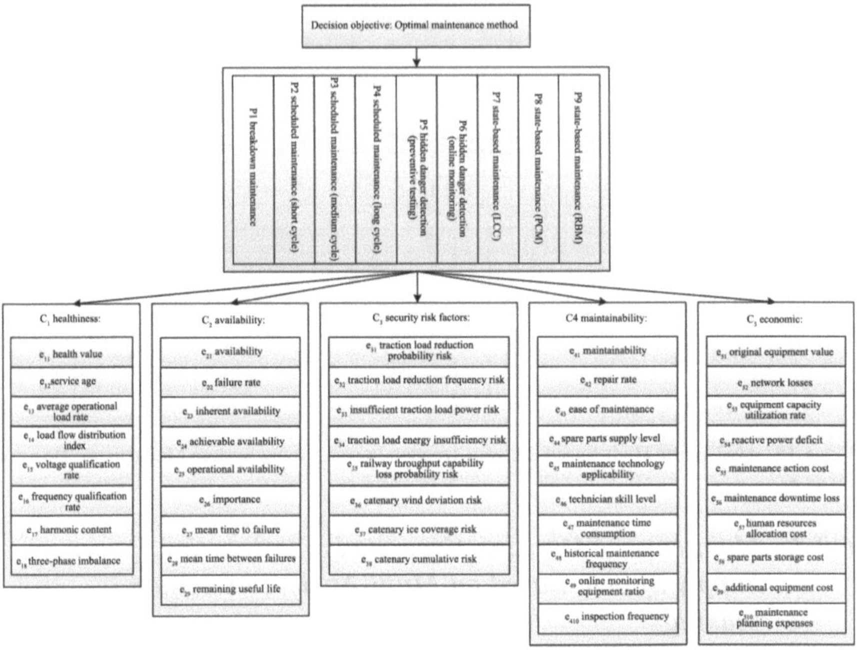

Fig. 6.16 ANP Decision model for maintenance strategy of the traction power supply system

weighted supra-matrices and limit supra-matrices, the normalized superiority degrees by element group and limit relative priorities are tabulated as shown in Table 6.5.

6.3.4.3 Calculation of Composite Weight for Maintenance Methods

Using the ANP comprehensive weighting method, the composite weighting of the model is performed for each maintenance method, resulting in the composite weights of the nine maintenance methods which are then sorted as shown in Table 6.6. It is evident that the three state-based maintenance methods and hidden danger detection (online monitoring) have composite weights (relative weights) above 0.5, thus they should be considered as priority options. On the other hand, hidden danger detection (preventive testing), the three types of scheduled maintenance, and breakdown maintenance may not be as suitable.

6.3.4.4 Sensitivity Analysis

To ensure the maintenance strategy adapts to the temporal and spatial differences of decision influencing factors, it is necessary to examine how the composite weight of

Table 6.5 Superiority Degree and Limit Relative Priority of Each Influencing Factor

Influencing factors	e_{11}	e_{12}	e_{13}	e_{14}	e_{15}	e_{16}	e_{17}	e_{18}		
Superiority degree	0.2490	0.1930	0.0973	0.1120	0.0903	0.0828	0.0894	0.0863		
Limit relative priority	0.0434	0.0336	0.0169	0.0195	0.0157	0.0144	0.0156	0.0150		
Influencing factors	e_{21}	e_{22}	e_{23}	e_{24}	e_{25}	e_{26}	e_{27}	e_{28}	e_{29}	
Superiority degree	0.1488	0.1422	0.0914	0.1364	0.1850	0.0829	0.0634	0.0785	0.0714	
Limit relative priority	0.0417	0.0398	0.0256	0.0382	0.0518	0.0232	0.0178	0.0220	0.0200	
Influencing factors	e_{31}	e_{32}	e_{33}	e_{34}	e_{35}	e_{36}	e_{37}	e_{38}		
Superiority degree	0.1013	0.1013	0.0996	0.0996	0.1577	0.1283	0.1312	0.1810		
Limit relative priority	0.0112	0.0112	0.0111	0.0111	0.0175	0.0142	0.0146	0.0201		
Influencing factors	e_{41}	e_{42}	e_{43}	e_{44}	e_{45}	e_{46}	e_{47}	e_{48}	e_{49}	e_{410}
Superiority degree	0.1454	0.1070	0.0560	0.1136	0.0811	0.0736	0.0875	0.1258	0.1274	0.0827
Limit relative priority	0.0146	0.0107	0.0056	0.0114	0.0081	0.0074	0.0088	0.0126	0.0128	0.0083
Influencing factors	e_{51}	e_{52}	e_{53}	e_{54}	e_{55}	e_{56}	e_{57}	e_{58}	e_{59}	e_{510}
Superiority degree	0.0690	0.0519	0.0457	0.0488	0.1234	0.1780	0.1654	0.1210	0.1124	0.0844
Limit relative priority	0.0077	0.0058	0.0051	0.0055	0.0138	0.0200	0.0185	0.0136	0.0126	0.0095

Table 6.6 Composite weight of 9 maintenance methods

Maintenance method	Absolute weight	Normalized weight	Relative weight
State-based maintenance (RBM)	0.0433	0.1948	1.0000
State-based maintenance (RCM)	0.0412	0.1853	0.9513
State-based maintenance (LCC)	0.0366	0.1646	0.8450
Hidden danger detection (online monitoring)	0.0323	0.1450	0.7445
Hidden danger detection (preventive testing)	0.0194	0.0870	0.4467
Scheduled maintenance (long cycle)	0.0146	0.0655	0.3360
Scheduled maintenance (short cycle)	0.0124	0.0556	0.2853
Scheduled maintenance (medium cycle)	0.0118	0.0531	0.2727
Breakdown maintenance	0.0109	0.0492	0.2524

each maintenance method changes with the limit relative weights of influencing factors. This is achieved using sensitivity analysis to adjust weights, thereby tuning and optimizing the priority order of maintenance methods. Specifically, sensitivity analysis is performed for maintenance method decision results using influencing factors such as health value (e_{11}), catenary lightning strike risk (e_{31}), and maintenance action cost (e_{55}) as a case study. The results are illustrated in Fig. 6.17, where the dashed line intersects the x-axis at the current limit relative weight value of the influencing factors.

From Fig. 6.17b, it can be seen that when the limit relative weights of health value (e_{11}) increase, the composite weight of state-based maintenance (RCM) increases significantly, while the composite weight of state-based maintenance (RBM) remains almost unchanged. When the limit relative weight of health value exceeds 0.2001, the priority order of state-based maintenance (RCM) will surpass that of state-based maintenance (RBM).

From Fig. 6.17c, it is observed that with the increase in the limit relative weights of catenary lightning strike risk (e_{38}), the composite weight of state-based maintenance (RBM) remains the highest. When the limit relative weights of e_{38} increase to 0.5300, hidden danger detection (online monitoring) will be prioritized over state-based maintenance (RCM). As the limit relative weights of e_{38} continue to increase to 0.8450, hidden danger detection (preventive testing) will also be prioritized over state-based maintenance (RCM).

From Fig. 6.17d, it can be observed that as the limit relative weights of maintenance action cost (e_{55}) increase, the composite weights of hidden danger detection (online monitoring) and scheduled maintenance (long cycle) will increase significantly. When the limit relative weight of e_{55} reaches 0.1323, hidden danger detection (online monitoring) will be prioritized over state-based maintenance (RBM). When the limit relative weight of e_{55} reaches 0.3379, scheduled maintenance (long cycle) will be prioritized over the three state-based maintenance methods. When the limit relative weight of e_{55} reaches 0.6560, hidden danger detection (preventive testing) will also be prioritized over the three state-based maintenance methods.

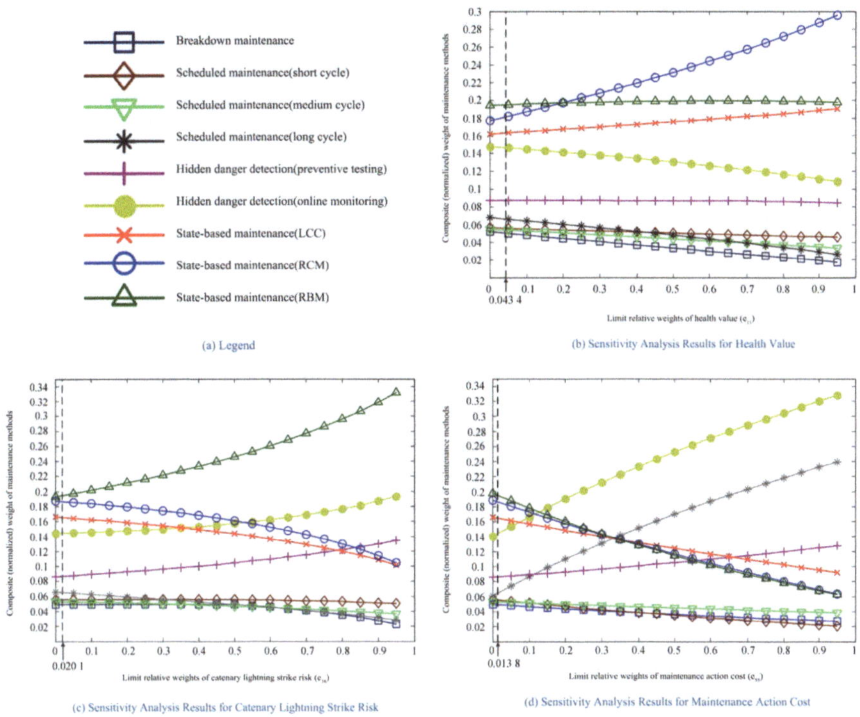

Fig. 6.17 Sensitivity analysis results for traction power supply system maintenance decision

From the above results, it can be seen that through maintenance decision-making, the overall optimal maintenance method for the traction power supply system can be determined before the maintenance activities are carried out. On one hand, the weight allocation can be adjusted according to the future development requirements of the same traction power supply system to modify the priority order of maintenance methods. On the other hand, the regional differences of different traction power supply systems can be flexibly taken into account to formulate their respective maintenance strategies. In the actual implementation of maintenance activities, it is also necessary to combine the health condition assessment results of each traction power supply equipment and the maintenance strategy to reasonably arrange and coordinate the development of specific maintenance plans and the allocation of maintenance resources.

6.4 Summary

This chapter primarily discusses the maintenance strategies for the traction power supply system of high-speed railways. It systematically introduces the concepts of maintenance and the evolution of maintenance technologies for traction power supply equipment. The chapter outlines the distribution patterns of the lifespan of traction power supply equipment and delves into the primary research methodologies for formulating maintenance strategies in high-speed railway contexts, closely aligned with practical engineering applications.

Building upon the principles outlined in Sect. 6.2, this chapter elaborates on the degradation of service performance and failure processes of traction power supply equipment. It details the availability and economic indicators associated with service performance degradation and advocates for a proactive maintenance strategy based on a *combined monitoring, inspection, and maintenance approach.* This strategy optimizes the proactive maintenance approach for equipment and systematically classifies and summarizes the maintenance methods of the traction power supply system along with their advantages and disadvantages. The maintenance decision-making process employs various methods such as application network analysis method, ANP, and multi-attribute decision making. This segment initiates research into decision-making and optimization from traction power supply equipment to maintenance strategies for traction power supply systems, offering practical case studies. Underpinned by the pillars of fault diagnosis, fault prognostics, availability, and risk assessment, and in conjunction with actual maintenance and repair models of traction power supply systems, this research is significantly relevant for devising maintenance strategies in high-speed railway traction power supply systems.

References

1. He, Z., & Cheng, H. (2012). Study on high-speed railway traction power supply system health management and fault early warning system [J]. *Power System Technology, 36*(10), 259–264.
2. He, Z., Feng, D., Lin, S., & Sun, X. (2016). Overview of security risk assessment studies for high-speed railway traction power supply systems [J]. *Journal of Southwest Jiaotong University, 51*(3), 418–429.
3. He, Z. (2012). Study on prognostics and health diagnosis of latent faults in complex distribution networks [J]. *Academic Dynamics, 02*, 7–10.
4. He, Z., Hu, H., Fang, L., et al. (2011). Study on harmonics and their transmission characteristics in high-speed railway traction power supply systems [J]. *Proceedings of the CSEE, 31*(16), 55–62.
5. Lin, S., Feng, D., & Sun, X. (2019). Traction power-supply system risk assessment for high-speed railways considering train timetable effects [J]. *IEEE Transactions on Reliability, 68*(3), 810–818.
6. Lin, S., Lin, X., & Feng, D. (2018). Availibility assessment of traction substation relay protection in urban rail transit [J]. *Journal of Southwest Jiaotong University, 53*(6), 1102–1109.
7. Wang, W., He, Z., Lin, S., et al. (2015). Research on PHM and proactive maintenance for high-speed railway traction power supply systems [J]. *Journal of Southwest Jiaotong University, 50*(5), 942–952.
8. Wang, W., He, Z., Cheng, H., et al. (2015). Health assessment method for metro traction power supply systems based on multi-layer immune principle and fuzzy statistics [J]. *Journal of the China Railway Society, 37*(12), 31–39.
9. Feng, D., He, Z. Y., Lin, S., Wang, Z., & Sun, X. J. (2017). Risk index system for catenary lines of high-speed railway considering the characteristics of time-space differences [J]. *IEEE Transactions on Transportation Electrification, 3*(3), 739–749.
10. Feng, D., Lin, S., He, Z. Y., & Sun, X. J. (2017). A technical framework of PHM and active maintenance for modem high-speed railway traction power supply systems [J]. *International Journal of Rail Transportation, 5*(3), 145–169.
11. Feng, D., Lin, S., He, Z. Y., Sun, X. J., & Wang, Z. (2018). Failure risk interval estimation of traction power supply equipment considering the impact of multiple factors [J]. *IEEE Transactions on Transportation Electrification, 4*(2), 389–398.
12. Feng, D., Lin, S., Zhang, A., Sun, X., Qin, N., & He, Z. (2017). Research on availibility prognostics methods for traction power supply equipment based on continuous-time Markov degradation processes [J]. *Proceedings of the CSEE, 37*(7), 1937–1945.

Z. He et al., *Prognostics and Health Management of High-Speed Railway Traction Power Substations*, https://doi.org/10.1007/978-981-95-4538-4

13. Feng, D., Lin, S., Sun, X., & He, Z. (2017). Availibility assessment of traction transformers considering high-speed rail load characteristics [J]. *Journal of the China Railway Society, 39*(8), 62–69.
14. Cheng, H., He, Z., Hu, H., et al. (2013). Lightning disaster risk assessment and early warning for high-speed railway traction power supply systems [J]. *Journal of the China Railway Society, 35*(5), 21–26.
15. Cheng, H., He, Z., & Mu, X. (2012). Health monitoring and assessment of high-speed railway traction power supply systems based on a multi-layer immune model [J]. *Power System Technology, 36*(9), 95–101.
16. Cheng, H., He, Z., Hu, H., et al. (2014). Comprehensive assessment of the health condition of high-speed train catenary networks using entropy weighting and multi-information sources [J]. *Journal of the China Railway Society, 36*(3), 19–24.
17. Hu, H., He, Z., Wang, J., et al. (2012). Load flow calculation of high-speed railway traction networks based on rail-network coupling [J]. *Proceedings of the CSEE, 32*(19), 101–108.
18. Sun, X., Lin, S., Feng, D., & He, Z. (2016). Short-term risk assessment of traction transformers based on load characteristics [J]. *Power System Technology, 40*(9), 2817–2823.
19. Zhang, A., Lin, S., Feng, D., & He, Z. (2017). Maintenance Decision model and its application for traction power supply equipment based on markov processes [J]. *Journal of the China Railway Society, 39*(11), 38–45.
20. Wang, Z., Lin, S., Feng, D., & Gao, S. (2018). Research on availibility assessment methods for contact networks considering weather conditions [J]. *Journal of the China Railway Society, 40*(10), 49–56.
21. Li, B. (2000). Foreign equipment management models and development trends (Part I) [J]. *Plant Maintenance Engineering, 07*, 36–37.
22. Xie, X. (2000). *Equipment state identification and maintenance decision [M]*. China Petrochemical Press Co. Ltd.
23. Fitch, E. C. (1992). Extending component service life through proactive maintenance [J]. *Noise and Vibration Worldwide, 23*, 38–38.
24. Fitch, E. C. (1992). *Proactive maintenance for mechanical systems [M]*. Elsevier.
25. Pecht, M., & Kang, R. (2010). *Diagnostics, prognostics and system's health management [M]*. City University of Hong Kong, PHM Centre.
26. Smith, G., Schroeder, J. B., Navarro, S., et al. (1997). Development of a prognostics & health management capability for the joint strike fighter [C]. *IEEE Autotestcon Proceedings, 1997*, 676–682.
27. Ferrell, B. L. (1999). JSF prognostics and health management [C]. In *1999 IEEE Aerospace Conference Proceedings* (Vol. 2, pp. 471–471).
28. Scanff, E., Feldman, K. L., Ghelam, S., et al. (2007). Life cycle cost impact of using prognostic health management (PHM) for helicopter avionics [J]. *Microelectronics Reliability, 47*(12), 1857–1864.
29. Mu, Z., Hu, H., & Hu, D. (2006). Weapon Equipment Prognostics and Health Management System Design [J]. *Ordnance Industry Automation, 25*(3), 20–21.
30. Pecht, M., & Jaai, R. (2010). A prognostics and health management roadmap for information and electronics-rich systems [J]. *Microelectronics Reliability, 50*(3), 317–323.
31. Murphy, B. P. (2000). *Machinery monitoring technology design methodology for determining the information and sensors required for reduced manning of ships [D]*. Master Dissertation of Massachusetts Institute of Technology.
32. Chen, Z. S., Yang, Y. M., Li, C., & el al. (2009). Model and algorithm of optimal embedded sensors placement for gearboxes based on signed directed graph of vibration propagation [J]. *Journal of Aerospace Power, 24*(10), 2384–2390.
33. Ma, S., Chen, G., & Fang, X. (2010). Preliminary study on prognostics and health management systems for complex equipment [J]. *Computer Measurement & Control, 18*(1), 1–4.

34. Sun, B., Kang, R., & Xie, J. (2007). Review of research and application status of prognostics and health management systems [J]. *Systems Engineering and Electronics, 29*(10), 1762–1767.
35. Kothamasu, R., Huang, S. H., & Verduin, W. H. (2006). System health monitoring and prognostics: A review of current paradigms and practices [J]. *Handbook of Maintenance Management and Engineering, 28*(9), 1012–1024.
36. Xu, L. (2009). *Study on prognostics and health management technology for electronic subsystems [D].* Thesis, University of Electronic Science and Technology of China.
37. Fan, J., Yung, K. C., & Pecht, M. (2011). Physics-of-failure-based prognostics and health management for high-power white light-emitting diode lighting [J]. *IEEE Transactions on Device and Materials Reliability, 11*(3), 407–416.
38. Climente-Alarcon, V., Antonino-Daviu, J. A., Strangas, E. G., et al. (2015). Rotor-bar breakage mechanism and prognosis in an induction motor [J]. *IEEE Transactions on Industrial Electronics, 62*(3), 1814–1825.
39. Pecht, M. (2008). *Prognostics and health management of electronics [M].* John Wiley & Sons, Ltd.
40. Cherry, G. A., & Qin, S. J. (2006). Multiblock principal component analysis based on a combined index for semiconductor fault detection and diagnosis [J]. *IEEE Transactions on Semiconductor Manufacturing, 19*(2), 159–172.
41. Tobon-Mejia, D. A., Medjaher, K., Zerhouni, N., et al. (2012). A data-driven failure prognostics method based on mixture of Gaussians hidden markov models [J]. *IEEE Transactions on Reliability, 61*(2), 491–503.
42. Poyhonen, S., Negrea, M., Jover, P., et al. (2003). Numerical magnetic field analysis and signal processing for fault diagnostics of electrical machines [J]. *The International Journal for Computation and Mathematics in Electrical and Electronic Engineering, 22*(4), 969–981.
43. Wang, P., & Vachtsevanos, G. (2001). Fault prognostics using dynamic wavelet neural networks [J]. *Artificial Intelligence for Engineering Design Analysis and Manufacturing, 15*(4), 349–365.
44. Hang, J., Zhang, J., & Cheng, M. (2014). Fault diagnosis of wind turbine based on multisensors information fusion technology [J]. *IET Renewable Power Generation, 8*(3), 289–298.
45. Pourali, M., & Mosleh, A. (2013). A functional sensor placement optimization method for power systems health monitoring [J]. *IEEE Transactions on Industry Applications, 49*(4), 1711–1719.
46. Zhao, Z., Quan, Q., & Cai, K. Y. (2014). A profust reliability based approach to prognostics and health management [J]. *IEEE Transactions on Reliability, 63*(1), 26–41.
47. Ju, P., Zuo, Y., Wen, F., et al. (2004). Power system health diagnosis [J]. *Electric Power Automation Equipment, 24*(6), 22–25.
48. Filev, D. P., & Tseng, F. (2006). Novelty detection based machine health prognostics [J]. *International Symposium on Evolving Fuzzy Systems, 1*, 193–199.
49. Young, J. C., Min, S. P., & Chong, N. C. (2008). Prediction of drill failure using features extraction in time and frequency domain of feed motor current [J]. *International Journal of Machine Tools & Manufacture, 48*(1), 29–39.
50. Sun, C., & Bi, W. (2003). New model of Grey prognostics parameters and its application in electrical insulation prognostics [J]. *Control Theory & Applications, 20*(05), 797–801.
51. Yang, N., Jiang, F., Meng, X., et al. (2003). Cable joint fault early warning system based on minimum variance temperature forecast [J]. *Automation of Electric Power Systems, 27*(24), 66–69.
52. Wu, X., Zhang, J., Zhao, T., et al. (2013). Power grid chain fault early warning method based on fuzzy clustering and fuzzy reasoning [J]. *Power System Technology, 37*(6), 1659–1665.
53. Xu, L., Zhu, P., Gu, H., et al. (2012). Early warning for current-carrying faults in power equipment based on variable-scale PCA [J]. *Electric Power Automation Equipment, 32*(5), 147–151.

54. Yao, J., Xiao, H., & Zhang, J. (2009). *Power safety assessment and management [M]*. China Electric Power Press.
55. Wei, Z., Xiao, X., Tong, X., et al. (2011). Disaster prevention analysis of catenary structure in railway traction power supply systems [J]. *Journal of the China Railway Society, 33*(6), 31–34.
56. Mor-Yaroslavtsev, A., & Levchenkov, A. (2011). Modeling the integration of expert systems into railway electric transport safety control [C]. In The *10th International Symposium Topical Problems in the Field Of Electrical and Power Engineering, Pamu, Estonia*, January 10–15.
57. Wang, B., You, D., Yin, X., et al. (2011). Security risk assessment system for complex power systems based on multi-factor Analysis [J]. *Power System Technology, 35*(1), 40–45.
58. Zhao, J. (2005). *Research on availability modeling based on performance degradation data and application [D]*. National University of Defense Technology.
59. Jin, G. (2014). *Degradation-based availibility technology [M]*. National Defense Industry Press.
60. Chen, W., Chai, X., & Sheng, J. (2001). Weibull availibility constant censoring test method for life-type products [J]. *Journal Of Zhejiang University (Engineering Science), 35*(2), 128–131.
61. Zhang, C., Wen, X., & Chen, X. (2004). Overview of accelerated life testing techniques [J]. *Acta Armamentarii, 25*(4), 485–490.
62. Chen, L., & Hu, C. (2009). Current status of research on availability analysis based on degradation modeling [J]. *Control and Decision, 24*(9), 1281–1293.
63. Meeker, W. Q., & Hamada, M. (1995). Statistical tools for the rapid development and evaluation of high-reliability products [J]. *IEEE Transactions on Reliability, 44*(2), 187–198.
64. Wang, H., & Gao, J. (2013). *Availibility analysis and evaluation of complex systems [M]*. China Science Publishing & Media, Ltd..
65. Lu, C. J., & Meeker, W. Q. (1993). Using degradation measures to estimate a time-to-failure distribution [J]. *Technometrics, 35*(2), 161–174.
66. Mawatari, H., Fukuda, M., Matsumoto, S., et al. (1997). Reliability and degradation behaviors of semi-insulating Fe-doped InP buried heterostructure lasers fabricated by RIE and MOVPE [J]. *Journal of Lightwave Technology, 15*(3), 534–537.
67. Lu, J., Park, J., & Yang, Q. (1997). Statistical inference of a time-to-failure distribution derived from linear degradation data [J]. *Technometrics, 39*(4), 391–400.
68. Meeker, W. Q., & Lu Valle, M. J. (1995). An accelerated life test model based on reliability kinetics [J]. *Technometrics, 37*(2), 133–146.
69. Yuan, X. X., & Pandey, M. D. (2009). A nonlinear mixed-effects model for degradation data obtained from in-service inspections [J]. *Reliability Engineering & System Safety, 94*(2), 509–519.
70. Guo, Q. (2015). *Research on availibility assessment method based on performance degradation data [D]*. South China University of Technology.
71. Zhuang, D., & Mao, S. (1996). Parameter estimation of mixed coefficient linear models [J]. *Chinese Journal of Applied Probability and Statistics, 12*(1), 81–87.
72. Freitas, M. A., Colosimo, E. A., Santos, T. R. D., et al. (2010). Reliability assessment using degradation models: Bayesian and classical approaches [J]. *Pesquisa Operacional, 30*(1), 194–219.
73. Mao, S., Tang, Y., & Wang, L. (2008). *availibility statistics [M]*. Higher Education Press.
74. Peng, B. (2010). *Research on availibility modeling method based on Wiener process [D]*. (Doctoral Dissertation). National University of Defense Technology.
75. van Noortwijk, J. M., van der Weide, J. A. M., Kallen, M. J., et al. (2007). Gamma processes and peaks-over-threshold distributions for time-dependent reliability [J]. *Reliability Engineering & System Safety, 92*(12), 1651–1658.
76. Tsai, C. C., Tseng, S. T., & Balakrishnan, N. (2012). Optimal design for degradation tests based on gamma processes with random effects [J]. *IEEE Transactions on Reliability, 61*(2), 604–613.

77. Zacks, S. (2005). Some recent results on the distributions of stopping times of compound Poisson processes with linear boundaries [J]. *Journal of Statistical Planning and Inference, 130*(1–2), 95–109.
78. Bocchetti, D., Giorgio, M., & Guida, M. (2009). A competing risk model for the reliability of cylinder liners in marine diesel engines [J]. *Reliability Engineering & System Safety, 94*(8), 1299–1307.
79. Li, C., Xu, M., Guo, S., et al. (2008). Current status and prospects of research on maintenance decisions [J]. *Thermal Turbine, 37*(1), 46–50.
80. Murthy, D. N. P., Atrens, A., & Eccleston, J. A. (2002). Strategic maintenance management [J]. *Journal of Quality in Maintenance Engineering, 8*(4), 287–305.
81. Jiang, R., & Murthy, D. N. P. (2008). *Maintenance: Decision models for management [M]*. Science Press.
82. Ivy, J. S., & Nembhard, H. B. (2005). A modeling approach to maintenance decisions using statistical quality control and optimization [J]. *Quality and Reliability Engineering International, 21*(4), 355–366.
83. Bertolini, M., & Bevilacqua, M. (2006). A combined goal programming-AHP approach to maintenance selection problem [J]. *Reliability Engineering & System Safety, 91*(7), 839–848.
84. Castanier, B., Grail, A., & Berenguer, C. (2005). A condition-based maintenance policy with non-periodic inspections for a two-unit series system [J]. *Reliability Engineering & System Safety, 87*(1), 109–120.
85. Wang, W. (2007). A two-stage prognosis model in condition based maintenance [J]. *European Journal of Operational Research, 182*(3), 1177–1187.
86. Wang, Y., Xu, H., Chen, W., et al. (2009). Grey system decision-making method for power transformer state-based maintenance strategy [J]. *Journal of Chongqing University, 32*(12), 1419–1424.
87. Xie, Q., Liang, J., & Zuo, H. (2006). Aircraft engine maintenance level decision based on variable precision rough sets [J]. *Systems Engineering and Electronics, 15*(4), 380–384.
88. Kallen, M. J., & van Noortwijk, J. M. (2005). Optimal maintenance decisions under imperfect Inspection [J]. *Reliability Engineering & System Safety, 90*(2), 177–185.
89. Chen, D., & Trivedi, K. S. (2005). Optimization for condition-based maintenance with semi-markov decision process [J]. *Reliability Engineering & System Safety, 90*(1), 25–29.
90. Song, G., Cao, J., Song, J., et al. (2007). Optimization of equipment maintenance support resources based on Rough set theory [J]. *Computer Engineering, 33*(4), 220–222.
91. Glazebrook, K. D., Mitchell, H. M., & Ansell, P. S. (2005). Index policies for the maintenance of a collection of machines by a set of repairmen [J]. *European Journal of Operational Research, 165*(1), 267–284.
92. Frangopol, D. M., & Liu, M. (2007). Maintenance and management of civil infrastructure based on condition, safety, optimization, and life-cycle cost [J]. *Structure and Infrastructure Engineering, 3*(1), 29–41.
93. Wang, J., Liu, W., Wei, F., et al. (2011). Study on transmission and transformation equipment state-based maintenance strategy based on life cycle cost management [J]. *Power System Protection and Control, 39*(5), 77–80.
94. Dehghanian, P., Fotuhi-Firuzabad, M., Aminifar, F., et al. (2013). A comprehensive scheme FBR reliability centered maintenance in power distribution systems—Part I: Methodology [J]. *IEEE Transactions on Power Delivery, 28*(2), 761–770.
95. He, Z., Xiao, C., & Ji, C. (2007). *Availibility-centered maintenance [M]*. International Cooperation Center of China Aerospace.
96. Khan, F. I., & Haddara, M. A. (2003). Risk-based maintenance (RBM): A quantitative approach fbr maintenance/inspection scheduling and planning [J]. *Journal of Loss Prevention in the Process Industries, 16*(6), 561–573.
97. Li, R., Wang, F., & Li, Z. (2013). Optimization of distribution network maintenance decision based on risk assessment [J]. *Electric Power Automation Equipment, 33*(11), 1–8.

98. Huang, J., Quan, L., & Quan, S. (2001). Current status and development of condition maintenance for high voltage electrical equipment in substations [J]. *Automation of Electric Power Systems, 8*(16), 56–61.
99. Yildirim, M., Sun, X. A., & Gebraeel, N. Z. (2016). Sensor-driven condition-based generator maintenance scheduling—Part I: Maintenance problem [J]. *IEEE Transactions on Power Systems, 31*(6), 4253–4262.
100. Zhang, H., Zhu, S., Zhang, Y., et al. (2009). Research and implementation of the condition maintenance technology system for transmission and distribution equipment [J]. *Power System Technology, 33*(13), 70–73.
101. Wang, H., Lin, D., Qiu, J., et al. (2015). Research on multiobjective group decision-making in condition-based maintenance for transmission and transformation equipment based on D-S. *Evidence Theory [J]., 6*(2), 1035–1045.
102. Guo, J., Xie, J., & Tang, G. (2003). Life cycle cost analysis in power equipment management [J]. *High Voltage Engineering, 29*(4), 13–15.
103. Mullerova, E., Hruza, J., Velek, J., et al. (2015). Life cycle management of power transformers: Results and discussion of case studies [J]. *IEEE Transactions on Dielectrics and Electrical Insulation, 22*(4), 2379–2389.
104. Ma, Z., Zhong, L., Lu, K., et al. (2011). Study on the optimal maintenance strategy based on the full life cycle cost of electrical equipment [J]. *Power System Protection and Control, 39*(16), 34–39.
105. Raghavan, S., & Chowdhury, B. (2015). State diagram-based life cycle management plans for power plant components [J]. *IEEE Transactions on Smart Grid, 6*(2), 965–972.
106. Grail, A., Dieulle, L., Berenguer, C., et al. (2002). Continuous-time predictive-maintenance scheduling fbr a deteriorating system [J]. *IEEE Transactions on Reliability, 51*(2), 141–150.
107. Conejo, A. J., Garcia-Bertrand, R., & Diaz-Salazar, M. (2005). Generation maintenance scheduling in restructured power systems [J]. *IEEE Transactions on Power Systems, 20*(2), 984–992.
108. Zhang, D., Li, W., & Xiong, X. (2012). Bidding based generator maintenance scheduling with triple-objective optimization [J]. *Electric Power Systems Research, 93*(10), 127–134.
109. Lu, G., Chung, C. Y., Wong, K. P., et al. (2008). Unit maintenance scheduling coordination mechanism in electricity market environment [J]. *IET Generation, Transmission & Distribution, 2*(5), 646–654.
110. Feng, C., & Wang, X. (2010). A competitive mechanism of unit maintenance scheduling in a deregulated environment [J]. *IEEE Transactions on Power Systems, 25*(1), 351–359.
111. Pan, L., Zhang, Y., Yu, G., et al. (2010). Calculation of electrical equipment failure rate in state maintenance decision [J]. *Electric Power Automation Equipment, 30*(2), 91–94.
112. Li, M., Han, X., Wang, Y., et al. (2012). Decision model and solution for condition-based maintenances at substations [J]. *Proceedings of the CSEE, 32*(25), 196–202.
113. Cui, X., Yin, L., Fan, C., et al. (2010). Research on transformer life cycle cost (LCC) model during substation renovation [J]. *Power System Protection and Control, 38*(7), 69–73.
114. Chen, S. (2011). *Research on condition maintenance strategy for substation equipment based on life cycle cost [D]*. North China Electric Power University.
115. Qiu, R. (2004). *availibility maintenance of electrical equipment [M]*. China Electric Power Press.
116. Shu, H. (2009). *availibility-centered maintenance of power systems [M]*. China Machine Press.
117. Endrenyi, J., & Anders, G. J. (2006). Aging, maintenance, and reliability—Approaches to preserving equipment health and extending equipment life [J]. *IEEE Power and Energy Magazine, 4*(03), 59–67.
118. Bertling, L., Allan, R., & Eriksson, R. (2005). A reliability-centered asset maintenance method for assessing the impact of maintenance in power distribution systems [J]. *IEEE Transactions on Power Systems, 20*(01), 75–82.

119. Tsai, Y. T., Wang, K. S., & Tsai, L. C. (2004). A study of availability-centered preventive maintenance for multi-component systems [J]. *Reliability Engineering & System Safety, 84*(03), 261–270.
120. Yang, F., & Chang, C. S. (2009). Multiobjective evolutionary optimization of maintenance schedules and extents for composite power systems [J]. *IEEE Transactions on Power System, 24*(4), 1694–1702.
121. Abiri-Jahromi, A., Parvania, M., Bouffard, F., et al. (2013). A two-stage framework for power transformer asset maintenance management—Part I: Models and formulations [J]. *IEEE Transactions on Power System, 28*(2), 1395–1403.
122. Yiimbe, Y., Hasegawa, T., & Furukawa, N. (2013). Optimization method for inspection scheduling of power distribution facilities [J]. *IEEE Transactions on Power Delivery, 28*(3), 1558–1565.
123. Yang, Y. (2011). *Study on RAMS evaluation of high-speed railway power supply system [D].* Beijing Jiaotong University.
124. Wu, J. (2013). *Theory and application of RAMS evaluation for high-speed railway power supply system [M].* Beijing Jiaotong University Press Limited Liability Company.
125. Qi, G. (2012). *Study on safety, availibility, and maintainability of high-speed railway overhead contact system [M].* Southwest Jiaotong University Press.
126. Li, X., Wu, J., Yang, Y., et al. (2010). Optimization study of maintenance plan for high-speed railway overhead contact system suspension system [J]. *Journal of the China Railway Society, 32*(2), 24–30.
127. Chen, S., Peng, H., Mao, B., et al. (2011). Optimization model for railway traction substation maintenance plan based on cost minimization [J]. *Journal of the China Railway Society, 33*(8), 39–44.
128. Chen, S., Wang, X., Bai, Y., et al. (2013). Optimization model for railway traction catenary maintenance plan based on cost minimization [J]. *Journal of the China Railway Society, 35*(12), 37–42.
129. Chen, M., Li, Q., & Xie, S. (2009). Optimization and simulation of maintenance plan for traction power supply system [J]. *Journal of South China University of Technology (Natural Science Edition), 37*(11), 100–106.
130. Wang, L., & Xu, S. (1990). *Introduction to analytic hierarchy process [M].* China Renmin University Press.
131. Saaty, T. (1980). *The analytic hierarchy process [M].* McGraw- Hilllnc.
132. Xiao, J., Wang, C., & Zhou, M. (2004). Comprehensive assessment and decision of urban power grid planning based on interval analytic hierarchy process [J]. *Proceedings of the CSEE, 24*(4), 50–57.
133. Wang, Z., & Wu, B. (2008). *Fuzzy data statistics [M].* Harbin Institute of Technology Press.
134. Tan, X. (2012). *Study on prognostics and health diagnosis of latent faults in complex distribution networks [D].* Southwest Jiaotong University.
135. Li, Q., Zhou, L., Zhang, F., et al. (2008). Comprehensive assessment of power system voltage state early-warning levels based on fuzzy theory and analytic hierarchy process (AHP) [J]. *Power System Technology, 32*(4), 40–45.
136. Zuo, M. J., Jiang, R., & Yam, R. C. M. (1999). Approaches for reliability modeling of continuous-state devices [J]. *IEEE Transactions on Reliability, 48*(1), 9–18.
137. He, S. (2008). *Random process [M].* Peking University Press.
138. Lin, Y. (2002). *Application of random processes [M].* Tsinghua University Press.
139. Li, Y., Liu, N., & Wang, Q. (2012). *Random process [M]* (2nd ed.). National Defense Industry Press.
140. Abdel-Hameed, M. (1975). A Gamma wear process [J]. *IEEE Transactions on Reliability, R-24*(2), 152–153.
141. van Noortwijk, J. M. (2009). A survey of the application of gamma processes in maintenance [J]. *Reliability Engineering and System Safety, 94*(1), 2–21.

142. Jia, X. (2007). *Availibility-centered maintenance decision model [M]*. National Defense Industry Press.
143. Nikulin, M. S., Nikolaos Limnios, N., Balakrishan, N., et al. (2010). *Advances in degradation modeling: Applications to reliability, survival analysis, and finance [M]*. Birkhauser.
144. Darling, R. W. R. (2002). Fluid limits of pure jump Markov processes: A practical guide [J]. *ArXiv Mathematics, 0210109*, 1–16.
145. Huynh, K. T., Castro, I. T., Barros, A., et al. (2012). Modeling age-based maintenance strategies with minimal repairs for systems subject to competing failure modes due to degradation and shocks [J]. *European Journal of Operational Research, 218*(1), 140–151.
146. Huynh, K. T., Barros, A., & Berenguer, C. (2015). Multi-level decision-making FBR the predictive maintenance of k-Out-of-n: F deteriorating systems [J]. *IEEE Transactions on Reliability, 64*(1), 94–117.
147. Li, W., & Pham, H. (2005). An inspection-maintenance model fbr systems with multiple competing processes [J]. *IEEE Transactions on Reliability, 54*(2), 318–327.
148. Peng, H., Feng, Q., & Coit, D. W. (2011). Reliability and maintenance modeling for systems subject to multiple dependent competing failure processes [J]. *HE Transactions, 43*(1), 12–22.
149. DeGroot, M. H., & Schervish, M. J. (2012). *Probability and statistics [M]* (4th ed.). Addison-Wesley.
150. Corder, G. W., & Foreman, D. I. (2014). *Nonparametric statistics: A step-by-step approach [M]* (2nd ed.). Wiley.
151. Rafiee, K., Feng, Q., & Coit, D. W. (2015). Condition-based maintenance for repairable deteriorating systems subject to a generalized mixed shock model [J]. *IEEE Transactions on Reliability, 64*(4), 1164–1174.
152. Li, Z., & Zhao, P. (2007). Reliability analysis on the 3-shock model of complex systems [J]. *IEEE Transactions on Reliability, 56*(2), 340–348.
153. Parvardeh, A., & Balakrishnan, N. (2015). On mixed δ-shock models [J]. *Statistics & Probability Letters, 102*(1), 51–60.
154. Lukacs, E. (1970). *Characteristic functions [M]* (2nd ed.). Griffin.
155. Gardiner, C. (2009). *Stochastic methods: A handbook fbr the natural and social sciences [M]* (4th ed.). Springer.
156. Wei, Z. (2008). *Tutorial on probability theory and mathematical statistics [M]* (2nd ed.). Higher Education Press.
157. Shephard, N. G. (1991). From characteristic function to distribution function: A simple framework for the theory [J]. *Econometric Theory, 7*(4), 519–529.
158. Finkelstein, M., Shafiee, M., & Kotchap, A. N. (2016). Classical optimal replacement strategies revisited [J]. *IEEE Transactions on Reliability, 65*(2), 540–546.
159. Elsayed, E. A. (2012). *Reliability engineering [M]* (2nd ed.). John Wiley & Sons.
160. Xu, J., Wang, J., Gao, F., et al. (2000). Review of research on condition maintenance technology of power equipment [J]. *Power System Technology, 24*(8), 48–52.
161. Duan, D., Wu, X., & Deng, H. (2012). Reliability evaluation in substations considering operating conditions and failure modes [J]. *IEEE Transactions on Power Delivery, 27*(1), 309–316.
162. Berdinyazow, A., Camci, F., Sevkli, M., et al. (2009). Economic analysis of maintenance policies for a system [C]. In *IEEE International Symposium on Diagnostics fbr Electric Machines, Power Electronics and Drivers, Cargese, France* (pp. 1–5).
163. Pan, L., Lu, G., Zhang, Y., et al. (2010). Optimization of equipment condition maintenance decision based on comprehensive risk assessment [J]. *Automation of Electric Power Systems, 34*(11), 28–32.
164. Zhang, X., Ge, S., Liu, H., et al. (2014). Comprehensive evaluation system and method for intelligent distribution networks [J]. *Power System Technology, 38*(1), 3642–3648.
165. Wang, B., He, G., Mei, S., et al. (2011). Construction method of evaluation index system for intelligent grids [J]. *Automation of Electric Power Systems, 35*(23), 1–5.

166. Yang, Y., Wu, J., Wu, Y., et al. (2008). Fuzzy assessment of electrified railway catenary availibility based on credibility theory [J]. *Journal of the China Railway Society, 30*(6), 115–119.
167. Xie, J., Wu, J., & Wu, Y. (2009). Availibility modeling of traction power supply system based on genetic algorithm [J]. *Journal of the China Railway Society, 31*(4), 47–51.
168. Wang, X. (2008). *Study on maintenance mode of railway traction power supply equipment based on RCM [D].* Southwest Jiaotong University.
169. Chen, S. (2007). *Optimization model and algorithm for maintenance plan of railway traction power supply system [D].* Beijing Jiaotong University, Doctoral Dissertation.
170. Saaty T. The analytic network process: Decision making with dependence and feedback [M], : , 2001.
171. Lin, Y. (2019). Review of Prognostics and health management technologies in the aerospace field [J]. *Aerospace China, 5*, 21–26.
172. Jing, B., Xu, G., Huang, Y., Jiao, X., & Liang, W. (2017). Analysis and study of the progress and issues in PHM technology for military aircraft [J]. *Journal of Electronic Measurement and Instrument, 31*(2), 161–169.
173. Zhang, B. (2008). Development and application of integrated diagnostic, predictive, and health management technologies abroad [J]. *Computer Measurement & Control, 5*, 591–594.
174. Lu, H., & Wang, Z. (2015). Development of integrated fault diagnosis and health management technology for aerospace electronic systems [J]. *Electronics Optics & Control, 22*(8), 60–65+86.
175. Pei, D., Wang, J., Zhou, P., & Luo, Q. (2016). Overview of PHM technology for ships [J]. *Journal of Electronic Measurement and Instrument, 30*(9), 1289–1297.
176. Xu, X. (2014). *Study on fault diagnosis methods and trend Prognostics technology for ship power systems [D].* Wuhan University of Technology.
177. Zhang, B., & Zeng, T. (2005). PHM: Key enabling technology to achieve the F-35's affordability goals [J]. *Aviation Maintenance & Engineering, 6*, 20–23.
178. Lü, C., Ma, J., & Wang, Z. (2016). Review of the development of PHM technology at home and abroad [J]. *Computer Measurement & Control, 24*(9), 1–4.
179. Qiu, L. (2012). Key technologies of Prognostics and health management systems for weapon equipment [J]. *Ship Electronic Engineering, 32*(5), 17–18.
180. Ju, P., Zuo, Y., Wen, F., Wei, Z., & Ma, H. (2004). Power system health diagnosis [J]. *Electric Power Automation Equipment, 6*, 22–25.
181. Hu, J., Zhang, W., Cao, D., Wen, Y., & Rong, Y. (2016). Research on the health status of power systems based on fuzzy theory [J]. *Power System Protection and Control, 44*(13), 61–66.
182. Mousavi, M. J., & Bulter-Purry, K. L. (2010). Detection of incipient faults in distribution underground cables [J]. *IEEE Transactions on Power Delivery., 25*(3), 1363–1371.

MIX
Papier aus verantwortungsvollen Quellen
Paper from responsible sources
FSC® C105338

If you have any concerns about our products,
you can contact us on
ProductSafety@springernature.com

In case Publisher is established outside the EU,
the EU authorized representative is:
Springer Nature Customer Service Center GmbH
Europaplatz 3, 69115 Heidelberg, Germany

Printed by Libri Plureos GmbH
in Hamburg, Germany